ARCTIC SHOREBIRDS *in* NORTH AMERICA

STUDIES IN AVIAN BIOLOGY

A Publication of the Cooper Ornithological Society

WWW.UCPRESS.EDU/GO/SAB

Studies in Avian Biology is a series of works published by the Cooper Ornithological Society since 1978. Volumes in the series address current topics in ornithology and can be organized as monographs or multi-authored collections of chapters. Authors are invited to contact the series editor to discuss project proposals and guidelines for preparation of manuscripts.

See complete series list on page 301.

ARCTIC SHOREBIRDS *in* NORTH AMERICA

A Decade of Monitoring

Jonathan Bart and Victoria Johnston, *Editors*
Paul A. Smith and Jennie Rausch, *Associate Editors*

Studies in Avian Biology No. 44

A PUBLICATION OF THE COOPER ORNITHOLOGICAL SOCIETY

University of California Press
Berkeley Los Angeles London

University of California Press, one of the most distinguished university presses in the United States, enriches lives around the world by advancing scholarship in the humanities, social sciences, and natural sciences. Its activities are supported by the UC Press Foundation and by philanthropic contributions from individuals and institutions. For more information, visit www.ucpress.edu.

Studies in Avian Biology, No. 44

University of California Press
Berkeley and Los Angeles, California

University of California Press, Ltd.
London, England

Library of Congress Cataloging-in-Publication Data

Arctic shorebirds in North America : a decade of monitoring /
Jonathan Bart and Victoria Johnston, editors.
p. cm. — (Studies in avian biology; No. 44)
"A Publication of the Cooper Ornithological Society."
Includes bibliographical references and index.
ISBN 978-0-520-27310-8 (cloth : alk. paper)

1. Shore birds—Canada, Northern. 2. Shore birds—Alaska.
3. Bird surveys—Canada, Northern. 4. Bird surveys—Alaska.
I. Bart, Jonathan. II. Johnston, Victoria Helen, 1962-

QL685.5.N63A73 2012
598.3'309719—dc23 2011048814

19 18 17 16 15 14 13 12
10 9 8 7 6 5 4 3 2 1

The paper used in this publication meets the minimum requirements of ANSI/NISO Z39.48-1992 (R 2002)(*Permanence of Paper*).

Cover image: White-rumped Sandpiper, in Nunavut, Canada. Photo by Jeffrey and Lisa-Jo van den Scott.

PERMISSION TO COPY

CONTENTS

Part IV • Synthesis

Online Content

Chapter 1 and abstracts for the remaining chapters are available online in French and Spanish from the University of California Press website, www.ucpress.edu/go/sab.

Chapitre 1 et résumés pour les chapitres suivants sont disponibles en français et espagnol à l'adresse internet de la Presse de l'Université de Californie, www.ucpress.edu/go/sab.

El capítulo 1 y resúmenes de los capítulos restantes están disponibles en línea en español y francés en la página web de la Prensa de la Universidad de California, www.ucpress.edu/go/sab.

CONTRIBUTORS

BRAD A. ANDRES
U.S. Fish and Wildlife Service
P.O. Box 25486, DFC
Denver, CO 80225-0486, USA
brad_andres@fws.gov

LINDSAY A. ARMER
Environment and Conservation
Indian and Northern Affairs Canada
4914–50 Street, 10th Floor Bellanca Building
P.O. Box 1500
Yellowknife, NT X1A 2R3, Canada
linds.armer@mac.com

JONATHAN BART
U.S. Geological Survey
Forest and Rangelands Ecosystem Science Center
970 Lusk Street
Boise, ID 83706, USA
jon_bart@usgs.gov

STEPHEN BROWN
Manomet Center for Conservation Sciences
P.O. Box 1770
Manomet, MA 02345, USA
sbrown@manomet.org

BRIAN T. COLLINS
National Wildlife Research Centre
Canadian Wildlife Service
Environment Canada
1125 Colonel By Drive
Ottawa, ON K1A 0H3, Canada
brianandhelencollins@sympatico.ca

GARRY DONALDSON
Canadian Wildlife Service
Environment Canada
351 St. Joseph Boulevard
Gatineau, QC K1A 0H3, Canada
garry.donaldson@ec.gc.ca

KYLE H. ELLIOTT
Department of Biology
University of Manitoba
Winnipeg, MB R3T 2N2, Canada
haliaeetus@gmail.com

CHARLES M. FRANCIS
National Wildlife Research Centre
Canadian Wildlife Service
Environment Canada
1125 Colonel By Drive
Ottawa, ON K1A 0H3, Canada
charles.francis@ec.gc.ca

VICTORIA JOHNSTON
Environment Canada
P.O. Box 2310
5019–52 Street, 4th Floor
Yellowknife, NT X1A 2P7, Canada
vicky.johnston@ec.gc.ca

DAVID J. KRUEPER
U.S. Fish and Wildlife Service
Migratory Bird Program
P. O. Box 1306
Albuquerque, NM 87103, USA
dave_krueper@fws.gov

CRAIG S. MACHTANS
Canadian Wildlife Service
Environment Canada
P.O. Box 2310
5019–52 Street, 4th Floor
Yellowknife, NT X1A 2P7, Canada
craig.machtans@ec.gc.ca

ANN MANNING
U.S. Geological Survey
Forest and Rangelands Ecosystem Science Center
970 Lusk Street
Boise, ID 83706, USA
(Current address:
520 East Ash Street,
Caldwell, ID 83605, USA,
manning_ann@yahoo.com)

BRIAN J. MCCAFFERY
U.S. Fish and Wildlife Service
Yukon Delta National Wildlife Refuge
P.O. Box 346
Bethel, AK 99559, USA
brian_mccaffery@fws.gov

R. I. G. MORRISON
National Wildlife Research Centre
Canadian Wildlife Service
Environment Canada
1125 Colonel By Drive
Ottawa, ON K1A 0H3, Canada
guy.morrison@ec.gc.ca

ELIN P. PIERCE
1134 Santa Luisa Drive
Solana Beach, CA 92075, USA
elpier@roadrunner.com

LISA PIRIE
Canadian Wildlife Service
Environment Canada
P.O. Box 1714
Qimugjuk Building 969
Iqaluit, NU X0A 0H0, Canada
lisa.pirie@ec.gc.ca

ROBERT PLATTE
U.S. Fish and Wildlife Service
Migratory Bird Management
1011 East Tudor Road
Anchorage, AK 99503-6199, USA
bob_platte@fws.gov

JENNIE RAUSCH
Canadian Wildlife Service
Environment Canada
P.O. Box 2310
5019–52 Street, 4th Floor
Yellowknife, NT X1A 2P7, Canada
jennie.rausch@ec.gc.ca

SUSAN K. SKAGEN
U.S. Geological Survey
Fort Collins Science Center
2150 Center Avenue, Building C
Fort Collins, CO 80526-8118, USA
susan_skagen@usgs.gov

PAUL A. SMITH
Smith and Associates Ecological
Research Ltd.
772 – 7th Conc South
Pakenham, ON K0A 2X0, Canada
paul_smith@smitheco.ca

CATHERINE WIGHTMAN
Montana Fish, Wildlife & Parks
1420 East Sixth Avenue
P.O. Box 200701
Helena, MT 59620-0701, USA
cwightman@mt.gov

FOREWORD

Contribution of Arctic PRISM to Monitoring Western Hemispheric Shorebirds

Susan K. Skagen, Paul A. Smith, Brad A. Andres, Garry Donaldson, and Stephen Brown

WHY MONITOR SHOREBIRDS?

Long-term monitoring of populations is of paramount importance to understanding responses of organisms to global environmental change and to evaluating whether conservation practices are yielding intended results through time (Wiens 2009). The population status of many shorebird species, the focus of this volume, remain poorly known. Long-distance migrant shorebirds have proven particularly difficult to monitor, in part because of their highly migratory nature and ranges that extend into highly inaccessible regions. As migrant shorebirds travel the length of the hemisphere, they congregate and disperse in ways that vary among species, locations, and years, presenting serious challenges to designing and implementing monitoring programs.

Rigorous field and quantitative methods that estimate population size and monitor trends are vitally needed to direct and evaluate effective conservation measures. Many management efforts depend on unbiased population size estimates; for example, the shorebird conservation plans for both Canada and the United States seek to restore populations to levels calculated for the 1970s based on the best information available from existing surveys. Further, federal wildlife agencies within the United States and Canada have mandates to understand the state of their nations' resources under various conventions for the protection of migratory birds. Accurate estimates of population size are vital statistics for a variety of conservation activities, such as prioritizing species for conservation action and setting management targets. Areas of essential habitat, such as those designated under the Western Hemisphere Shorebird Reserve Network, the Important Bird Areas program of BirdLife International and the National Audubon Society, or Canada's National Wildlife Areas program, are all evaluated on the basis of proportions of species' populations which they contain. The size, and trends in size, of a species' population are considered key information for assessing its vulnerability and subsequent listing under the U.S. *Endangered Species Act* and the Canadian *Species at Risk Act.*

To meet the need for information on population size and trends, shorebird biologists from Canada and the United States proposed a shared blueprint for shorebird monitoring across the Western Hemisphere in the late 1990s; this effort was undertaken in concert with the development of the Canadian and U.S. Shorebird Conservation Plans (Donaldson et al. 2000, Brown et al. 2001). Soon thereafter, partners in the monitoring effort adopted the name "Program for Regional and International Shorebird Monitoring" (PRISM). Among the primary objectives of PRISM were to estimate the population sizes and trends of breeding North American shorebirds and describe their distributions (Bart et al. 2002). PRISM members evaluated ongoing and potential monitoring approaches to address 74 taxa (including subspecies) and proposed a combination of arctic and boreal breeding surveys, temperate breeding and non-breeding surveys, and neotropical surveys.

CHALLENGES WITH MONITORING SHOREBIRDS

Despite their importance for conservation and management of shorebirds, accurate estimates of population size have proven difficult to obtain for many species; some species disperse widely during migration, vary in their lengths of stay at stop-over sites, and differ in their ratios of imperfect detection. Estimates provided in Morrison et al. (2006) represent the best information currently available, and although the authors have devoted substantial effort to refining the estimates, 47 of 75 (63%) of the taxa described have population estimates that are considered only within one or more orders of magnitude. Trend data, too, are in many cases not sufficiently robust to support management action. Despite the apparent widespread declines in shorebird populations, imprecision and potential bias in the trend estimates mean that some species in need of conservation attention do not have the basic level of information necessary to support unequivocally sound management actions, such as listing for protection under the Canadian *Species at Risk Act*.

BENEFITS OF ARCTIC MONITORING

Arctic PRISM was designed to monitor the status of shorebird populations by estimating population size across the entire North American arctic at regular intervals using data collected at a regional scale. This multiscale geographic focus offers a number of advantages from a management perspective. Local and regional density information can be applied directly in an impact assessment context (as was the case for the proposed Mackenzie Valley Oil and Gas project; see Rausch and Johnston, Chapter 5, this volume) or could be applied indirectly to estimate flyway populations for management efforts elsewhere. Partitioning population sizes among smaller-scale geographies allows local managers to set management targets for their area of interest. The range-wide nature of the population information provided by PRISM also will allow managers to better understand national and international responsibilities for a species' protection.

Efforts prior to Arctic PRISM to estimate population size based primarily on wintering and migration surveys were unable to determine the relative distribution of breeding shorebirds between Alaska versus Canada for widely dispersed species. Establishing the relative responsibility borne by Canada and the United States for a particular species' conservation was therefore difficult. Interpretation of results from surveys that varied in proportional coverage of birds along different flyways may have provided misleading perspectives for shorebirds such as the Semipalmated Sandpiper, a species currently under consideration for listing by the Committee on the Status of Endangered Wildlife in Canada. Historically, 75% of all Semipalmated Sandpipers were thought to migrate through the Bay of Fundy based on comparisons of total counts at this site during migration with counts on the wintering grounds. It was assumed that these birds bred in the eastern arctic, suggesting that Canada had the lead responsibility for conservation of both breeding and migrating Semipalmated Sandpipers.

Recent surveys in areas that previously lacked coverage are now revealing a new perspective on Semipalmated Sandpiper distribution. At sites surveyed to date, densities are much higher in the western regions than in eastern areas. Although much of the species' range remains to be surveyed, it seems that a large fraction of the population likely breeds in Alaska and may therefore use inland migration routes where they will not be exposed to environmental threats operating in the Bay of Fundy. Moreover, results to date suggest that the population size of Semipalmated Sandpipers is much larger than once believed. The Bay of Fundy may be used by a smaller fraction of the species than once thought, yet it is still of critical importance to southbound populations traveling along the Atlantic coast.

The regional information on population trends obtained through Arctic PRISM surveys will provide insights valuable to conservation efforts. This information will help to determine the geographic areas or flyways where conservation action is most urgently needed, but without some of the problems associated with counting birds at migratory stop-over locations. Arctic surveys could also complement migration surveys by helping to determine if some apparent declines are due in part to shifts in migratory stopover locations. Consider, for example, the dramatic decline of Red-necked Phalaropes in the Bay of Fundy. Range-wide counts on the breeding grounds would provide evidence to help discriminate between explanations of population decline or geographic shift.

Arctic surveys may also help to identify changes in non-wetland arctic habitats. Shorebirds are often found at highest densities in wetland habitats, and these habitats are where monitoring efforts have traditionally been focused. In PRISM surveys to date, significant numbers of shorebirds have been found in more upland habitats such as upland heath tundra, although densities are usually significantly lower. GIS-based estimates of available habitat suggest that these habitats are nearly eight times more extensive than wetlands, and support an important fraction of shorebird populations across the arctic. Declines may occur first in marginal habitats, and would go unnoticed if monitoring targeted only a restricted number of high-density sites.

THE ROLE OF ARCTIC PRISM SURVEYS IN CONTINENTAL SHOREBIRD MONITORING

When the PRISM planning document was drafted in the early 2000s, information on population size and trends of 46 of the 74 taxa (32 of 49 species) was anticipated from arctic and boreal breeding surveys (ABBS), to be supplemented by temperate migration and winter single-species surveys (Bart et al. 2002). As pilot efforts revealed the low feasibility of implementing extensive boreal surveys (see Surveying the Boreal Fractions of Northern Species below), this list of focal species with potential for population size and trend estimates was modified to the 26 species covered by arctic breeding ground surveys (Johnston and Bart, chapter 1, this volume).

To identify the species for which arctic breeding ground surveys clearly yield the best approach for estimation of populations sizes and trends, we distinguished among the 26 focal species according to whether at least 70% of a species' range falls within the North American arctic, and the likelihood of reaching PRISM accuracy targets based on the analyses in this volume (Bart and Smith, chapter 13, this volume). The target would be met by a CV of the population estimates of 0.42 for future surveys revisiting the same plots or 0.31 for future surveys of new independently chosen plots. Based on these criteria, we classified species into four categories of concern, including: (1) species with the primary range in the arctic for which the accuracy target could easily be met; (2) species primarily breeding in the arctic for which the accuracy target could be met by adding additional elements to the survey design; (3) species with more than 30% of their range falling outside of the arctic study area but for which estimates of the arctic component would satisfy the PRISM target; and (4) species with more than 30% of their range falling outside of the arctic study area that require substantial additional design elements to adequate survey the arctic portion of their populations (Table F.1).

We consider the arctic breeding surveys to clearly be a viable and valuable approach to population size and trend estimation for all of the focal species in category 1, including seven species ranked as Highly Imperiled or High Concern (Bar-tailed Godwit, Ruddy Turnstone, Rock Sandpiper, Sanderling, Red Knot, Dunlin, and Buff-breasted Sandpiper), three species of moderate concern (Black-bellied Plover, Semipalmated Sandpiper, and Red Phalarope), and four species of low concern (White-rumped Sandpiper, Baird's Sandpiper, Pectoral Sandpiper, and Long-billed Dowitcher). This declaration is based on the assumption of 50 crew years and repeat surveys of plots. If the number of crew years dropped to 40, estimates of Buff-breasted Sandpipers and Long-billed Dowitchers would no longer meet the accuracy target.

The two species in category 2, Black Turnstone (High Concern) and Purple Sandpiper (Moderate Concern) breed primarily in the arctic but have a lower certainty of reaching the accuracy target, primarily because they are rare and patchily distributed. Additional coverage within their ranges is needed to approach the accuracy target. Conservation priority ranks for these species, from High Concern to Low Concern, are derived from the Canadian and United States shorebird conservation plans and subsequent updates (U.S. Shorebird Conservation Plan 2004).

For several category 3 species (those with less than 70% of their range in the arctic), the predicted accuracy of counts in the arctic is good; thus accurate trend estimates of the arctic component of their populations can feasibly come from the arctic surveys. This is true for American Golden-Plover and Whimbrel (High Concern), Pacific Golden-Plover, Wilson's Snipe, Least

TABLE F.1

Feasibility of reaching desired accuracy target (CV < 0.42) for 26 focal species with arctic breeding ground surveys, assuming repeat surveys of plots and effort of 50 crew-years in future surveys.

Breeding distribution	Focal species	
More than 70 % of breeding range in arctic	*Category 1:* Species for which accuracy target will likely be obtained (predicted CV; Table 12.2)	
	Black-bellied Plover (0.20)	White-rumped Sandpiper (0.31)
	Bar-tailed Godwit (0.36)	Baird's Sandpiper (0.24)
	Ruddy Turnstone (0.28)	Pectoral Sandpiper (0.23)
	Red Knot (0.33)	Dunlin (0.33)
	Rock Sandpiper (0.38)	Buff-breasted Sandpiper (0.29)
	Sanderling (0.26)	Long-billed Dowitcher (0.38)
	Semipalmated Sandpiper (0.21)	Red Phalarope (0.29)
	Category 2: Species for which accuracy target will likely be obtained only with additional sampling within range (predicted CV)	
	Black Turnstone (0.46)	Purple Sandpiper (1.20)
Less than 70 % of breeding range in arctic	*Category 3:* Species for which accuracy target will likely be obtained for arctic component (predicted CV; % range in arctic)	
	American Golden-Plover (0.21; 59)	Least Sandpiper (0.39; 21)
	Pacific Golden-Plover[a] (0.34; 66)	Stilt Sandpiper (0.30; 68)
	Semipalmated Plover (0.40; 39)	Wilson's Snipe (0.26; 6)
	Whimbrel (0.29; 30)	Red-necked Phalarope (0.28; 43)
	Category 4: Species that cannot be adequately surveyed in the arctic (predicted CV; % range in arctic)	
	Hudsonian Godwit (1.27; 26)	Western Sandpiper[a] (0.47; 64)

[a] Likelihood of reaching accuracy target may be improved with additional survey design elements.

Sandpiper, Stilt Sandpiper, Red-necked Phalarope (all of Moderate Concern), and Semipalmated Plover (Low Concern). For these eight species, additional breeding surveys in other biomes or during migration or winter will be necessary to capture trends of the populations that breed outside the arctic. The remaining two focal species with less than 70% of their range in the arctic, Hudsonian Godwit and Western Sandpiper (High Concern), are in category 4 and will require both targeted design elements within the arctic and additional survey efforts outside the arctic to adequately assess trends.

Generally, the species that have proven most difficult to survey in the arctic are relatively rare and/or have a restricted distribution. While rarity can make a species difficult to survey throughout the annual cycle, a restricted distribution may in some cases be an asset for targeted, single-species surveys. High fidelity of Pacific Golden-Plovers to their Pacific island wintering sites suggests that information on changes in their population size could be obtained there (Johnson et al. 2006). Virtually all of the Hudsonian Godwits wintering along the Pacific Coast do so in the vicinity of Chiloé Island, Chile (Andres et al. 2009), and systematic ground counts could provide information on population size and trends. Recent analyses of Christmas Bird Count data (Butcher and Niven 2007) may prove useful for tracking changes in a select group of shorebirds that winter in North America, such as the Purple Sandpiper. In the Pacific Flyway, design is under way for a program to estimate trends of shorebirds during winter. A thorough review of alternative methods for species not surveyed well by Arctic PRISM should be undertaken.

SURVEYING THE BOREAL FRACTIONS OF NORTHERN SPECIES

Of the 26 taxa breeding in boreal and arctic regions, ten species have more than 30% of their range in the boreal ecozones (see Table F.1), and shorebird surveys in boreal habitats could therefore play an important role in a continental monitoring program for shorebirds. Initially, it was hoped that a ground-based survey similar to Arctic PRISM, with double sampling and stratified random plot selection, would also be effective in boreal habitats. Pilot efforts revealed that foot travel across boreal wetlands was difficult and unsafe, and methods were reconsidered. Sinclair et al. (2004) reviewed the potential for a variety of air- and ground-based survey methods, and research to evaluate some of these methods is ongoing (Elliott et al. 2010). Still, a strategy for monitoring shorebirds within challenging boreal habitats remains elusive. Surveys for boreal species may be best addressed by well-designed migration or winter surveys.

MONITORING AND CLIMATE CHANGE

Predicted changes in precipitation patterns and increases in temperature will be intensified in the arctic and will have dramatic effects on the distribution, abundance, and viability of many arctic bird species. Changes will be manifested through alterations in surface hydrology, increases in vegetation height, and shifts in lifecycle phenology of bird food resources (ACIA 2005). Continued monitoring of arctic bird populations to determine their response to changing environmental factors influenced by climate change will aid scientists, policymakers, and society in developing cost-effective mitigation actions that will maintain arctic avifaunal diversity. Anticipated changes in arctic marine and terrestrial systems will likely include increased industrial mining, oil and gas development, and international shipping, all of which can contribute additive stresses to arctic birds and their habitats. Understanding differences in shorebird abundance across the arctic will allow for development of effective land protection strategies to maintain populations of shorebirds and other tundra birds. Identification and protection of sedge–grass tundra refugia may become increasingly important in maintaining arctic shorebird diversity if current tundra habitats are altered by climate change.

CONCLUSION

The arctic plays a key role in the life cycle of many Western Hemisphere shorebird species, yet at present neither the Canadian nor the United States government has committed to a long-term program to monitor populations there. The suspected population declines occurring already, coupled with the impending threat of dramatic changes to climate and habitats, increase the need for such a program. Moreover, results to date demonstrate that the Arctic PRISM surveys, if carried out with the recommended sampling intensity, will achieve their objective of delivering reliable information on distributions, population sizes, and population trends. Continued implementation of Arctic PRISM will meet the monitoring needs for most arctic-breeding shorebirds; measuring population trends of shorebirds on their arctic breeding grounds is a critical step toward evaluating ongoing conservation efforts for this group of birds.

ACKNOWLEDGMENTS

C. M. Francis, C. Hickey, and D. B. Lank provided valuable comments on an earlier draft of this foreword.

SUSAN K. SKAGEN
Fort Collins Science Center
Fort Collins, Colorado

PAUL A. SMITH
Smith and Associates Ecological Research Ltd.
Pakenham, Ontario

BRAD A. ANDRES
U.S. Fish and Wildlife Service
Denver, Colorado

GARRY DONALDSON
Environment Canada
Gatineau, Québec

STEPHEN BROWN
Manomet Center for Conservation Sciences
Manomet, Massachusetts

PART ONE

Introduction

CHAPTER ONE

Goals and Objectives

Victoria Johnston and Jonathan Bart

We report results from shorebird surveys in the North American arctic, defined here as Bird Conservation Regions 2 and 3 of the North American Bird Conservation Initiative (http://www.nabci.net/International/English/bcrmap.html). The surveys estimate population size and trend and provide information on habitat relationships at the regional and arctic-wide scale (Table 1.1, Fig. 1.1). Of the 53 species of shorebirds that breed in the United States and Canada, 26 (47%) breed in the arctic in sufficient numbers that arctic surveys are an important part of monitoring programs for them (Donaldson et al. 2000, Brown et al. 2001; Table 1.1).

Arctic-breeding shorebirds are a diverse group that exhibit a wide range of migration, reproductive, and wintering strategies (Table 1.1). Some species migrate a short distance to the northern United States and southern Canada (e.g., Purple Sandpiper; for scientific names see Appendix C), while others undertake epic migrations to West Africa (e.g., Red Phalarope) or southern South America (e.g., Hudsonian Godwit, Red Knot). Some species migrate in huge flocks, while others trickle south singly or in small groups. There are monogamous, polygamous, and polyandrous breeders, and most habitats in the arctic provide nesting opportunities for shorebird species. Despite their different life history characteristics, all arctic shorebird species share two traits: (1) They are all migrants (none inhabit the arctic year-round) and (2) because of their migratory behavior, all are exposed to anthropogenic hazards at some point(s) in their life cycle.

For more than a decade, concern has been expressed that shorebird populations around the world may be in serious decline. For example, a major review concluded that 48% of 200 populations with known trends were in decline, whereas only 16% were increasing (International Wader Study Group 2003). Since this review, further evidence has accumulated indicating broad declines among shorebird populations. Nebel et al. (2008) reported declines during the past 24 years of 81% for shorebirds breeding in Australia and 75% for shorebirds present there during migration. Many studies have shown that migration counts of shorebirds in eastern and central North America have declined, especially along the eastern seaboard (Howe et al. 1989, Morrison et al. 1994, 2001a, Morrison and Hicklin 2001, Bart et al. 2007). Niles et al. (2008, 2009) reported that *rufa* Red Knot populations have declined more than 75% and that populations of other species using the Delaware Bay may have declined. Numerous studies at arctic breeding locations have also recorded declines in some species (Gratto-Trevor

Johnston, V., and J. Bart. 2012. Goals and objectives. Pp. 3–8 *in* J. Bart and V. Johnston (editors). Arctic shorebirds in North America: a decade of monitoring. Studies in Avian Biology (no. 44), University of California Press, Berkeley, CA.

TABLE 1.1

Arctic-breeding shorebird species whose populations will be estimated by the Arctic PRISM program and a summary of their life history traits.

Species	Species code	Breeding location[a]		Preferred breeding habitat[b]	Migration distance[c]	Wintering location[d]	Reproductive strategy[e]	N.A. population estimate[f]	Population trend[g,h]	Shorebird conservation plans priority species?[g,i]
		East–west	South–north							
Black-bellied Plover	BBPL	E, C, W	Mid-Hi	U	L	US, SA	Monog	200,000	↓↓	MC
American Golden-Plover	AMGP	E, C, W	Lo-Hi	U	L	SA	Monog	200,000	↓↓	HC
Pacific Golden-Plover	PGPL	W	Lo	U	S	US	Monog	50,000	↔	MC
Semipalmated Plover	SEPL	E, C, W	Lo-Hi	B	L	CA, SA	Monog	150,000	↓?	LC
Whimbrel	WHIM	C, W	Lo-Mi	W, U	L	US, SA	Monog	66,000	↓?	HC
Hudsonian Godwit	HUGO	C, W	Lo	W, U	L	SA	Monog	70,000	↓	HC
Bar-tailed Godwit	BARG	W	Lo	W, B	L	Australasia	Monog	90,000	↔	HC
Ruddy Turnstone	RUTU	E, C, W	Hi	B	L	US, SA	Monog	245,000	↓↓	HC
Black Turnstone	BLTU	W	Lo	B	S	AK to Mex	Monog	95,000	↔	HC
Rock Sandpiper	ROSA	W	Lo	B	S	AK to California	Monog	150,000	↓	HC
Purple Sandpiper	PUSA	E	Hi	B	S	US	Monog	25,000	↓?	MC
Red Knot	REKN	E, C, W	Hi	B	L	US, SA	Monog	120,000	↓↓	HI
Sanderling	SAND	E, C	Hi	B	L	US, SA	Polyan	300,000	↓↓	HC
Dunlin	DUNL	E, C, W	Mid	W	M	US, CA, East Asia	Monog	1,525,000	↓↓	HC
Semipalmated Sandpiper	SESA	E, C, W	Lo-Hi	W, U	L	SA	Monog	2,000,000	↓↓	MC
Western Sandpiper	WESA	W	Lo-Mid	C, U	L	US, CA	Monog	3,500,000	↓↓	HC
Least Sandpiper	LESA	E, C,	Lo-Mid	W, U	L	US, SA	Monog	700,000	↓↓	MC
White-rumped Sandpiper	WRSA	E, C, W	Mid-Hi	W	L	SA	Polyg	1,120,000	↔	LC

Baird's Sandpiper	BASA	E, C, W	Lo-Hi	C, U	L	SA	Monog	300,000	↓?	LC
Pectoral Sandpiper	PESA	E, C, W	Lo-Hi	W, U	L	SA	Polyg	500,000	↔	LC
Buff-breasted Sandpiper	BBSA	C, W	Lo-Hi	W, U	L	SA	Polyg	30,000	↓	HI
Long-billed Dowitcher	LBDO	W	Lo-Mid	W, U	M	US, Mex	Monog	400,000	↑	LC
Stilt Sandpiper	STSA	C, W	Lo-Mid	W	L	SA	Monog	820,00	↔	MC
Wilson's Snipe	WISN	E, C, W	Lo	W	S	US to SA	Monog	2,000,000	↓↓	MC
Red-necked Phalarope	RNPH	E, C, W	Lo-High	W	L	SA	Polyan	2,500,000	↓↓	MC
Red Phalarope	REPH	E, C, W	Lo-High	W	L	US	Polyan	1,250,000	↓↓	MC

[a] Breeding location: east (E), central (C), west (W), low arctic (Lo), middle arctic (Mid), high arctic (Hi).

[b] Breeding habitat: wetlands (W), uplands (U), barren (B), coast (C).

[c] Migration distance: long (L), medium (M), short (S).

[d] Wintering location: United States (US), Central America (CA), South America (SA), Mexico (Mex), Alaska (AK).

[e] Reproductive strategy: monogamous (Monog), polygamous (Polyg), polyandrous (Polyan).

[f] Morrison et al. 2006; but also see Bart and Smith, chapter 14, this volume.

[g] Donaldson et al. 2000; Brown et al. 2001; USSCP 2004.

[h] Population trend: significant population decline (↓↓), probable or nonsignificant population decline (↓), apparently stable populations or unknown (↔), conflicting information (↓?), probable or nonsignificant population increase (↑).

[i] Conservation concern: low concern (LC), moderate concern (MC), high concern (HC), highly imperiled (HI).

[j] From Arctic PRISM Tier 1 surveys.

References: Gratto-Trevor 1992, Parmalee 1992b, Lanctot and Laredo 1994, Wilson 1994, Paulson 1995, Johnson et al. 1996a, 1996b, Skeel and Mallory 1996, Warnock and Gill 1996, Holmes and Pitelka 1998, Klima and Jehl 1998, Mueller 1999, Nol and Blanken 1999, Donaldson et al. 2000, Nettleship 2000, Rubega et al. 2000, Takekawa and Warnock 2000, Brown et al. 2001, Handel and Gill 2001, Harrington 2001, McCaffery and Gill 2001, Elphick and Klima 2002, Gill et al. 2002, Macwhirter et al. 2002, Moskoff and Montgomerie 2002, Payne and Pierce 2002, Tracy et al. 2002, USSCP 2004, Morrison et al. 2006, Nebel and Cooper 2008.

Figure 1.1. Study area and study species (see Appendix C, this volume, for scientific names of all species).

1993, Gratto-Trevor et al. 1998, Jehl and Lin 2001, McCaffery et al. 2006, Jehl 2007, Johnston and Pepper 2009, Smith 2009).

Shorebirds are difficult to monitor because few species are well surveyed on their breeding grounds (due largely to access problems) and few migration or wintering regions of the world have rigorously designed surveys. When the North American shorebird initiative was initiated in the late 1990s, few survey data were available to use in estimating trends (Brown et al. 2001, Morrison et al. 2001a, 2001b). These factors, combined with other concerns about impacts of climate change, toxic pollutants, and habitat loss throughout the range of arctic-breeding shorebirds led American and Canadian shorebird biologists to create the Program for International and Regional Shorebird Monitoring (PRISM) in the late 1990s (Skagen et al. 2003). The goals of PRISM are to:

1. Estimate the size of breeding populations of shorebirds in North America;
2. Describe shorebirds' distribution, abundance, and habitat relationships;
3. Monitor trends in shorebird population size;
4. Monitor shorebird numbers at stopover locations; and
5. Assist local managers in meeting their shorebird conservation goals.

PRISM members proposed an arctic-wide program whose objectives were to produce estimates of population size and trends, along with information on habitat relationships, at regional and arctic-wide scales (Table 1.1). Obtaining population estimates, rather than just indices, was the most difficult, yet important, aspect of the Arctic PRISM sampling design. Most index surveys have numerous sources of potential bias, so it is difficult to know whether changes in survey results indicate changes in population size or might be due to changes in detection ratios (Williams et al. 2005). Knowing population size may also help managers judge how quickly a declining trend will put a species in conservation danger.

When fully implemented, the Arctic PRISM program will have three "tiers": Tier 1 includes the arctic-wide surveys to estimate population size and trends for each of the 26 focal species. The accuracy target for trend estimation adopted by the shorebird initiative, 80% power to detect a decline of 50% occurring in 20 years, will be adopted for Tier 1 surveys (accuracy targets for the estimates of population size have not been adopted by the shorebird initiative, Skagen et al. 2003). Tier 2 includes long-term studies of demographic rates and other aspects of natural history at permanent sites widely spaced across the arctic in Alaska and Canada. Tier 3 is a checklist program that includes annual collection of information on the distribution and abundance of shorebirds at sites visited throughout the arctic. Taken together, these programs will provide (1) continuous, suggestive—although not conclusive—indications of population change (Tier 3 checklists supplemented by counts outside the arctic during migration and wintering periods); (2) highly accurate—although expensive—estimates of trends, to be undertaken every 20 years or more often if other, less conclusive, evidence suggests serious declines (Tier 1 surveys); and (3) investigations of causes of declines and how they might be reversed (Tier 2 surveys).

In this monograph, we present results from the first ten years of Arctic PRISM surveys. The monograph is organized in four parts. The Introduction includes this chapter and a detailed discussion of the methods including the sampling plan, field methods, and analysis (Bart et al., chapter 2, this volume). Methods are not described in later chapters unless they differed from the methods described in Bart et al. (chapter 2, this volume).

The second part, Regional Reports, includes six chapters reporting results from all of the regions surveyed to date. The areas surveyed include four sites in western Alaska (McCaffery et al., chapter 3, this volume); the North Slope of Alaska (Bart et al., chapter 4, this volume), Yukon North Slope and Mackenzie Delta (Rausch and Johnston, chapter 5, this volume), Southampton and Coats Islands (Smith et al., chapter 6, this volume), Prince Charles Island, and Air Force Island, and western Baffin Island (Johnston and Smith, chapter 7, this volume), and briefer surveys in several widely distributed areas (Bart et al., chapter 8, this volume). The surveys in all of these areas used the same general approach, and all provided estimates of population size (number of shorebirds present in the surveyed area prior to hatch) that had little if any bias and that were accompanied by standard errors (SEs) and coefficients of variation (CVs).

The third part, Methodology, includes five chapters that cover topics not discussed in Bart et al. (chapter 2, this volume). During the past few years, we have realized that aerial surveys, flown in helicopters at low elevation and slow speeds, can be used to obtain extensive information on shorebird distribution and abundance. While the results cannot be used to estimate population size, they are of great value in developing and refining the sampling plans used to select plots. This method is discussed in Elliott and Smith (chapter 9, this volume). We have always known that a few species—especially those that are uncommon or have large home ranges—can be difficult to survey using our ground-based methods. Pirie and Johnston (chapter 10, this volume) addresses this issue for one such species, the Whimbrel, by comparing results from aerial and ground surveys. Pirie et al. (chapter 11, this volume) and Armer et al. (chapter 12, this volume) discuss the current status of Tiers 2 and 3 and suggest how these components of the overall Arctic PRISM approach may best be enhanced. Bart and Smith (chapter 13, this volume) discuss allocation of effort to parts of the sampling plan and present results from a simulation to estimate how much effort will be required to achieve the Arctic PRISM accuracy target for each species.

Part Four, Synthesis, includes three chapters beginning with a detailed summary of the results presented in regional chapters (Bart and Smith, chapter 14, this volume). This chapter will be an appropriate starting point for readers wishing the "big picture." Bart et al. (chapter 15, this volume) discusses priorities for the future, including both methodology and regions to survey.

The final section of the monograph contains more detailed information including an explanation of why the framers of Arctic PRISM did not believe that migration or wintering counts could achieve the PRISM accuracy target of 80% power to detect a 50% decline in 20 years (Appendix A). Extensive tables of estimates with measures of precision are provided in Appendix B. The entire data set has been posted on the Coordinated Bird Monitoring website (http://cbmdms.dev4.fsr.com/default.aspx) along with the comprehensive analytic program DS (for double sampling), prepared during the course of this study.

Arctic PRISM is one of the most ambitious monitoring projects ever undertaken for nongame species. It has involved dozens of principal investigators, hundreds of cooperators, and thousands of hours of field and analysis time. Controversies, particularly over the methods, have occurred (Arctic PRISM Peer Review Committee 2010), as is inevitable in large, novel programs. However, despite the difficulties, data have been collected throughout the arctic in North America, and as this volume goes to press, Canada has committed in principle to complete the first round of Arctic PRISM surveys and shorebird specialists in Alaska are discussing plans to continue the surveys in western Alaska, perhaps in 2012. We are therefore hopeful that the entire North American arctic will be surveyed using the PRISM methods, after which another round of evaluation should be carried out.

ACKNOWLEDGMENTS

Acknowledgments are included in each of the subsequent chapters. Here we wish to thank our agencies, the U.S. Geological Survey and the Canadian Wildlife Service, for their support of this work during the past decade. Also, we pay tribute to early support from the U.S. Fish and Wildlife Service, and especially to K. Wohl and B. A. Andres, for helping J. Bart and others develop the basic survey methods. When that work began, no one knew that within a few years a major shorebird monitoring initiative would begin and that, ten years later, the shorebird community would be able to produce this volume.

ONLINE CONTENT

Abstracts are available in French and Spanish from www.ucpress.edu/go/sab.
Une traduction du résumé est disponible en français.
Una traducción del resumen está disponible en español.

CHAPTER TWO

Methods

Jonathan Bart, Victoria Johnston, Paul A. Smith, Ann Manning, Jennie Rausch, and Stephen Brown

Abstract. Detecting declines in population size is one of the highest priorities of the shorebird initiatives in Canada and the United States. The quantitative goal is 80% power to detect a 50% decline, occurring during no more than 20 years, with a significance level of 0.15, using a two-tailed test, and incorporating effects of potential bias into the estimator. The Arctic PRISM program was designed to achieve this goal for arctic-nesting shorebird populations. The survey methods are an application of double sampling. Rapid surveys were made on a large number of plots selected from throughout arctic Alaska and Canada using stratified random sampling. Intensive surveys were made on a subsample of the plots to obtain detection rates, which were used to calibrate results from rapidly surveyed plots. Surveys will be made of the entire arctic region, each lasting several years and producing an estimate of average population size during the survey period. Results from two or more survey periods will be used to estimate change, or trend, in population size.

Key Words: arctic, monitoring, population size, PRISM, shorebirds.

A major goal of Arctic PRISM is to estimate change in shorebird population size occurring during 20 years with power of 80% to detect a 50% decline occurring in no more than 20 years, using a significance level of 0.15 and a two-tailed test, and acknowledging effects of potential bias (Skagen et al. 2003, Bart et al. 2005). This chapter describes the methods being used to achieve the desired power. We discuss the delineation of plots and strata, selection of plots to be surveyed, how the surveys were conducted, and analysis of the resulting data. Methods for analyzing habitat data varied between regions and are described in region-specific chapters (chapters 3–8, this volume). The methods described in this chapter were first developed in northern Alaska during 1997–2001. The basic approach was described by Bart and Earnst (2002). A detailed description of the field methods, including training, was provided by Bart and Earnst (2005) and has been expanded into a comprehensive Arctic PRISM manual (Rausch, Canadian

Bart, J., V. Johnston, P. A. Smith, A. Manning, J. Rausch, and S. Brown. 2012. Methods. Pp. 9–16 *in* J. Bart and V. Johnston (editors). Arctic shorebirds in North America: a decade of monitoring. Studies in Avian Biology (no. 44), University of California Press, Berkeley, CA.

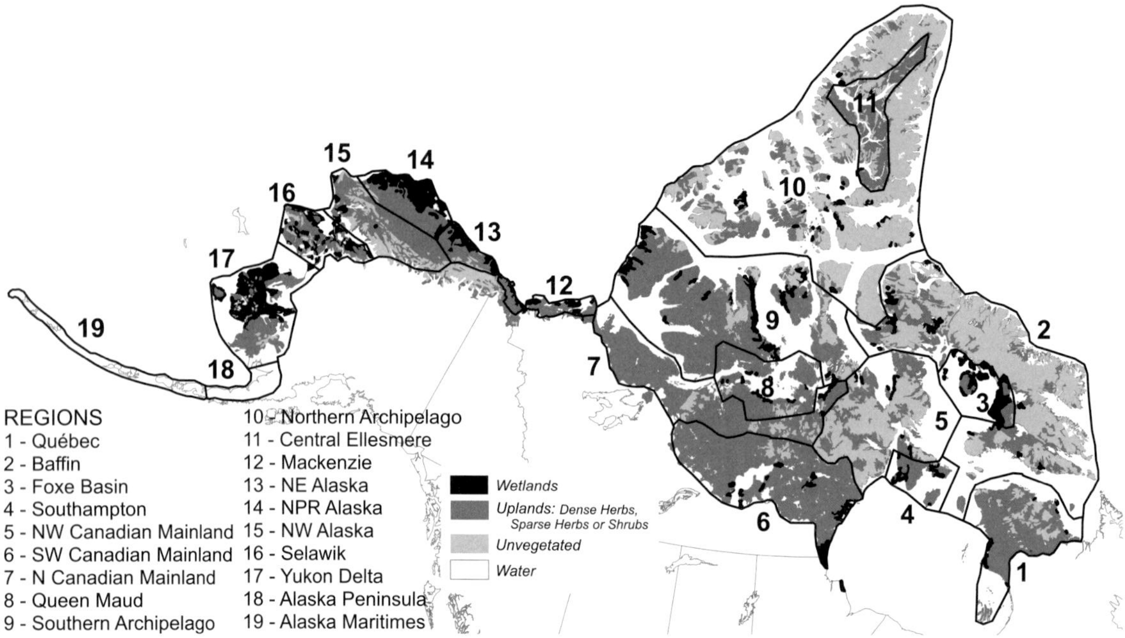

Figure 2.1. Study area, regions, and major habitats as depicted on the Circumpolar Arctic Vegetation Map (CAVM; CAVM Team 2003).

Wildlife Service, unpubl. report). In this chapter, we concentrate on the quantitative methods involving the use of double sampling, a topic which has not previously been fully described.

The study area is the arctic portion of North America, as delineated on the Circumpolar Arctic Vegetation Map (CAVM; CAVM Team 2003), with modifications to exclude some mountainous areas (Fig. 2.1). The study area was partitioned into 19 regions (Fig. 2.1) based on logistic considerations. Prior to conducting surveys in each region, the region was further divided into subregions on the basis of how much area could be covered by field crews and anticipated density of shorebirds. Large areas not suitable for nesting shorebirds, such as oceans, lakes, and barren areas, were excluded. Care was taken not to exclude barren areas unless we had convincing evidence that birds did not occur in them.

DOUBLE SAMPLING

A form of double sampling (Cochran 1977) was used to estimate density and population size. A large sample of plots was surveyed with a rapid method of unknown accuracy, and a subsample of these plots was surveyed intensively to determine actual numbers present. The ratio of the results from the rapid and intensive surveys was used to estimate the "detection ratio" and to adjust the results from the rapid surveys (Bart et al. 1998).

Compared to many bird survey methods, double sampling requires only a few assumptions. Specifically, double sampling produces unbiased estimates of population size, and thus trend in population size, if the nominal sampling plan is followed and if the counts on intensive plots yield unbiased estimates of numbers present. No other assumptions are required. Note that counts on the intensive plots do not each have to be completely accurate; they only have to be unbiased (overestimates must just balance underestimates). Note, too, that failure to follow the nominal sampling does not necessarily cause any bias in estimates; it only makes such bias possible. Finally, bias in population size estimates also does not necessarily cause any bias in trend estimates; it only makes such bias possible. Bias in the trend estimate is algebraically equal to trend in the detection ratio, 'expected value of the estimated population size/actual population size'. Thus, if the expected value of the estimated population size was 15% too low during the first set of surveys and was 10% too low during the second set of surveys, then bias in the trend estimate would be 5%. If the population actually declined by 50%, then the expected value of the estimated change would be a 45% decline. Thus, small changes in the detection ratio, due either to non-random selection of intensive plots or to errors on the intensive plot surveys, cause little bias in the trend estimate in relation to the large change (50% decline) that the surveys were designed to detect. Stated a different way, a bias of 5–10% causes little change in the power to detect a 50% decline. See Bart et al. (2004) for a more complete discussion of this issue and quantification of how bias affects power.

PLOT SELECTION AND SURVEYS

Each subregion was partitioned into plots, most of which covered 0.12–0.16 km^2 and were designed to be surveyed in 1.0–1.5 hours. Plots were substantially larger in some early years of the study. We defined "wetland," "moist," and "upland" habitats in each subregion (see region-specific chapters, this volume, for definitions) and calculated the proportion of each plot covered by each habitat type. Plots with no habitat suitable for nesting birds were deleted. Plots with only small amounts of wetland, moist, or upland habitat that were primarily covered by water were unsatisfactory due to high travel costs (it did not make sense to spend hundreds or thousands of dollars to visit a very small plot). We therefore combined small plots with surrounding plots to reduce the variance in plot size.

Plots were assigned to wetland, moist, and upland habitat types. Habitat types were used (along with region) to define strata; thus each plot had to be assigned to a single habitat type. The rules used to make these assignments varied across the arctic because the extent of different habitats varied substantially. In many cases, plots were assigned to the type corresponding to the habitat that covered the largest fraction of the plot (e.g., if wetland habitat covered more than 50% of the plot, the plot was assigned to the wetland plot type). If wetland habitat was rare within a subregion, a different rule was used; for example, 'if wetlands habitat covers more than 20% of the plot, then assign the plot to the wetland plot type; otherwise, assign the plot to the type corresponding to the habitat that covers the largest fraction of the plot'. The rules used in each region are described in more detail in the regional reports (chapters 3–8, this volume).

The sampling plan for selecting plots to survey involved stratification using subregion and plot type (wetland, moist, upland), followed by selection of clusters of plots and then selection of plots. We selected plots to survey in groups to reduce distances between plots being surveyed at the same time. These groups usually included plots in different habitats, and thus in different strata. Selection of plots was thus not independent in different strata (i.e., plots in different strata were close together much more often than if we had used independent selection). We referred to the groups of plots as "zones" (Fig. 2.2) to distinguish them from clusters, which, by definition, are plots in the same strata. Most zones covered 4–36 km^2 and comprised 25 to a few hundred plots. We acknowledged the lack of independence in selecting plots in different strata by modifying the standard formulas for cluster sampling (see below). Some reviewers have had difficulty grasping why we had to define zones, but had we ignored the lack of independence caused by selecting plots in different strata within zones, our variance estimates would have had substantial negative bias. Zones to be surveyed were selected systematically to ensure even coverage across the subregion. Simple random sampling was used to select plots within clusters.

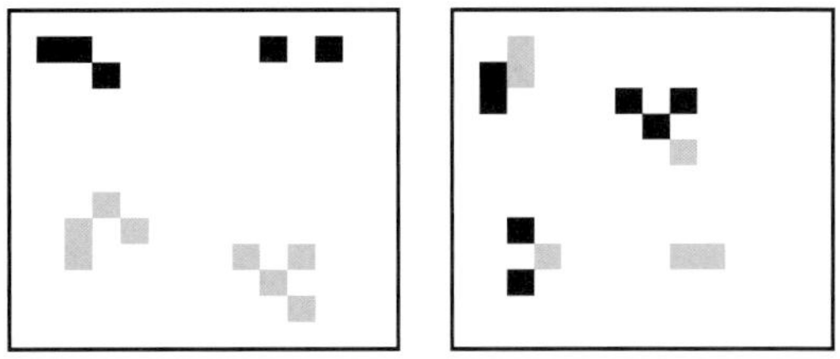

Figure 2.2. Standard cluster sampling (left) in which all plots in each cluster are in one stratum (indicated by shading) and our sampling plan, in which plots in a cluster were often in different strata (right). We termed the groups of plots (right) "zones." Plots of the same habitat type, within a zone, were a cluster.

In the early years of the study, we attempted to carry out the steps above by hand. With large subregions, this was not possible and we were forced to use short-cut methods which inevitably caused us problems later in the analysis. We therefore prepared a series of ArcGIS tools, collectively referred to as the Arctic PRISM ArcGIS extension (Table 2.1), to automate delineation of plots and assignment of plots to clusters, zones, and strata. This tool was essential for partitioning large regions into plots. It is available free from the senior author and may be useful to others who need to define a rigorous sampling frame for large, heterogeneous areas.

Random selection of locations for intensive surveys, which is part of the double sampling

TABLE 2.1

Description of the tools in the Arctic PRISM extension of Arcview.

Tool	Function
Create zones and plots	Generates sampling plots contained within larger zones (in polygon shapefile format).
Merge zones and plots	Merges plots and/or zones that contain less than a specified area of suitable habitat.
Plot habitat summary	Adds fields to the plot attribute table representing the proportion of each plot occupied by each habitat type.
Assign plot types	Codes a field in the plot attribute table using habitat types within each plot.
Cluster area	Produces a comma-delimited text file summary of the plots.
Plot selection	Creates a random sample of plots, stratified by plot type.

protocol, proved to be impossible. We found that many selected zones (groups of plots that we might have used for intensive surveys) lacked a suitable campsite close enough to the plots or the selected plots had too few birds to be able to compute detection ratios. These problems eventually forced us to select intensive camp locations nonrandomly on the basis of logistics and expectations about shorebird density. Selection of plots around camps followed a similar method. Areas thought likely to have shorebirds were first identified and, when possible, a random selection was made of locations for the plots. When the habitat was variable, an effort was made to distribute plots across habitats. When initial surveys failed to reveal territorial shorebirds, the plots were moved, though this happened only occasionally. Because intensive plots were not selected randomly, results of rapid surveys on intensive plots were not combined with results from other rapid surveys. The results from rapid surveys and the estimated detection rates were thus independent, a fact that had implications for variance calculations, as discussed below. Although we were disappointed not to be able to select intensive plots in a fully random manner, analyses of detection rates showed little variation in either habitat or density, suggesting that the selection bias, if any, in our estimates of detection rates was small.

Field work was conducted during 1998–2006 at numerous locations widely distributed across the study area. Rapid surveys used time-constrained, area search methods. Total time per plot varied among regions but was usually about 1 hour in Alaska and 2 hours in Canada. Observers covered each plot thoroughly, walking transects 25 m apart when the habitat was uniform and following irregular paths when waterbodies or other obstacles were common. When habitats were variable, more time was spent in areas where birds were thought to be more common, but all parts of each plot were covered. Surveyors recorded "*indicated pairs*" (nests, probable nests, single birds by sex) on plot maps and summarized their observations in tabular form immediately after each survey (before continuing to the next plot). In the analyses, we assumed that each indicated pair represented two birds.

During most years and in most survey areas, 1–2 camps were established for intensive surveys (at previously selected sites for random surveys when feasible). At each camp, 4–6 plots were established. Two to four surveyors spent 3–5 weeks visiting these plots every 1–2 days, searching for nests and non-nesting birds. Search effort was usually more than 30 hours per plot and was greater than 50 hours per plot in a few areas where shorebird density was extremely high. The intensive searches began as birds returned to the plots and continued past hatch. Surveyors attempted to find all active shorebird nests and to identify all other territorial shorebirds.

The issue of how much effort, on intensive surveys, is required to find nearly all territorial birds was investigated in our study area by Smith et al. (2009). Surveyors searched for shorebird nests at four locations, widely distributed across the arctic. At each location, two or more teams, working independently, conducted surveys of the same plots. At the conclusion of the surveys, the teams compared results to determine how many nests each had missed. Results were then used to model the likelihood of missing a nest as a function of number and timing of visits, species, and other factors. After five visits, the estimated proportion of nests found was 0.88 for the most difficult species to detect and 1.00 for the easiest species to detect. These results show that nearly all nests surviving for at least 7–10 days are found, but nests that survive only a few days may not be found. The study by Smith et al. (2009) did not assess the overall likelihood of detecting territorial birds because they only recorded nests found. During all Arctic PRISM surveys made through 2006, nests were not found for 22% of all birds recorded. This fact, in combination with the results obtained by Smith et al. (2009), suggests that the total detection rate of territorial birds was close to 1.0. As noted above, even if the intensive surveys were biased, this does not necessarily cause any bias in trend estimates (though bias in population size estimates does occur).

ESTIMATORS

For notational simplicity, we assume that population size is the same in each year during each survey period. Population declines, or other changes, during the survey period reduce precision but cause no bias. This can be seen by adding a subscript to indicate year and then showing that random selection of which areas to survey each year guarantees that the estimate, assuming it is unbiased within the year, also yields an unbiased estimate across years.

For a given species and period, the estimate of population density (d; birds per km^2) was

$$d = \frac{\hat{\bar{X}}}{\hat{R}} \tag{1}$$

where $\hat{\bar{X}}$ is an estimate of the mean density of birds that would have been recorded if an indefinitely large sample of rapid surveys had been conducted and $\hat{R}$ is an estimate of the detection ratio (birds recorded/birds present) on the rapid surveys. $\hat{\bar{X}}$ was obtained from the rapid surveys; $\hat{R}$ was obtained from the intensive surveys. From the standard equation for the estimated variance of a ratio of independent random variables (Cochran 1997),

$$\hat{V}(d) = d^2\left(\frac{\hat{V}(\hat{\bar{X}})}{\hat{\bar{X}}^2} + \frac{\hat{V}(\hat{R})}{\hat{R}^2}\right) \tag{2}$$

The estimated population size was

$$\hat{Y} = Ad \tag{3}$$

where A is the size of the study area. The variance of $\hat{Y}$ was estimated as

$$\hat{V}(\hat{Y}) = A^2\hat{V}(d). \tag{4}$$

To derive estimators for the terms in expression (2), let

$\bar{z}_{uhi}$ = mean number of birds recorded per plot in the ith cluster of type h plots in region u

$\bar{b}_{uhi}$ = mean area covered per surveyed plots in the ith cluster of type h plots in region u

a_{uhi} = area covered by all plots in the ith cluster of type h plots in region u

n_{uh} = number of clusters of type h plots surveyed in region u

N_{uh} = number of clusters of type h plots in region u

$\hat{\bar{X}}$ was estimated using the "combined approach" (Cochran 1977) for ratios with stratification:

$$\hat{\bar{X}} = \frac{\hat{Z}}{\hat{a}} = \frac{\sum_u^U \sum_h^H \frac{N_{uh}}{n_{uh}} \sum_i^{n_{uh}} a_{uhi}(\bar{z}_{uhi}/\bar{b}_{uhi})}{\sum_u^U \sum_h^H \frac{N_{uh}}{n_{uh}} \sum_i^{n_{uh}} a_{uhi}} = \frac{\sum_u^U \sum_h^H \frac{N_{uh}}{n_{uh}} \sum_i^{n_{uh}} \hat{Z}_{uhi}}{\sum_u^U \sum_h^H \frac{N_{uh}}{n_{uh}} \sum_i^{n_{uh}} a_{uhi}} \tag{5}$$

The quantity $\hat{Z}_{uhi} = a_{uhi}\ (\bar{z}_{uhi}/\bar{b}_{uhi})$ is the estimated number of birds that would be recorded if all the plots were surveyed in the ith cluster of type h plots in region u. The numerator in expression (5) is the mean of the $\hat{Z}_{uhi}$ times the number, N_{uh}, of clusters in the stratum. The numerator may thus be viewed as an estimate of the number of birds that would be recorded on rapid surveys if all plots in all regions were surveyed. The denominator is an estimate of the total area, based on the surveyed plots. The ratio is thus an estimate of density (uncorrected for the detection rate). The rationale for this estimator may be explained as follows. If the mean area of the surveyed zones is larger than the mean area of all zones, then the numerator will tend to exceed the true number of birds present, but the denominator will also tend to be greater than the true area, so the ratio will tend to be closer to the actual density.

It will be convenient in deriving the variance estimator to express $\hat{\hat{X}}$ as

$$\hat{\hat{X}} = \frac{\hat{Z}}{a} = \frac{\sum_u^U \sum_h^H N_{uh} \hat{\bar{Z}}_{uh}}{\sum_u^U \sum_h^H N_{uh} \bar{a}_{uh}} \quad (6)$$

where $\hat{\bar{Z}}_{uh}$ and $\bar{a}_{uh}$ are the means of the $\hat{Z}_{uhi}$ and a_{uhi}, respectively, in stratum u-h (i.e., region u and type h). $\hat{V}(\hat{\hat{X}})$ was estimated using the standard formula for the estimated variance of a ratio of correlated random variables (Cochran 1997),

$$\hat{V}(\hat{\hat{X}}) = \left(\frac{\hat{Z}}{a}\right)^2 \left(\frac{\hat{V}(\hat{Z})}{\hat{Z}^2} + \frac{\hat{V}(a)}{a^2} - \frac{2\hat{Cov}(\hat{Z},a)}{\hat{Z}a}\right). \quad (7)$$

As noted above, sampling in different strata was not independent. To acknowledge this dependence, let the subscripts g and h indicate type, n_{ugh} = the number of zones in region u in which at least one type g plot and one type h plot were surveyed, and $\hat{\bar{Z}}_{ug|b}$ = the mean of the $\hat{Z}_{ugi}$ among the n_{ugh} plots. The "$|b$" (for "both") notation means the sum is restricted to zones in which both types g and h were surveyed. Let $\hat{\bar{Z}}_{uh|b}$, $\bar{a}_{ug|b}$ and $\bar{a}_{uh|b}$ be defined in a similar manner. With this notation, and using standard survey sampling methods, it may be shown that

$$\hat{V}(\hat{Z}) = \sum_u^U \sum_g^H \sum_h^H N_{ug} N_{uh} \frac{n_{ugh}}{n_{ug} n_{uh}} \sum_i^{n_{ugh}} (\hat{Z}_{ugi} - \hat{\bar{Z}}_{ug|b}) \times (\hat{Z}_{uhi} - \hat{\bar{Z}}_{uh|b}) / (n_{ugh} - 1), \quad (8)$$

$$\hat{V}(a) = \sum_u^U \sum_g^H \sum_h^H N_{ug} N_{uh} \frac{n_{ugh}}{n_{ug} n_{uh}} \sum_i^{n_{ugh}} (a_{ugi} - \bar{a}_{ug|b}) \times (a_{uhi} - \bar{a}_{uh|b}) / (n_{ugh} - 1), \quad (9)$$

$$\hat{Cov}(\hat{Z}, a) = \sum_u^U \sum_g^H \sum_h^H N_{ug} N_{uh} \frac{n_{ugh}}{n_{ug} n_{uh}} \sum_i^{n_{ugh}} (\hat{Z}_{ugi} - \hat{\bar{Z}}_{ug|b}) \times (a_{uhi} - \bar{a}_{uh|b}) / (n_{ugh} - 1). \quad (10)$$

When $n_{ugh} = 1$, the corresponding term in expressions (8)–(10) cannot be evaluated even though the variance or covariance the term represents does exist. Models could be used to estimate the missing terms (e.g., using the mean of the terms for which $n_{ugh} > 1$), but we have not investigated this approach and therefore omitted terms in expressions (8)–(10) when n_{ugh} was 1.

The detection ratio, R, was estimated as

$$\hat{R} = \frac{\bar{x}}{\bar{y}} \quad (11)$$

where $\bar{x}$ is the mean number of birds recorded on rapid surveys of the intensive plots and $\bar{y}$ is the mean number of birds determined to be present on these plots through intensive surveys.

Let m = the number of camps, $\bar{x}_i$ = the mean number recorded per rapid survey at camp i (>1 rapid survey was made at intensive plots to increase precision), and $\bar{y}_i$ = the mean number actually present at all plots in camp i, then $\bar{x} = \Sigma\bar{x}_i/m$ and $\bar{y} = \Sigma\bar{y}_i/m$. The $\bar{x}_i$ were calculated as the simple means of the means/plot because sometimes plots at a camp were not all surveyed the same number of times by rapid surveyors. Camps were widely distributed across the study area, so they were treated as a simple random sample (rather than a stratified random sample). Under this assumption, the variance of $\hat{R}$ was estimated as

$$\hat{V}(\hat{R}) = \hat{R}^2 \left(\frac{\hat{V}(\bar{x})}{\bar{x}^2} + \frac{\hat{V}(\bar{y})}{\bar{y}^2} - \frac{2\hat{Cov}(\bar{x}, \bar{y})}{\bar{x}\bar{y}}\right) \quad (12)$$

where

$$\hat{V}(\bar{x}) = \frac{1}{m} s^2(\bar{x}_i), \hat{V}(\bar{y}) = \frac{1}{m} s^2(\bar{y}_i), \hat{Cov}(\bar{x}, \bar{y}) = \frac{1}{m} cov(\bar{x}_i, \bar{y}_i)$$

and s^2 and $\hat{cov}$ indicate the sample variances and covariances, respectively.

Estimates will often be needed within a stratum, habitat, or region. The point and interval estimators for such cases are easily derived from the expressions above. Within a single stratum we have simple random sampling. From (6), $\hat{\hat{X}}_{uh} = \hat{\bar{Z}}_{uh}/\bar{a}_{uh}$. The estimated variance has the same structure as expression (7), but $\hat{V}(\hat{\bar{Z}}_{uh}) = s^2(\hat{Z}_{uhi})/n_{uh}$, $\hat{V}(\bar{a}_{uh}) = s^2(a_{uhi})/n_{uh}$. For the mean within one habitat, across >1 region, expressions (6)–(10) apply, except that g and h are equal and constant, so expressions (8)–(10) simplify:

$$\hat{V}(\hat{Z}) = \sum_u^U N_{uh}^2 s^2(\hat{Z}_{uhi})/n_{uh}, \quad (13)$$

$$\hat{V}(a) = \sum_u^U N_{uh}^2 s^2(a_{uhi})/n_{uh}, \quad (14)$$

$$\hat{Cov}(\hat{Z}, a) = \sum_u^U N_{uh}^2 cov(\hat{Z}_{uhi}, a_{uhi})/n_{uh}. \quad (15)$$

The estimate for >1 habitat within a single region is provided directly by expressions (6)–(10), with the sums in u having a single value. For all three cases, the equations for the detection rate and its variance do not change except that the expressions may have fewer terms.

A comprehensive Windows-based program, DS, was written to carry out all of the calculations described above. DS produces estimated densities and population totals, along with standard measures of precision, by habitat, region, and species. It has many features to facilitate analysis. For example, any set of species can be used to estimate the detection ratios for any species. It is available free from the senior author along with a detailed user's manual.

DETECTION RATIOS

As noted immediately above, program DS allows the user to specify which species will be used in estimating the detection ratio for each species. It might seem that species-specific ratios should be estimated whenever feasible, but how do we define "feasible"? If intensive plots did not contain any pairs of a species, then we obviously cannot obtain a species-specific detection ratio. If the intensive plots only contained one or two pairs, then even though we could obtain a species-specific estimate, it would not be very useful because it would have a huge SE. Furthermore, species identity is only one factor that might affect detection ratios. When we compare densities across regions or habitats, we may be misled if detection ratios differ, but we use a combined rate. This problem is no different from those analysts face with any method for estimating detection ratios. For example, distance methods require 70 or more detections, yet, for many species (especially in specific habitats and regions), the number detected may be far smaller, so data must be combined across species (or regions and habitats) to obtain detection ratios. More generally, this is an example of a common statistical problem: how many "parameters" (in our case, detection ratios) to define. A common approach, and the one we largely follow in this monograph, is to perform a comprehensive test such as an analysis of variance (ANOVA) to determine whether we have good reasons for rejecting the null hypothesis that the rates are equal. Failing to reject the null hypothesis doesn't mean we conclude that the rates are equal; it just means we conclude that the data set is not large enough to justify calculation of separate rates. As a practical matter, if all of the confidence intervals overlap, then it is very unlikely that the ANOVA and subsequent pairwise tests will support calculation of any species-specific rates. In subsequent chapters of this volume we often report whether any of the detection rates are significantly different and, if they are not (and especially if they are not even close to being significantly different), then we generally use the combined detection ratio. One exception, however, is that if we believe on the basis on biological information that the true detection ratios are quite different, then we often report the densities and population sizes using both the combined ratio and the species-specific ratio.

WHY DOESN'T POPULATION SIZE EQUAL THE SUM OF STRATUM-SPECIFIC ESTIMATES?

Readers will notice that estimated population sizes for two or more regions do not usually equal the sum of the region-specific estimates. Often the two figures are quite different. This is a standard problem in survey sampling that arises when a ratio estimator is used in each of several strata. The presence of a random variable in the denominator causes the estimate to be biased. As a simple example, suppose we were trying to estimate the inverse of mean plot size, as is the case in expression (6). To keep the example simple, suppose plots were actually of size 1 or size 9 (the units do not matter), and that the two sizes were equally common. The mean size is thus 5, so the inverse is 0.2. Now suppose we tried to estimate the inverse with a sample of size 1. We might get a plot of size 1, in which case our estimate of the inverse would be 1, or we might get a plot of size 9 in which case our estimate would be 0.11. Since the two sizes are equally common, the mean of the possible estimates (i.e., its expected value) would be $(1.0 + 0.11)/2 = 0.55$, which is substantially different from the true value, 0.2. If we made the plot sizes 1 and 2, then the difference between the expected value and true value (i.e., the bias) would be much smaller. If we used plot sizes of 1 and 9 but increased the sample size to 2, then the bias would also be smaller. More generally, the bias in a ratio estimate decreases with sample size and when the variance of the random variable in the denominator decreases.

With ratio estimates in stratified sampling, and either small sample sizes or large variance of the random variable in the denominator, the bias in each stratum can be considerable. If an overall estimate is computed by adding up stratum-specific estimates, the bias also sums and can become quite large when estimates are summed over several strata. To avoid this problem, estimates are usually made by summing the numerators and dividing the result by the sum of the denominators. Thus, we might compute ("sum of birds")/("sum of areas"), known as the "combined estimate," rather than "sum of birds/areas," known as the separate estimate.

With our estimator (e.g., expression 6), the important random variable in the denominator is a_{uhi}, the area covered by all plots in the *i*th cluster of habitat *h* plots in region *u*. This quantity often varies substantially across clusters in a given habitat and region, and this variation would generate substantial bias if we used the separate approach in estimating population sizes. We therefore used the combined approach to reduce the bias, even though this means that the estimated population size for a region with several strata does not equal the sum of the stratum-specific estimates. The combined approach yields somewhat larger SEs than using the separate approach but much smaller bias. We consider the approach "conservative" in that sense.

CONCLUSION

When we began the study we expected to use distance, double-observer, or some other well-developed method for estimating detection rates and obtaining unbiased estimates of population size. We found, however, that some species responded to us strongly (which violates distance assumptions); in some plots the best habitat covered a small proportion of the area, so random selection did not seem feasible (which violates all methods that require random selection of points); and during parts of the season some birds had not arrived or some had already left (which meant that no single-visit survey could produce an accurate count). These problems forced us to adopt double sampling, a method that includes intensive efforts to find all the birds but requires few assumptions and thus is unaffected by the problems mentioned above. As noted in this chapter, the only assumptions are that the sampling plan is followed and that the intensive estimates are unbiased. The one remaining statistical problem was that standard cluster sampling was not very efficient; we needed to have plots in different habitats in close proximity. This necessitated the modification to typical cluster sampling described above in expressions (8)–(10). Use of double sampling also allowed us to avoid the use of index methods, which, in turn, means that the surveys can be modified as new analytic and field methods appear. We believe this is a major strength of the approach and will be appreciated by those who design the survey in the coming decades.

ACKNOWLEDGMENTS

Many people helped develop the Arctic PRISM methods. Initial, critical support was provided by K. Wohl and B. A. Andres, U.S. Fish and Wildlife Service. S. Earnst was involved for the first several years and played a key role, especially in developing methods for the intensive surveys. B. T. Collins, C. Elphick, P. Geissler, C. Handel, R. Stehn, and two anonymous reviewers provided detailed comments on the method in 2005 in a peer review plan organized by B. Peterjohn. Their comments led to many improvements in the method. Discussions about how best to conduct the surveys were held more or less continuously in the camps and helped identify problems, especially logistic ones that we later addressed through changes in the design. These people are too numerous to list here but are acknowledged in later chapters. We thank S. Schulte, S. Brown, and an anonymous reviewer for comments on the penultimate draft. Finally, we acknowledge B. A. Andres, S. K. Skagen, and G. Donaldson, in their roles as U.S. Shorebird Coordinator, PRISM Coordinator, and Canada Shorebird Coordinator, for their support through the many years that Arctic PRISM has taken to develop.

ONLINE CONTENT

Abstracts are available in French and Spanish from www.ucpress.edu/go/sab.
Une traduction du résumé est disponible en français.
Una traducción del resumen está disponible en español.

PART TWO

Regional Reports

CHAPTER THREE

Shorebird Surveys in Western Alaska

Brian J. McCaffery, Jonathan Bart, Catherine Wightman, and David J. Krueper

Abstract. Surveys for breeding shorebirds were conducted during 2001–2002 in four National Wildlife Refuges (NWRs) in western Alaska—Alaska Maritime, Alaska Peninsula, Yukon Delta, and Selawik. The sizes of our study areas on and adjacent to these four refuges were 9,243 km^2, 24,493 km^2, 853 km^2, and 15,170 km^2, respectively. Eleven sites were selected non-randomly: three in the Alaska Maritime NWR, six in the Alaska Peninsula, and one each in the other two NWRs. Survey and analytic methods are described in Bart et al. (chapter 2, this volume). Rapid surveys were conducted on 224 plots; 2,163 indicated pairs of shorebirds were recorded, of which 1,485 were judged to be nesting in the surveyed plots. Detection ratios were estimated using intensive plot data from northern Alaska as well as from two plots on the Yukon Delta NWR. The highest estimated densities (shorebirds/km^2) were on the Yukon Delta Study Area: 416 in wetlands and 300 in moist areas. The estimated densities on the Alaska Peninsula study area were 118 in wetlands and 62 in uplands. Other densities were markedly lower. Estimated numbers of shorebirds were 62,000 (CV = 0.58), 1,804,000 (CV = 0.32), 310,000 (CV = 0.11), and 390,000 (CV = 0.35) in the Alaska Maritime, Alaska Peninsula, Yukon Delta, and Selawik study areas, respectively. The former two estimates were affected by selection bias of unknown magnitude and so should be regarded with caution. A small estimate was generated for the Yukon Delta study area because it covered only about 1% of the Yukon Delta NWR. We identify several species-specific estimates from our study which appear inconsistent with previous continental estimates. This pilot study provides preliminary estimates of species composition and density in the surveyed areas. By incorporating several region-specific modifications to the sampling protocols for future surveys, we believe that the Arctic PRISM method is suitable for covering large areas in western Alaska.

Key Words: Alaska, Alaska Maritime, Alaska Peninsula, Density, Dunlin, monitoring, National Wildlife Refuge, phalaropes, population size, PRISM, sandpipers, Selawik, shorebirds, turnstones, Yukon Delta, Yukon–Kuskokwim Delta.

McCaffery, B. J., J. Bart, C. Wightman, and D. J. Krueper. 2012. Shorebird surveys in western Alaska. Pp. 19–36 *in* J. Bart and V. Johnston (editors). Arctic shorebirds in North America: a decade of monitoring. Studies in Avian Biology (no. 44), University of California Press, Berkeley, CA.

Western Alaska supports one of the richest tundra shorebird faunas in the world. In western North America, shorebird species richness peaks in Alaska between 60° and 65° north latitude, roughly the zone from the mouth of the Kuskokwim River north to the Seward Peninsula (Pitelka 1979). Not only does the region host a diverse suite of breeding species and populations that are endemic to Beringia (Alaska Shorebird Group 2008), it also supports some of the highest local breeding densities of shorebirds in the world (Meltofte et al. 2007b).

Most estimates of breeding shorebird densities in western Alaska have been derived from two types of studies. The first type includes site-specific studies, often characterized by small plots that were not randomly selected, wet and/or moist habitats, and shorebird communities numerically dominated by the smaller species (particularly phalaropes and *Calidris* sandpipers; Holmes 1970, 1971; Schamel et al. 1979; Gill and Handel 1990; McCaffery and Ruthrauff 2004). The methods in these studies included transect surveys, plot inventories, and area searches. Some consisted of simple counts, while others incorporated information on breeding behavior and re-sightings of marked birds. Only one such study was expanded to generate a larger-scale population estimate that incorporated a regional sampling design, stratification, and double sampling (Handel and Gill 1992).

A second class of studies with the potential for generating estimates of breeding shorebird densities and/or regional population sizes in western Alaska (i.e., in and adjacent to Bird Conservation Region 2; NABCI 1998) has been conducted by the U.S. Geological Survey, primarily on federal lands (including National Parks, National Wildlife Refuges, and on Bureau of Land Management holdings). Most of these studies were characterized by the use of 10 km × 10 km sample units within large geographical areas, count points with detection probabilities estimated through distance sampling, stratified random designs, and a primary focus on upland and montane habitats (Tibbitts et al. 2005, Ruthrauff et al. 2007, Ruthrauff and Tibbitts 2009). These habitats typically supported very different shorebird communities than the small, local plot studies described in the preceding paragraph.

In 2001 and 2002, we and our colleagues at four National Wildlife Refuges (NWRs) in western Alaska initiated surveys that combined these two general approaches to estimate shorebird numbers and describe distribution. We used a double-sampling design, with small plots selected from relatively large geographic areas. Our goals were to (1) derive statistically defensible regional density and population estimates for shorebirds in the areas we surveyed, (2) obtain new information about the distribution and habitat associations of shorebirds, and (3) develop methods that could be used to conduct more comprehensive shorebird surveys in western Alaska. In the text below, we refer to the study areas as the Alaska Maritime study area (AMSA), the Alaska Peninsula study area (APSA), the Yukon Delta study area (YDSA), and the Selawik study area (SSA). Geographic regions and NWRs are referred to using their descriptive names (e.g., Yukon Delta, Yukon Delta NWR).

METHODS

In each study area, one or more sites were selected non-randomly (Fig. 3.1). In each site, one or more clusters of plots were selected. We attempted to select clusters randomly at each site, but logistic constraints prevented us from doing so on the AMSA and APSA. Plots were then randomly selected within each cluster and were surveyed using the rapid method described in Bart et al. (chapter 2, this volume).

The AMSA included only the Aleutian Islands. Two observers traveling on a U.S. Fish and Wildlife Service vessel were left ashore for short periods at three sites (Fig. 3.1), from west to east: Adak Island, Amlia Island, and Avatanak Island. Sites were selected based on the route and schedule followed by the ship, and by accessibility. Completely random selection of sites was not feasible due to logistic constraints, but we tried to make the sites representative of low- and mid-elevation areas. At each site, areas for the surveys were delineated and one (Adak and Avatanak Islands) or two (Amlia Island) clusters of 16-ha plots were randomly selected with the constraint that plots had to be separated by at least 500 m. Sixty plots were each surveyed once during 15 May–2 June 2002. We tried to survey Black Oystercatchers (for scientific names, see Appendix C) on the shore from high vantage points but found that fog and rugged topography made accurate counts impossible. We believe that ocean-based surveys using small boats would be the most effective method for surveying Black Oystercatchers in this study area.

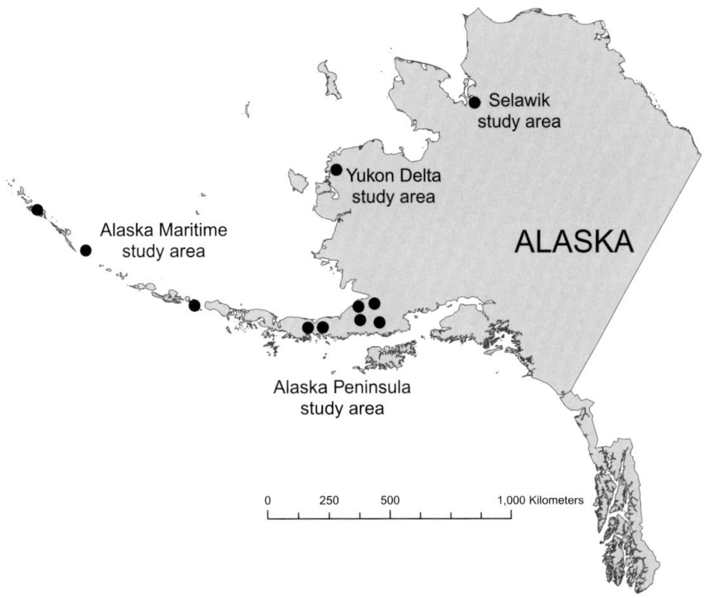

Figure 3.1. Four study areas (names) and 11 study sites (dots).

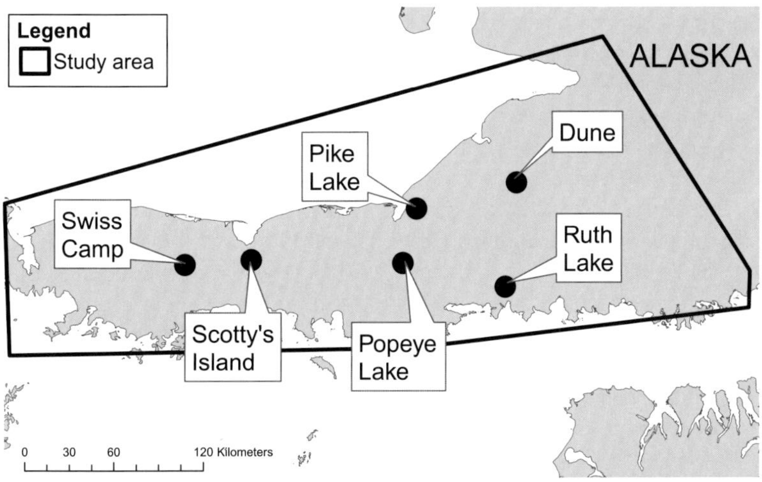

Figure 3.2. Sites in the Alaska Peninsula study area.

The AMSA encompassed 13,696 km^2 but we estimated that one-third of this area was unsuitable for shorebirds because it was covered by snow, ice, or water or because the slopes were too steep to be used for nesting. We used the remaining area, 9,243 km^2, in estimating population sizes throughout the AMSA. We acknowledge that selection bias of unknown magnitude may have affected our estimates of density and that the actual amount of area suitable for shorebirds may be quite different from the value we used. In addition, because of the complexity of habitats in the AMSA, particularly in terms of the rugged topography and its effect on hydrology, we were unable to readily assign plots to the habitat categories used in our other study areas. In the tables summarizing our findings across the four study areas, AMSA data are entered into the "moist" category, but this represents a default category only and should not be construed as reflecting actual shorebird habitat preferences in the AMSA.

In APSA, six sites were selected between North Becharof Lake and Chignik based on their accessibility by fixed-wing aircraft (Fig. 3.2). We

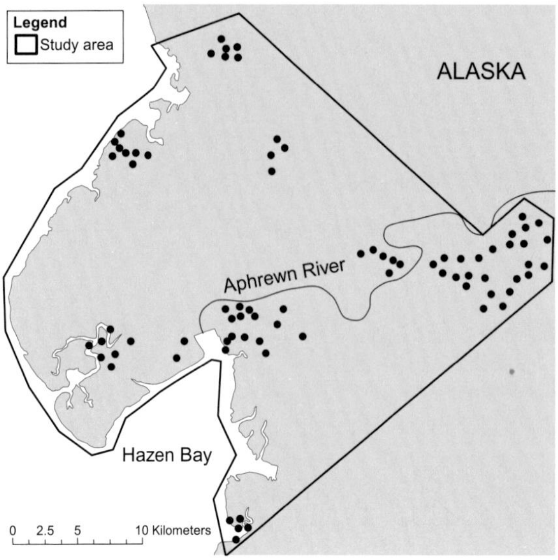

Figure 3.3. Site and plots (dots) in Yukon Delta study area.

randomly selected a total of 51 points within these sites and used each point as the northwest corner for a 16-ha plot. If the point fell in deep water, an alternate point was selected. We used a cover map prepared by Wibbenmeyer et al. (1982) to delineate habitats as follows: marsh = wetland (4,487 km^2), wet meadow = moist (6,678 km^2), and shrub = upland (13,328 km^2). Pixels classified as unsuitable, including deep water, snow, and ice, were excluded. Plot assignment was by the "dominant type rule" (each plot was assigned to the most common habitat). Although we extrapolate to the entire study area, the sites were not randomly selected, and we were not able to estimate the magnitude of the selection bias. Surveys were conducted during 15 May–5 June 2002.

In the YDSA, we selected a single site along the coast next to Hazen Bay (Fig. 3.3). The site covered 853 km^2, which is about 1% of the Yukon Delta NWR. In a landcover classification scheme for the Hazen Bay area, moisture was quantified to generate initial high-order categories (wet, moist, and dry); lower-level habitat categories were then classified within each moisture category (Tande and Jennings 1986). Therefore, we divided our study area into two strata, wet (665 km^2) and moist (188 km^2), as described by Tande and Jennings (1986). We selected 78 plots, in nine clusters. Eight of nine clusters were randomly selected; the ninth was selected due to its proximity to a permanent field station. Plot borders followed natural features; they varied in size from 9- to 14-ha. The eastern group of plots in Figure 3.3 was subdivided into two clusters. Plots were surveyed a single time during the first two weeks of June in 2001 or 2002. Two locations were also selected for intensive surveys (see Bart et al., chapter 2, this volume). At one of the locations, the surrounding area was classified as wetland or upland and two 10-ha plots were randomly selected in each stratum. At the other location, a single 16-ha plot was selected. These plots were surveyed using the intensive methods described in Bart et al. (chapter 2, this volume).

In the SSA, a single large site was selected. It covered 15,170 km^2 including most of the Selawik NWR and some additional areas (Fig. 3.4). We used a landcover map of the Selawik area to identify habitats prior to stratification (Kirk and Markon 1989). We then combined floodplain, wet graminoid, wet mosaic, and dwarf scrub lichen habitats into a wetland stratum (6,494 km^2), and moist graminoid and dwarf scrub tussock habitats into an upland stratum (8,676 km^2). Thirty-five randomly selected plots in 21 one- or two-plot clusters were surveyed a single time during 11–27 June 2002. Plot borders followed natural features and were located completely within habitat strata. Their size varied from 14- to 22-ha. We initiated intensive

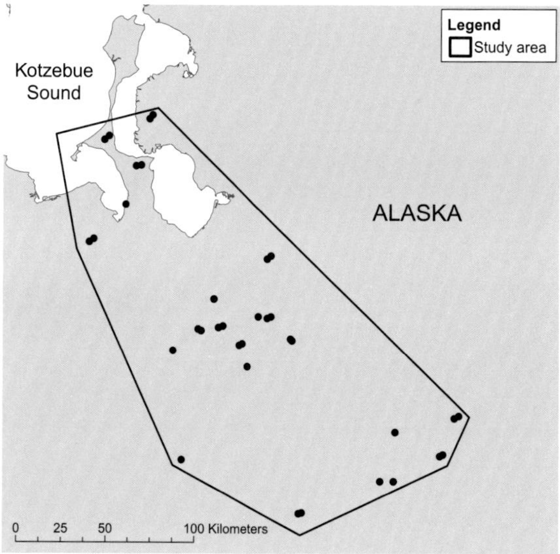

Figure 3.4. Site and plots (dots) in the Selawik study area.

surveys, but no shorebird nests were found and the surveyor could not confirm that any of the shorebirds observed were nesting on the plot. Because of this uncertainty, we did not use intensive data from Selawik in estimating detection rates.

Analytical methods are described in Bart et al. (chapter 2, this volume). To estimate densities in birds/km^2, we doubled the number of indicated pairs/km^2. We provide a table summarizing densities and population sizes for all species combined and those detected in at least three of our study areas. Another summarizes observations by study area and habitat for most species with at least one estimated pair on plots in at least two study areas. We report additional results in the text.

RESULTS

Detection Ratios

We had only two intensive camps in western Alaska, both in the YDSA. This is far too small a sample size for accurate estimates, and we therefore used data from the North Slope of Alaska in estimating detection ratios. However, we compared the estimated detection ratios from our limited sample in western Alaska with that from the North Slope and found that they were nearly identical (0.79 and 0.81, respectively). The combined value, used here for analyses, was 0.81 with a SE of 0.06.

All Shorebirds Combined

Surveyors recorded 2,163 shorebird observations (nests, pairs, and single birds) during our surveys and estimated that 1,485 pairs (or unpaired but territorial males) were attempting to breed within the surveyed plots (Table 3.1). Surveyors found 69 nests, but most records were of single birds, and most birds were recorded in wetland plots.

Estimated densities (birds/km^2) for all shorebirds combined were highest on the YDSA (416 birds/km^2 in wetlands and 300 birds/km^2 in moist areas; Table 3.2). Densities were also high on wetlands in the APSA (118 birds/km^2), but were markedly lower in other regions and habitats. The estimated population size on the APSA was just over 1,800,000 shorebirds. The estimate for the much smaller YDSA was 309,000. Table 3.2 shows that even with a modest effort, estimates with respectable precision can be obtained. The species accounts, presented next, provide a much more detailed examination of the survey results.

Species Accounts

Black-bellied Plover

This species was fairly common in the YDSA, and four sightings were recorded at the Dune and Pike Lake sites in the APSA (Table 3.3). No

TABLE 3.1

Numbers of shorebird observations recorded on rapid surveys.

	Observations by type			Observations by habitat				
Location	Nest	Pairs	Single	Wet	Moist	Upland	Total recorded	Estimated *n* of pairs
SSA	5	40	104	135		14	149	67
YDSA	48	426	1,199	1,353	320		1,673	1,193
APSA	13	94	193	238	4	58	300	203
AMSA	3	8	30		41		41	22
All	69	568	1,526	1,726	365	72	2,163	1,485

TABLE 3.2

Estimated densities (birds/km²) and population sizes (and CVs).

		Density				
Species	Location	Wetlands	Moist	Uplands	All habitats	Population size (CV)
All shorebirds	SSA	41 (0.37)		9 (0.81)		390,247 (0.35)
	YDSA	416 (0.12)	300 (0.12)			309,371 (0.11)
	APSA	118 (0.20)	17 (0.79)	62 (0.54)		1,803,925 (0.32)
	AMSA		7 (0.58)			61,602 (0.58)
Dunlin	SSA	3.2 (0.70)		0 (0)	1.7 (0.72)	25,279 (0.72)
	YDSA	198 (0.19)	8.9 (0.42)		112 (0.23)	95,250 (0.23)
	APSA	66 (0.56)	0 (0)	32 (0.58)	38 (0.49)	932,554 (0.49)
	AMSA		0 (0)		0 (0)	0 (0)
Western Sandpiper	SSA	4.4 (0.61)		0 (0)	2.3 (0.62)	34,636 (0.62)
	YDSA	14 (0.60)	172 (0.17)		86 (0.21)	73,677 (0.21)
	APSA	1.8 (0.79)	0 (0)	1.8 (0.66)	1.5 (0.58)	35,742 (0.58)
	AMSA		0 (0)			0 (0)
Red-necked Phalarope	SSA	2.6 (0.77)		0.29 (1.18)	1.5 (0.72)	22,688 (0.72)
	YDSA	64 (0.17)	55 (0.21)		60 (0.16)	50,961 (0.16)
	APSA	9.0 (0.11)	0 (0)	0 (0)	3.2 (0.28)	77,096 (0.28)
	AMSA		0 (0)		0 (0)	0 (0)

other records were obtained. More pairs than single birds were recorded. The species was found primarily in wetlands and moist habitats. This latter finding is not surprising; in the YDSA, most foraging and brood-rearing is carried out in wetland habitats, but most ground courtship and virtually all nesting occurs in moist habitats (B. J. McCaffery, pers. obs.). The estimated number of pairs breeding in surveyed plots was 14.

TABLE 3.3

Number of observations for each species recorded on rapid surveys.

Species	Location	Observations by type			Observations by habitat			Total recorded	Estimated *n* of pairs
		Nests	Pairs	Singles	Wet	Moist	Upland		
Black-bellied Plover	SSA	0	0	0	0		0	0	0
	YDSA	0	8	7	7	8		15	12
	APSA	0	3	1		2	2	4	2
	AMSA	0	0	0		0		0	0
	Total	0	11	8	7	10	2	19	14
Pacific Golden-Plover	SSA	1	5	15	18		3	21	6
	YDSA	0	0	0	0	0		0	0
	APSA	0	1	1	1	0	1	2	2
	AMSA	0	0	0		0		0	0
	Total	1	6	16	19	0	4	23	8
Bar-tailed Godwit	SSA	0	0	4	4		0	4	1
	YDSA	0	4	70	69	5		74	9
	APSA	0	0	0	0	0	0	0	0
	AMSA	0	0	0		0		0	0
	Total	0	4	74	73	5	0	78	10
Black Turnstone	SSA	0	1	0	1		0	1	1
	YDSA	7	77	60	141	3		144	117
	APSA	0	0	0	0	0	0	0	0
	AMSA	0	0	0		0		0	0
	Total	7	78	60	142	3	0	145	118
Rock Sandpiper	SSA	0	0	0	0		0	0	0
	YDSA	1	5	12	2	16		18	15
	APSA	0	0	0	0	0	0	0	0
	AMSA	3	8	22		33		33	20
	Total	4	13	34	2	49	0	51	35
Dunlin	SSA	0	0	3	3		0	3	3
	YDSA	22	107	479	601	7		608	475
	APSA	10	47	68	101	0	24	125	99
	AMSA	0	0	0		0		0	0
	Total	32	154	550	705	7	24	736	577

TABLE 3.3 (*continued*)

TABLE 3.3 (CONTINUED)

Species	Location	Observations by type: Nests	Pairs	Singles	Observations by habitat: Wet	Moist	Upland	Total recorded	Estimated *n* of pairs
Semipalmated Sandpiper	SSA	2	9	38	49		0	49	23
	YDSA	4	33	173	208	2		210	168
	APSA	0	0	0	0	0	0	0	0
	AMSA	0	0	0		0		0	0
	Total	6	42	211	257	2	0	259	191
Western Sandpiper	SSA	2	7	11	20		0	20	12
	YDSA	8	47	147	29	173		202	163
	APSA	0	1	4	4	0	1	5	4
	AMSA	0	0	0		0		0	0
	Total	10	55	162	53	173	1	227	179
Wilson's Snipe	SSA	0	0	0	0		0	0	0
	YDSA	0	2	10	2	10		12	9
	APSA	0	2	27	21	0	8	29	19
	AMSA	0	0	0		0		0	0
	Total	0	4	37	23	10	8	41	28
Red-necked Phalarope	SSA	0	2	4	5		1	6	5
	YDSA	5	96	158	188	71		259	179
	APSA	0	12	6	17	0	1	18	13
	AMSA	0	0	0		0		0	0
	Total	5	110	168	210	71	2	283	197
All Species	SSA	5	24	75	100	0	4	104	51
	YDSA	47	379	1,116	1,247	295	0	1,542	1,147
	APSA	10	66	107	144	2	37	183	139
	AMSA	3	8	22	0	33	0	33	20
	Total	65	477	1,320	1,491	330	41	1,862	1,357

The estimated densities in birds/km^2 (and CVs) on the YDSA were 2.8 (0.80), 7.1 (0.59), and 4.8 (0.41) in wetlands, moist areas, and both habitats combined, respectively. The estimated population size (and CV) was 4,070 (0.41). Estimated densities in birds/km^2 averaged lower on the APSA: 5.8 (1.44), 1.75 (0.66), 1.84 (0.89) in moist habitats, dry habitats, and across the entire APSA, respectively. The distribution of this species among our study areas was not entirely consistent with the Birds of North America range map (BNA; Paulson 1995). As expected, Black-bellied Plovers were present on the YDSA, but we did not find them in the SSA, although that area is within the depicted range. In addition, the birds at Dune and Pike Lake were slightly outside that range; only recently has the species been confirmed as a breeder on the Alaska Peninsula, hundreds of kilometers from the nearest previously known nesting areas (Savage and Johnson 2005).

Pacific Golden-Plover

Pacific Golden-Plovers were recorded commonly in the SSA and on two occasions in the APSA (Table 3.3). Single birds in wet areas were the most common record. The estimated population sizes (and CVs) were 31,333 (0.58) in the SSA and 24,789 (0.77) in the APSA. Pacific Golden-Plovers have only recently been confirmed as a breeding species on the Alaska Peninsula (Savage and Johnson 2005). They are also widely distributed across much of the interior of the Yukon–Kuskokwim Delta in upland habitats, but they do not occur as a breeding species within the YDSA (Connors et al. 1993, Johnson et al. 2001, B. J. McCaffery, pers. obs.).

Semipalmated Plover

This species was seen only in the APSA but at four different sites there. A pair and three singles were recorded in all three habitats, and two pairs were judged to be breeding on the surveyed plots, one each in wet and moist habitats. The estimated population size (and CV) was 28,198 (1.00). Elsewhere on the Alaska Peninsula, Semipalmated Plovers occupy dunes, river banks, and disturbed areas (S. E. Savage, pers. comm.). On the Yukon–Kuskokwim Delta, this species is a fairly common nester among coastal dunes, along sandy beaches north of Hazen Bay, in disturbed habitats in the Askinuk Mountains, and along mountain rivers farther inland (Brandt 1943, Petersen et al. 1991, McCaffery and Harwood 1997), but it does not occupy the coastal wet and moist meadows surveyed in this study. It is an uncommon breeder in the eastern Aleutians and has been recorded as a breeding species over the last quarter century in the Andreanof Islands, including Adak Island (Gibson and Byrd 2007).

Greater Yellowlegs

This species was recorded only in the APSA, near the edge of its known breeding range (Elphick and Tibbitts 1998). Single birds in wetlands were the most common sighting. Surveyors recorded 16 nests, pairs, or single birds and judged ten pairs to be nesting on surveyed plots. The estimated population size (and CV) was 49,565 (0.62). On the Alaska Peninsula, Greater Yellowlegs breed well beyond tree-line, but on the Yukon–Kuskokwim Delta to the north, this species has been found breeding only on the moist tundra adjacent to the spruce tree-line more than 100 km inland (McCaffery, pers. obs.).

Whimbrel

Whimbrel were recorded only in the SSA but were common there. They were found mainly in wetlands ($n = 33$ observations), but also in uplands ($n = 9$). Surveyors recorded 42 pairs or single birds and judged 13 pairs to be breeding in surveyed plots, seven and six pairs in wetland and uplands, respectively. The estimated population size (and CV) was 90,278 (0.46). The BNA range map shows Whimbrel occurring throughout the YDSA, whereas we did not record any breeding birds there (Skeel and Mallory 1996). Although the species breeds patchily on inland moist and upland tundra elsewhere on the Yukon–Kuskokwim Delta (McCaffery 1996), and is fairly common during migration on the delta's coastal meadows, it does not breed in the YDSA (Handel and Dau 1988, B. J. McCaffery, pers. obs.).

Hudsonian Godwit

Hudsonian Godwits were recorded in the APSA (two males) and in the SSA (one pair). The surveyors were uncertain about the status of the birds in the APSA, which is about 150 km from the nearest known breeding population (S. E. Savage, pers. comm.). The SSA is close to the known distribution, and the pair recorded there was judged to be breeding on the plot, but the nest was not found. Because we consider the status of all three records to be uncertain, we do not present density or population estimates for this species in either study area. Hudsonian Godwits do not breed in the coastal zone of the Yukon–Kuskokwim Delta where our study area was located, although they do breed further inland (McCaffery and Harwood 2000).

Bar-tailed Godwit

Bar-tailed Godwits were frequently detected in the YDSA, uncommon in the SSA, and absent in the APSA and AMSA (Table 3.3). In both the SSA and YDSA, however, most birds (68 of 78) were judged to be breeding off the surveyed plots. Most sightings were of single birds in wetlands.

The number of pairs estimated to be breeding on surveyed plots was ten, and these breeding records were distributed more evenly between wetland and moist habitats (six and four, respectively), with all pairs in moist habitats located on the YDSA. In that area this species forages most frequently in wetland habitats, while nests can be found in both wet and moist habitats (McCaffery and Gill 2001, B. J. McCaffery pers. obs.). The estimated densities in birds/km^2 (and CVs) on the YDSA were 2.8 (0.56) in wetlands and 4.7 (0.52) in moist areas. The estimated population size (and CV) in the YDSA was 3,112 (0.36). On the SSA, the estimated densities in birds/km^2 (and CVs) were 0.6 (0.99) in wetlands, and the estimated population size (and CV) was 5,071 (1.00).

Marbled Godwit

Marbled Godwits were observed at two sites in the APSA (Pike and Popeye Lakes). Surveyors recorded two pairs and 13 single birds on the surveyed plots, all in wetlands, plus five birds off the plots. They estimated that two pairs were breeding on their plots. Their estimated population size (and CV) was 9,913 (0.62), considerably higher than other recent estimates (Tibbitts et al. 2005, Morrison et al. 2006).

Ruddy Turnstone

Ruddy Turnstones were recorded only in the YDSA. One pair and five singles were recorded in wetlands and moist areas. The estimated population size (and CV) was 1,189 (0.27), but that estimate should be interpreted with caution. Near the coast of the Yukon–Kuskokwim Delta, nests are almost exclusively situated in the drier and higher patches of moist tundra (B. J. McCaffery, pers. obs.), so the birds recorded as breeding in wetlands were probably nesting outside of the plots where they were detected. Although the exact value of our point estimate may therefore be suspect, the most important inference to be drawn from our Ruddy Turnstone data is that, relative to other breeding shorebirds in the YDSA, the species is quite rare.

Black Turnstone

Black Turnstones were abundant in the YDSA but rare or absent in the other study areas (Table 3.3). Seven nests or probable nests, 77 pairs, and 60 singles were recorded on the YDSA, almost exclusively in wetlands, while only a single pair was observed on the SSA. The estimated number of pairs breeding on the surveyed plots was 118.

In the YDSA, estimated densities in birds/km^2 (and CVs) in wetlands and moist areas were 48 (0.19) and 2.4 (0.24), respectively. Prior estimates in the same area ranged from > 100/km^2 in wet coastal saltgrass meadows to 4/km^2 > 2 km inland in moist dwarf shrub mat tundra (Handel and Gill 1992). The estimated population size (and CV) in the YDSA was 22,924 (0.25). This estimate is a significant fraction of the estimated global population of 95,000 (Handel and Gill 1992, Morrison et al. 2006).

Rock Sandpiper

Rock Sandpipers were the only widespread shorebird in the AMSA and were also recorded commonly in the YDSA (Table 3.3). On the YDSA, where we distinguished between wet and moist habitats, most sightings were in moist areas. These data are consistent with previous observations of breeding birds on the Yukon–Kuskokwim Delta (Gill et al. 2002, Johnson and McCaffery 2004, Johnson et al. 2009). The estimated number of pairs breeding on surveyed plots was 35.

The estimated densities (and CVs) were 16.0 (0.30) in moist areas of the YDSA and 6.2 (0.63) on the AMSA. These values virtually bracket the range of densities reported in most other studies throughout this species' range (Gill et al. 2002). White et al. (1977), however, estimated densities of 40–80 birds/km^2 on Amchitka Island in the central Aleutians. Estimated population sizes (and CVs) were 6,656 (0.34) in the YDSA and 56,997 (0.63) in the AMSA. It must be recalled, as described in the Methods section, that the sites in the AMSA were not randomly selected and we have little way to estimate selection bias. The estimated population size is therefore only a first approximation. In addition, an arctic fox removal program was in its third year on Amlia Island but no such program had been initiated on Avatanak Island. Because predation pressure may be lower on Amlia Island, Rock Sandpiper populations may be increasing, whereas on Avatanak Island, predation pressure may still be a factor limiting nesting success and/or abundance of Rock Sandpipers. We did not find Rock Sandpipers on plots in the APSA, although they are known to breed on the Alaska Peninsula, and in some areas (e.g., Izembek NWR), they are one of the most abundant tundra-nesting bird species (Gill et al. 2002, K. Sowl, pers. comm.).

Dunlin

Dunlin were abundant in the YDSA and APSA and were rare or absent elsewhere (Table 3.3). Surveyors recorded 32 nests or probable nests and 154 pairs, but single birds were by far the most common type of observation. This species was recorded almost exclusively in wetlands on YDSA, but nearly one-fifth of all observations on the APSA were in uplands. The estimated number of pairs nesting in surveyed plots was 577.

The estimated density of Dunlin in wetlands in the YDSA was 198 birds/km^2 (Table 3.2). Although this value seems remarkable, it is virtually identical to Gill and Handel's (1990) estimate from a site within our study area, and not markedly higher than densities found by Holmes on the central coastal YKD just north of our study area (Holmes 1970, Holmes and Black 1973). Dunlin densities drop off dramatically, however, just inland of the YDSA (B. J. McCaffery, pers. obs.). To the north, among wetlands on the Seward Peninsula, estimated densities ranged from 28 to 80 birds/km^2 (Kessel 1989). Overall, densities in the YDSA and APSA were much higher in wetlands than in other habitat strata. The estimated population sizes (and CVs) were 95,250 (0.23) in the relatively small YDSA, and 932,554 (0.49) in the APSA. Although the density of Dunlin on the SSA was low compared to the YDSA and APSA, the population estimate for the SSA exceeded 25,000 (Table 3.2).

Semipalmated Sandpiper

Semipalmated Sandpipers were common in the YDSA and were regularly observed in the SSA, but were not recorded in the other study areas (Table 3.3). A few dozen pairs and a handful of nests were recorded, but most observations were of single birds and nearly all records were from wetlands. The estimated number of pairs breeding in surveyed plots was 191. In the central part of the Yukon–Kuskokwim Delta, Semipalmated Sandpipers only occur regularly in the lower coastal meadows (R. E. Gill, Jr., and C. R. Ely, pers. comm., B. J. McCaffery pers. obs.). A few isolated pairs were reported from the sand dunes in the vicinity of Dall Point north of Hooper Bay (Brandt 1943).

The estimated density (and CV) in wetlands in the YDSA was 70 birds/km^2 (0.20), somewhat higher than a site-specific estimate within our study area of 40/km^2 from the mouth of the Tutakoke River, but lower than estimates of local densities from multiple wetland sites on the Seward Peninsula to the north of the Yukon–Kuskokwim Delta, which ranged from 101 to 268 birds/km^2 (Kessel 1989, Gill and Handel 1990). Although Semipalmated Sandpipers were the most common shorebird in the SSA, their wetland density of 19 birds/km^2 (0.60) was still lower than that reported from either the Seward Peninsula or the YDSA (this study). The estimated population sizes (and CVs) in the YDSA and SSA were 33,315 (0.33) and 148,467 (0.62), respectively.

Western Sandpiper

Western Sandpipers were recorded in all study areas except the AMSA, but they were common only in the YDSA (Table 3.3). Over a quarter of the observations were nests or pairs, mainly on moist areas though often also in wetlands. The estimated number of pairs breeding on surveyed plots was 179.

As with Dunlin, estimated densities of Western Sandpipers were extremely high in the YDSA, although for Western Sandpipers, moist areas, rather than wetlands, were preferred (Table 3.2). Such high densities were not unexpected. Our estimate of 172 birds/km^2 in moist habitats compares with site-specific estimates of 680–980 birds/km^2, 68 birds/km^2, and 290–363 birds/km^2 elsewhere in the central Yukon–Kuskokwim Delta (Holmes 1971, McCaffery et al. 1998, McCaffery and Ruthrauff 2004). Local densities at multiple sites on the Seward Peninsula ranged from 17 to 200 birds/km^2 (Kessel 1989). The estimated population size (and CV) in the YDSA was 73,677 (0.21).

Although the BNA account does not show the breeding range of the Western Sandpiper extending onto the Alaska Peninsula, our observations of them were not surprising (Wilson 1994). In fact, they were the second most abundant shorebird on Gill et al.'s (1981) 2000-km^2 study area on the north-central Alaska Peninsula, where breeding was confirmed. We estimated a population size (and CV) of 35,742 birds/km^2 (0.58) on the APSA, very similar to the population estimate of 34,636 (0.62) on the considerably smaller SSA.

Least Sandpiper

Least Sandpipers were common on the APSA, occurring at all sites except Ruth Lake, the high-elevation site. One nest, 13 pairs, and 38 single birds were recorded on the APSA, 39 in wetlands and 13 in uplands, for a total of 52 observations. The estimated number of pairs breeding on surveyed plots was 35. The estimated density in birds/km^2 (and CV) on the APSA was 19 (0.24) in wetlands, 0 (0) in moist areas, and 9.5 (0.30) in uplands. The estimated population size (and CV) was 273,751 (0.22). On Avatanak Island, the easternmost site in the AMSA, three single birds, judged to indicate two breeding pairs, were recorded on the surveyed plots. These observations yielded a population estimate (and CV) for the AMSA of 4,438 (0.96). Least Sandpipers are considered to be uncommon breeders in the eastern Aleutians (Gibson and Byrd 2007). The species was not recorded in the YDSA or SSA. Although they are spottily distributed in the interior of the Yukon–Kuskokwim Delta, they do not breed in the coastal fringe where we sampled (B. J. McCaffery, pers. obs.).

Pectoral Sandpiper

Pectoral Sandpipers were recorded only on the YDSA, where they are a common spring migrant in late May, but only a rare breeder (McCaffery, pers. obs.). Ten pairs and 19 single birds were recorded, mainly in wetlands. The number of pairs estimated to be breeding on surveyed plots was 18, but we note that courtship displays are common among birds in migrant flocks. The estimated density (and CV) in all habitats was 6.5 birds/km^2 (0.35). The estimated population size (and CV) was 5,539 (0.35).

Short-billed Dowitcher

Short-billed Dowitchers were recorded only on the APSA. Surveyors recorded one nest, five pairs, and 20 single birds, mainly in wetlands. The number of pairs estimated to be breeding on surveyed plots was 14. The estimated density in birds/km^2 was highest in wetlands (6.9) and lowest in uplands (3.5), but the difference was not significant. The estimated overall density (and CV) in the APSA was 5.1 (0.45). The estimated population size (and CV) was 124,124 (0.45). Gabrielson and Lincoln (1959) summarized multiple breeding season records of this species from the Alaska Peninsula, and Gill et al. (1981) also reported this species breeding in the vicinity of Nelson Lagoon. The species' breeding range is now known to extend north to the Yukon–Kuskokwim Delta, but at that latitude, they are limited to inland riparian meadows and muskeg habitats, and are not found breeding along the coast in or near the YDSA (B. J. McCaffery, pers. obs.).

Long-billed Dowitcher

Long-billed Dowitchers were recorded mainly on the YDSA, though two single birds were recorded on the SSA. Surveyors in the YDSA recorded 21 pairs and ten single birds, mainly in wetlands but sometimes in moist areas (ten observations). The estimated number of pairs breeding on surveyed plots was ten. In the YDSA, the estimated density in birds/km^2 (and CV) was 4.4 (0.73). The estimated population size (and CV) was 3,754 (0.73). In the SSA, the estimated density was 0.67 birds/km^2, and the estimated population size (and CV) was 10,143 (1.00).

Like the Pectoral Sandpiper, fairly large numbers of Long-billed Dowitchers are present on the coastal meadows of the Yukon–Kuskokwim Delta in late May. Many apparently continue north to breed, but some remain to nest. Based on observations of nests and broods, Long-billed Dowitchers are uncommon breeders on the central Yukon–Kuskokwim Delta. Dowitcher nests have been found most frequently in meadows and small wetlands within moist tundra areas (Brandt 1943, B. J. McCaffery, pers. obs.).

Wilson's Snipe

Wilson's Snipe were recorded mainly on the YDSA and APSA (Table 3.3), although four birds were recorded outside the surveyed plots on the SSA. Most records were of single birds in wetlands. Surveyors recorded 41 total observations, and 28 pairs were judged to be breeding on the surveyed plots.

The estimated density (and CV) in birds/km^2 in all habitats was 5.3 (0.54) in the YDSA and 7.3 (0.49) in the APSA. The estimated population size was 4,485 in the YDSA and 177,781 in the APSA. Although we did not detect this species on the AMSA, it is a fairly common breeder locally in the eastern Aleutians (Gibson and Byrd 2007). Nesting has been confirmed on islands both east (Unimak) and west (Unalaska) of Avatanak Island, and birds have been detected in spring at Umnak even farther west.

Red-necked Phalarope

Red-necked Phalaropes were abundant on the YDSA and were uncommon or absent elsewhere (Table 3.3). The most frequent records were of single birds in wetlands, but pairs and birds in moist areas were also common. The estimated number of pairs breeding on surveyed plots was 197.

Estimated densities on the YDSA included 64 birds/km^2 in wetlands and 55 birds/km^2 in moist areas (Table 3.2). Densities in the APSA and SSA were much lower. The estimated population sizes in the SSA, YDSA, and APSA (with CVs) were 22,688 (0.72), 50,961 (0.16), and 77,096 (0.28), respectively.

Our estimated density within the YDSA is about half the density previously estimated for this species at a coastal site within our study area, and is at the low end of the range of densities (70–200 birds/km^2) found at sites on the Seward Peninsula (Kessel 1989, Gill and Handel 1990). Although Red-necked Phalaropes are locally uncommon to fairly common breeders in the eastern and central Aleutians (including Adak Island; Gibson and Byrd 2007), we failed to detect them on any of the three islands we surveyed.

Red Phalarope

Red Phalaropes were recorded only on the YDSA. A nest, 15 pairs, and 49 singles were recorded, almost exclusively in wetlands. The estimated number of birds breeding on surveyed plots was only 15, 12 in wet habitats and three in moist habitats. The estimated density and population size (and CVs) in the YDSA were 4.2 birds/km^2 (0.53) and 3,599 (0.53). This low density is consistent with previous findings in our area (Gill and Handel 1990, B. J. McCaffery, pers. obs.), but markedly lower than densities ranging from 88 to 144 birds/km^2 at two sites north of the Yukon–Kuskokwim Delta on the Seward Peninsula (Kessel 1989).

DISCUSSION

Methodological Concerns

A number of factors limit the inferences about bird densities and population sizes that can be drawn from the pilot PRISM surveys conducted in western Alaska in 2001–2002. Small sample sizes and non-random site selection are perhaps the most important factors, particularly in the AMSA and APSA. Because sites in those study areas were not randomly selected, estimates from the AMSA and APSA are subject to selection bias, which we have not attempted to quantify. For example, our estimate of Marbled Godwits on the APSA is nearly 10,000 birds, several times larger than that generated by a recent focused effort to derive a population estimate for this subspecies; the discrepancy is probably at least partially a result of selection bias in our study (Tibbitts et al. 2005). In addition, the sample size in the AMSA (three sites) and APSA (six sites) were too small to capture some important shorebird nesting habitats. For example, Rock Sandpipers are fairly common in open, low ericaceous shrub habitats along the north coast of the Alaska Peninsula, but this habitat did not occur at our six survey sites (Gill et al. 2002, S. E. Savage pers. comm.). As a result, one of the regularly nesting shorebirds on the peninsula was not detected in our surveys. Semipalmated Plovers, Wilson's Snipes, and Red-necked Phalaropes are known to breed on the AMSA, but we did not encounter them, presumably because they either do not breed, or breed only at very low densities, at our three Aleutian study sites (Mueller 1999, Nol and Blanken 1999, Rubega et al. 2000, Gibson and Byrd 2007).

Several issues regarding our habitat classifications also constrain interpretation of our data. We partitioned our observations into three habitat strata—wetlands, moist areas, and uplands. Because we based these stratifications on land-cover maps specific to each study area and with different classification criteria, however, the three strata as defined in one study area may not correspond directly to those same strata in another study area. As a result, comparisons of shorebird–habitat associations should probably not be made among our study areas. Second, as noted earlier, the complexity of habitats within the AMSA precluded effectively assigning plots and observations to specific habitats that corresponded with the categories we used elsewhere. Even in study areas where habitat assignment was more straightforward, a plot might be assigned to a stratum based on the dominant habitat type(s) within that plot. As a result, the ecological texture produced by fine-grain habitat mosaics is lost at the landscape-level analysis that we provide here. Last, in the APSA, the Ruth River site was located in alpine tundra. We have included this site (and its lone shorebird observation, a Semipalmated Plover) within our upland stratum, but it was ecologically quite distinct from the habitats classified as upland elsewhere on the APSA. Given these limitations, we acknowledge that this pilot study made only modest progress toward our goal of obtaining information about shorebird habitat associations. We encourage other researchers to adopt a more nuanced approach to habitat stratification and analysis in subsequent investigations.

Another concern is survey timing, which can be critical in the arctic (Meltofte 2001). Studies of survey timing that have included data from western Alaska have focused on seasonal declines in breeding vocalizations and the probability of nest detection, respectively (Nebel and McCaffery 2003, Smith et al. 2009). Timing also affects anti-predator behavior of adult shorebirds. During courtship, egg-laying, and early incubation, breeding shorebirds vary in their tendency to approach observers from beyond the range of high detectability. For example, once a nest site has been selected, both members of a Whimbrel pair (including the incubating bird) may alarm-call and approach an observer, while in other species, a cryptic strategy is employed. Even among the latter species, however, their behavior can change radically when eggs are hatching and when adults are tending broods; alarm-calling adults may converge on observers from dozens, or even hundreds, of meters away.

This seasonal pattern of behavior may have affected our results, at least at one site. On the SSA, the failure to find nesting birds on intensive plots was tentatively attributed to sampling too late in the season. Late surveys might also explain the high densities of Whimbrels. A species already prone to approaching observers during the incubation period, brood-tending Whimbrels are also known to jointly mob human observers and predators; territorial borders are not respected during brood-rearing, so that multiple pairs can rapidly converge on a predator or human observer from hundreds of meters away. Our point estimate for the Whimbrel population on the Selawik refuge exceeded 90,000. This estimate is over three times the previous *continental* estimate for this subspecies (Morrison et al. 2006). Although the accuracy of, and procedures used for deriving, those earlier estimates have been challenged recently (Farmer 2008), our estimate for the number of Whimbrels on the Selawik refuge is certainly questionable as well. We suggest that caution is required before placing too much confidence in any of the estimates derived to date.

In western Alaska, at least on the YDSA, two aspects of spring shorebird ecology complicate inferences about the status of birds on rapid plots. First, on the outer delta several species of shorebirds exhibit courtship behavior even though they occur primarily (Pectoral Sandpipers and Long-billed Dowitchers) or exclusively (Red Knots) as migrants (Gill and Handel 1981, 1990; McCaffery et al. 2008; B. J. McCaffery, unpubl. data). Small numbers of the former two species do remain to breed on the delta, but recording all or most encountered birds of these species as breeders would result in a substantial overestimate. Furthermore, in our study, the estimated detection ratio may not have captured these errors because they were calculated mainly using other species and in areas where these errors did not occur. In future work, it will be important to have enough intensive plots that detection ratios for western Alaska, or even each study area, can be calculated.

The second aspect of shorebird ecology in western Alaska that complicates interpretation of the data from rapid plots is the tendency for many species that nest primarily in upland or moist tundra to forage extensively in wetland habitats. Such behavior is regularly exhibited by Black-bellied Plovers, Bar-tailed Godwits, Ruddy Turnstones, Rock Sandpipers, and Western Sandpipers. In addition, the few nests of Pectoral Sandpipers and Long-billed Dowitchers located in the vicinity of our intensive plots have also been in moist tundra. Both Bar-tailed Godwits and Western Sandpipers regularly display over, and occasionally fight in, wetland habitats where no pairs are actually nesting (B. J. McCaffery, pers. obs.). Such

TABLE 3.4

Examples of marked discrepancies in population estimates between this study and Morrison et al. 2006.

Species	Study area	Population estimate	Morrison et al. continental estimate
Pacific Golden-Plover	SSA	31,000	35,000–50,000
Pacific Golden-Plover	APSA	25,000	35,000–50,000
Greater Yellowlegs	APSA	50,000	100,000
Whimbrel (*N. p. rufiventris*)	SSA	90,000	26,000
Dunlin (*C. a. pacifica*)	APSA	933,000	550,000
Least Sandpiper	APSA	274,000	700,000
Short-billed Dowitcher (*L. g. caurinus*)	APSA	124,000	75,000

behaviors can render an assessment of breeding status in wetland habitats problematic. It thus seems likely that our habitat-specific estimates of breeding density for these species have positive bias in wetland habitats and negative bias in their actual nesting habitat (because of the frequent absence of the breeding birds from that habitat). Our current study does not permit us to estimate the magnitude of either the potential error or the loss of precision caused by this complication.

Two corollaries are apparent from the preceding discussion. First, we are reasonably confident about our density estimates for those species that concentrate the majority of their activities, including nesting, in wetland habitats (e.g., Black Turnstone, Dunlin, Semipalmated Sandpiper, and Red Phalarope). Second, in areas such as the outer Yukon–Kuskokwim Delta, a more accurate assessment of regional densities for species that regularly move between habitats may be derived by (1) *not* stratifying by habitat, and (2) relaxing the requirement for ascertaining breeding status. A simple random selection of plots could take into account inter-habitat movements by the common local breeders; in effect, if there is some probability that they are missed in their primary habitat, they also have some probability of being detected in the secondary habitat. As long as breeding status did not have to be tied to the particular plot on which the bird was detected, the bird could be counted as part of the sampled breeding population by virtue of its presence in a particular season and region. Moreover, all of the problems mentioned above would be reduced by having a sufficient sample of intensive plots.

Densities and Population Estimates

Reasonably precise ($CV < 0.35$) estimates of regional populations were achieved for several of the smaller, more abundant species in the YDSA and APSA. In the YDSA, these included Ruddy and Black Turnstones, all four species of regularly breeding *Calidris*, and the Red-necked Phalarope. On the APSA, comparably precise estimates were generated for Least Sandpiper and Red-necked Phalarope. In most cases, the density estimates for these species were also within, or at least close to, the range of site-specific density estimates from western Alaska derived for these species from other studies using different methods.

Given continental population estimates provided by Morrison et al. (2006), however, some of our regional population estimates are quite surprising (Table 3.4). For Pacific Golden-Plover, Greater Yellowlegs, and Least Sandpiper, the population estimates for our study areas equaled a large proportion of the Morrison estimate despite our study areas covering only a small proportion of the species' ranges. In fact, for Pacific Golden-Plover, the combined estimate for Selawik and Alaska Peninsula exceeds the Morrison estimate for the continental population. For Whimbrel, Dunlin, and Short-billed Dowitcher, the differences between estimates is even more striking, with our limited regional estimates being higher than Morrison et al.'s estimates by 65–200%.

It is difficult to evaluate the significance of these discrepancies, and a number of factors might be involved. We have already noted that, at least for some species, high estimates may be a function of birds approaching human observers, resulting in high apparent plot densities. This is particularly likely when there are no local intensive plots for calibrating this behavior on a species-specific basis. The Whimbrel estimate on the SSA and the Greater Yellowlegs estimate on the APSA may both be inflated as a result of this phenomenon. In addition, with the exception of the Least Sandpiper estimate on the Alaska Peninsula, the CVs of the species noted in Table 3.4 were all $\geq$0.45. The lack of precision in our estimates may contribute to the differences with Morrison et al. (2006), but it is important to note that in *all* of the examples with marked discrepancies relative to Morrison et al. (2006), our estimates are higher. This suggests that there may be a systematic difference between estimates generated by the two approaches. A more detailed discussion of our results in relation to those of Morrison et al. (2006) is provided in Bart and Smith (chapter 14, this volume). Here we simply note that in many areas and for many species, including cases in which the problems described above were much reduced, our surveys suggested larger populations than estimated by Morrison et al. (2006).

Importance of Western Alaska for Breeding Shorebirds

Even a relatively conservative assessment of our results shows the significance of western Alaska to continental, and indeed global, shorebird populations. Both the Alaska Peninsula and the Yukon–Kuskokwim Delta support extremely large numbers of breeding shorebirds. We estimate that slightly more than 1.8 million shorebirds breed in the APSA. Approximately half of those are Dunlin. As noted previously, our Dunlin estimate on the Alaska Peninsula alone is nearly double the current estimate for the *pacifica* race. Admittedly, the CV of our estimate is large (0.49), but we can identify nothing about Dunlin biology that would result in a consistent positive bias in our point estimate. In fact, if anything, we suspect that our estimate might be biased low because of our non-random site-selection process. High densities of Dunlin can be found in the inland regions we sampled on the APSA, but high densities also occur along the coast in habitats we did not survey (S. E. Savage, pers. comm.). Elsewhere in western Alaska, coastal habitats regularly support the highest densities of Dunlin (Holmes 1970, Kessel 1989, Gill and Handel 1990), so our failure to sample in coastal habitats of the APSA may have led to an underestimate of Dunlin numbers.

The CVs of our APSA population estimates for Least Sandpipers and Red-necked Phalarope were reasonably good; in fact, the CV of 0.22 for Least Sandpiper was the third-lowest of any species at any site in our study. Although our estimate of 77,000 Red-necked Phalaropes is not a striking proportion of the estimated continental population of 2.5 million, our Least Sandpiper estimate of 274,000 is nearly 40% of the most current continental estimate for this species (Morrison et al. 2006). Given that Least Sandpipers range across the entire continent as a breeding species, our results indicate that there may be many more of this species than previously suspected. Regardless of the actual overall North American population, it is clear that the Alaska Peninsula is a very important region for Least Sandpipers.

For the Yukon–Kuskokwim Delta, our data corroborate previous studies that have found extremely high local densities of breeding shorebirds on the outer central portion of the Delta. Significantly, our study demonstrates that those high densities are not limited to a handful of small study sites, but rather occur over hundreds of km^2. Overall densities on the YDSA in wetlands were 416 shorebirds/km^2, which are the highest recorded in North America (Bart and Smith, chapter 14, this volume), and are among the highest in the world in either arctic or subarctic habitats (Meltofte et al. 2007b).

Because the YDSA at Hazen Bay was small (853 km^2) compared to our other study areas in western Alaska, the population estimates for this area are relatively modest. Although it is tempting to generate a preliminary refuge-wide estimate of shorebird densities based on our work near Hazen Bay, we do not feel it is appropriate to extrapolate these densities across the entirety of Yukon Delta NWR. Habitats change dramatically north, east, and south of our study area, and our limited experience in these other areas suggests that shorebird densities and/or species composition are markedly different there.

On the central Yukon–Kuskokwim Delta, however, there are broad ecological similarities across coastal habitats (i.e., those within ~20 km of the Bering Sea) between Nelson Island to the south and the Askinuk Mountains 120 km to the north. Within this area (which includes our YDSA near Hazen Bay), there are vast expanses of additional low coastal meadows that are similar to our wet habitat stratum, and limited observations of shorebirds in this larger area suggest shorebird densities at least comparable to those that we found in the YDSA (Holmes 1970, 1971; B. J. McCaffery pers. obs.). As noted previously, we are most confident in our estimates for wetland-nesting species. Using a recently derived landcover classification system for the central coastal Yukon–Kuskokwim Delta, we have identified an area of 2,200 km^2 of low coastal meadows that we believe is comparable to our wet habitat stratum (T. Jorgensen, pers. comm.). As a speculative exercise to consider the potential population sizes of the common shorebirds in this area, we extrapolated our Hazen Bay density estimates (for those species with CVs < 0.35) to this expanded area. This extrapolation results in an estimate of nearly 840,000 breeding individuals of our four most abundant wetland-nesting species, including 436,000 Dunlin, 154,000 Semipalmated Sandpipers, 141,000 Red-necked Phalaropes, and 106,000 Black Turnstones.

The estimated Dunlin and Black Turnstone numbers merit comment. The Dunlin estimate on just this limited section of the Yukon–Kuskokwim Delta equals 80% of the previously published population estimate of 550,000 for *C. a. pacifica* (Morrison et al. 2006). When we consider (1) that Dunlin nest on the Yukon–Kuskokwim Delta both north and south of this expanded area, (2) that some *pacifica* nest north of the Yukon–Kuskokwim Delta, and (3) that we estimated nearly a million on the Alaska Peninsula, it seems appropriate to reevaluate the population estimate for this population.

For the Black Turnstone, the current global population estimate of 95,000 is one of the relatively few rigorously generated species-specific population estimates for northern breeding shorebirds in North America (Handel and Gill 1992). As such, this estimate warrants a careful comparison with the estimate generated in our study. Handel and Gill's point estimate for the number of Black Turnstones on the central Yukon–Kuskokwim Delta is 80,000 (±19,000 95% CI). Our estimate of 106,000 within this same region just exceeds the upper 95% confidence limit of 99,000 generated by Handel and Gill (1992). Do these results suggest relative stability in this population, or do they reflect a population that has changed dramatically in size over the two decades between estimates (as suggested by a point estimate nearly a third higher in our survey)? Because most northern-breeding shorebird populations are thought to be declining, an increase in population size of the magnitude suggested by our turnstone data seems counterintuitive (Morrison et al. 2006, Bart et al. 2007). On the other hand, gradually warming conditions over the last several decades could theoretically increase some shorebird population growth rates, at least in the short term, so the possibility for significant population growth cannot be ruled out (Meltofte et al. 2007b). Given this uncertainty, the potential sources of bias in our estimate, and the variation in survey methodology, we are unable to draw a definitive conclusion about the population trajectory of this population, even when comparing two rigorously derived estimates. We do note, however, that the two Black Turnstone population estimates approach the same order of magnitude (i.e., 100,000); our estimate corroborates Handel and Gill's (1992) conclusion about the importance of the Yukon–Kuskokwim Delta for breeding Black Turnstones.

Last, expanding our overall density estimate for all shorebirds detected in wet habitats in the YDSA to the entire 2,200 km^2 of low wetlands on the central coastal Yukon–Kuskokwim Delta yields an estimate of 915,200 shorebirds. This area may support the highest shorebird densities on the delta, but it still makes up less than 5% of the entire region. In particular, it does not include those moist tundra areas that apparently support the highest densities of Western Sandpipers (Holmes 1971, McCaffery and Ruthrauff 2004), a species estimated to occur in the millions (Morrison et al. 2006). Based on this and other studies, we conclude that the entire Yukon–Kuskokwim Delta probably supports several million nesting shorebirds.

Design of Additional Shorebird Surveys in Western Alaska

We believe our study demonstrates the potential value of widespread shorebird surveys following the basic Arctic PRISM double sampling method, but it also shows that some modifications of this method may be appropriate for western Alaska, especially on the Yukon–Kuskokwim Delta. In this final section, we briefly identify some of the modifications that should be considered.

As noted above, surveying shorebirds in western Alaska raises three problems that rarely occur in other areas: local densities are often extremely high, birds often nest and forage in different habitats, and migrants are often present during the survey period. All of these factors suggest that detection rates may be more variable on the Yukon–Kuskokwim Delta than in other, more northerly areas. If so, this would suggest that some resources should be shifted from rapid surveys to intensive surveys. In the rest of the arctic, we have expended roughly equal resources on rapid and intensive surveys. In western Alaska, perhaps two-thirds of the effort should be devoted to intensive surveys. Note that we are not saying surveys here will be more expensive. To the contrary, it seems likely that the higher bird densities will make it easier to obtain the target CV. The opportunity for adjusting allocation of effort to take account of local conditions is one of the great advantages of the double sampling method.

The problems mentioned above, along with the large number of species encountered, also suggest the need for more extensive training prior to the surveys. Such training, for example, should identify species that may exhibit courtship but still be migrants, and it should emphasize that habitats used for nesting and foraging may differ. Another option to consider is surveying rapid plots more than once. This is being done in a study of riparian birds in Arizona due to the long breeding season there (E. Juarez, pers. comm.). A single estimate is derived from the two (or three) rapid surveys. This obviously necessitates an increased survey effort, but there is no reason that plots have to be surveyed only once. The increased correlation between rapid and intensive survey results may more than compensate for the reduction in sample size for a given level of effort. As a qualitative guideline, if rapid surveyors feel they are largely guessing about numbers present (e.g., because some birds may be migrants), then consideration should be given to repeated visits on rapid surveys. One caution about this approach, however, is that repeating rapid surveys will surely change the detection ratio, so a large enough sample of intensive plots must be available to calculate detection ratios separately for these surveys.

Given these modifications, and probably others yet to be suggested, we believe the basic double sampling method will provide estimates with acceptable precision and bias, in both the point and interval estimates, for a reasonable total cost. More detailed suggestions about needed sample sizes for surveys in western Alaska are contained in Bart and Smith (chapter 13, this volume) and Bart et al. (chapter 15, this volume).

ACKNOWLEDGMENTS

In the Alaska Maritimes study area, T. Godfrey assisted with surveys. We relied heavily on the advice of J. Williams, S. Ebbert, and V. Byrd to determine whether surveys could be conducted. We also appreciate the support provided by the captain and crew of the M/V *Tiglax*. In the Alaska Peninsula study area, C. Adler, B. Blush, R. Kaler, A. Leppold, and S. E. Savage conducted the surveys. The Alaska Peninsula/Becharof NWR provided funding. In the Yukon Delta study area, the surveys were conducted by T. Booms, F. Broerman, C. Fitzpatrick, M. Johnson, S. Nebel, A. Niehaus, J. Prather (deceased), D. Ruthrauff, M. Sardy, and M. Spies. The U.S. Fish and Wildlife Service provided funding and logistic support. Yukon Delta NWR pilot G. Walters and wildlife biologist D. Gillikin were particularly helpful. In the Selawik study area, L. A. Ayers, G. Doney, T. Moran, M. Prehoda, C. Villa, and G. Wiles assisted with the surveys. The U.S. Fish and Wildlife Service, Selawik NWR, provided funding and logistic support. We thank S. E. Savage and R. Ydenberg for their constructive reviews of an earlier draft of this paper. The U.S. Geological Survey provided funds for survey design, analysis, and preparing this chapter. Although the U.S. Geological Survey and U.S. Fish and Wildlife Service provided significant financial and logistical support for this study, the views expressed herein may not reflect the views of either agency.

ONLINE CONTENT

Abstracts are available in French and Spanish from www.ucpress.edu/go/sab.

Une traduction du résumé est disponible en français.

Una traducción del resumen está disponible en español.

CHAPTER FOUR

North Slope of Alaska

Jonathan Bart, Stephen Brown, Brad A. Andres, Robert Platte, and Ann Manning

Abstract. We used a combination of ground and aerial surveys to characterize the abundance and distribution of shorebirds and other birds on the North Slope of Alaska. The double sampling method, which we used for the ground surveys, is described in Bart et al. (chapter 2, this volume). The aerial surveys were conducted during 1992–2005 and covered most of the study area. We present numbers recorded, estimated densities and population sizes, and habitat relationships for shorebirds and other species. Most species occurred in higher density, and had much larger populations, in the National Petroleum Reserve–Alaska than west of the Colville River. The most abundant shorebirds were Semipalmated Sandpipers and Pectoral Sandpipers, with estimated populations of about 1.2 million birds each; and Long-billed Dowitchers, Red and Red-necked Phalaropes, and Dunlin, with estimated populations of 500,000–700,000 each. The most abundant waterfowl were Northern Pintails, Greater White-fronted Geese, and Long-tailed Ducks, with estimated populations of 200,000–300,000 each. Glaucous and Sabine's Gulls, Arctic Terns, and Parasitic and Long-tailed Jaegers each had estimated populations of about 30,000–100,000. Lapland Longspurs, Savannah Sparrows, and Willow Ptarmigan were the most common landbirds, with estimated populations of about 1 million each. All but a few species were most common in wetlands and least common in uplands.

Key Words: Alaska, landbirds, North Slope, population estimates, shorebirds, surveys, waterbirds, waterfowl.

The North Slope of Alaska is a major nesting area for a wide variety of bird species, including shorebirds, waterfowl, gulls and terns, and landbirds (National Research Council 2003, Johnson et al. 2007). However, there has never been an analysis of the density of these species across the entire range of habitats in the region. Previous studies have provided either descriptions of the ranges of these species within the region or densities for some species in some parts of this larger area (Mallek et al. 2004, Larned et al. 2005, Brown et al. 2007, Johnson et al. 2007,

Bart, J., S. Brown, B. A. Andres, R. Platte, and A. Manning. 2012. North Slope of Alaska. Pp. 37–96 *in* J. Bart and V. Johnston (editors). Arctic shorebirds in North America: a decade of monitoring. Studies in Avian Biology (no. 44), University of California Press, Berkeley, CA.

unpublished U.S. Fish and Wildlife Service reports). This monograph provides the results of the first such analysis, for all arctic-nesting birds that occur on the North Slope.

The primary goals of this study were (1) to collate and present results from bird surveys conducted during many years, and provide graphic displays of bird numbers recorded; (2) to estimate the density of each species, and calculate estimates of total population size in the study area; and (3) to examine interrelationships among habitat types and nesting densities by species to determine which habitats are most important for each species.

Measurements of density and total population are important for understanding both the biology of these species and the relative importance of the region for their global populations. Many conservation systems use the relative contribution of a specific area to total population size to determine importance for conservation. For example, the Western Hemisphere Shorebird Reserve Network (WHSRN; www.whsrn.org) and the Ramsar Convention on Wetlands of International Importance, especially as waterfowl habitat (Ramsar Convention; www.ramsar.org), both use the percentages of a flyway population as a conservation threshold. Similar criteria have been developed by the Important Bird Areas Program of Birdlife International (www.birdlife.org/action/science/sites). All of these programs rely on having both accurate overall population size estimates for each species and estimates of the proportion of the population using particular areas of interest. The development of accurate population estimates for the North Slope region of Alaska will contribute significantly to both of these goals by providing the first regional population estimates and by providing a context against which to evaluate total population size estimates published for various species.

Major threats to breeding birds in the North Slope region include oil and gas development and climate change. Understanding the current densities of breeding birds in the region is a prerequisite to determining impacts of any future changes in the wetland and tundra habitats supporting these birds.

Oil and gas development have been under way on the North Slope since 1977, and development has expanded to the point that assessments of cumulative impacts on the biology of arctic birds and mammals are warranted (Gilders and Cronin 2000, National Research Council 2003). Significant oil and gas development has been proposed for the Arctic National Wildlife Refuge and is already under way in parts of the National Petroleum Reserve–Alaska and the state lands in between (National Research Council 2003). Impacts from these developments on nesting birds may occur from a variety of factors, including direct loss of habitat from development of roads, drilling pads, pipelines and related infrastructure, and supporting activities such as gravel mining. In addition, indirect impacts may occur from increases in dust or changes in hydrology including snow accumulation patterns (Auerbach et al. 1997, National Research Council 2003). Other indirect impacts may include changes in predator populations that could negatively affect nesting birds (Eberhardt et al. 1983, National Research Council 2003, Liebezeit et al. 2009).

Climate change impacts are expected to be severe in the arctic compared to other regions (ACIA 2005). A variety of different types of impacts may be important. Northward expansion of shrub habitats has already been documented in the arctic, and is expected to increase with the rising temperature in the region (Sturm et al. 2001, ACIA 2005). The resulting loss of tundra habitats may reduce breeding areas for many species dependent on graminoid tundra. Sea level rise projected to occur may reduce coastal habitat availability or quality for species that nest or forage in coastal areas (Jorgensen and Ely 2001, Rehfisch and Crick 2003). Changes in the seasonal availability of food sources such as invertebrates may also change, affecting the ability of each species to time their life history to coincide with peaks in available food (the "match/mismatch" hypothesis; Durant et al. 2007).

The population size estimates presented here are the first for the entire North Slope region, and will be critical in determining future impacts from ongoing development and climate change.

METHODS

Regions

We distinguished between the Arctic Coastal Plain (ACP) and the Foothills to the south (Fig. 4.1). The ACP was the area covered by U.S. Fish and Wildlife

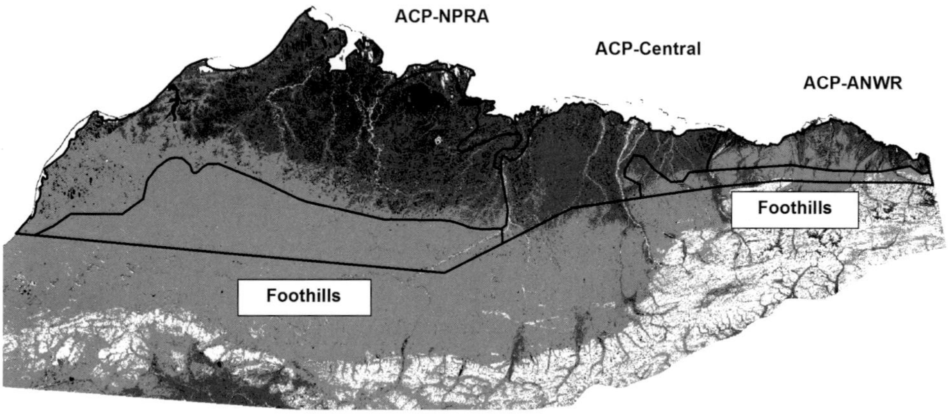

Figure 4.1. Regions used in the study. Light gray indicates drier areas (source for vegetation: Muller et al. 1999). Sampling intensity in the ACP-NPRA was much higher in the northeast in the delta of the Colville River and along Fish Creek. This area is delineated above (and in Fig. 4.2), but separate estimates for the area are not reported in the tables because it did not have ecological significance. A fifth area for which estimates are reported, the Colville River, includes the riparian areas along the Colville River upstream from the delta. It is too small to show on the figure.

Service aerial surveys; the Foothills included our study area south of these regions. We also subdivided the ACP into the National Petroleum Reserve of Alaska (NPRA), the Central region, and the Arctic NWR. A small region in the northeastern part of the NPRA was also distinguished because sampling intensity was particularly high there. The areas of the regions were ACP-NPRA (34,489 km^2), ACP-Central (10,825 km^2), ACP-Arctic NWR (4,793 km^2), Foothills (23,254 km^2), and Colville River (30 km^2). In the analyses below, results from this region are combined with the rest of the ACP-NPRA region.

Results from both ground and aerial surveys were depicted using a grid of square cells, 6 km on a side (Fig. 4.2). For both data sets, we simply calculated the overall density of observations. For the aerial surveys, this provided an unbiased estimate because locations for the aerial surveys were selected without regard to habitat. For the ground surveys, the simple average ignored habitat within the cell, whereas the plots were usually concentrated in wetlands and moist habitat. However, the purpose of producing these maps was simply to show the general distribution of sightings using a method that involved the least possible manipulation of the data and that produced a comprehensible image (which displaying the counts in each plot did not due to the large number of plots). Three density categories were defined for each map using zero for one category and an intermediate threshold to define the other two categories. The threshold was chosen so that about one-third of the cells were in the high-density category. This approach produced maps that highlighted the areas of highest

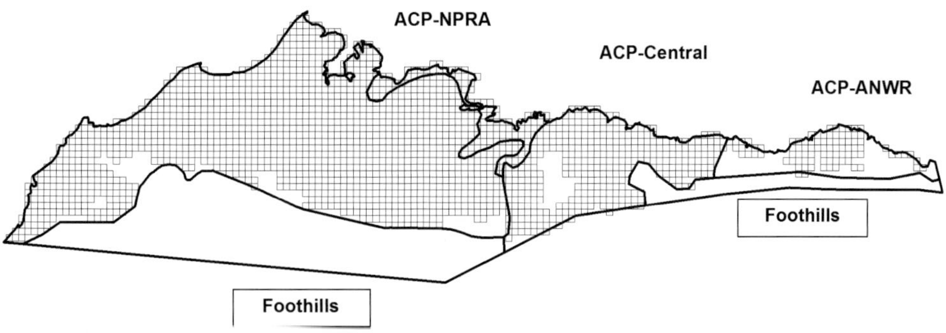

Figure 4.2. Cells used to display results from ground and aerial surveys.

density. We used these maps for various purposes, including a description of the distribution of each common species within our study areas. Our descriptions update those of Johnson et al. (2007), which were based solely on our ground surveys and on presence (rather than density) data.

Ground Surveys

Bart et al. (chapter 2, this volume) discuss methods used to select and survey the plots and analyze the resulting data. Here, we discuss only the methods that are specific to this chapter.

Habitats (wetland, moist, upland, unsuitable) were defined using the landcover map by Muller et al. (1999). At the time the study was initiated, the Muller et al. (1999) landcover map was the only GIS layer available that covered the entire North Slope region. The pixel size was 100 m; eight classes were identified, which we combined to four classes. More recently, other landcover maps have become available with 30-m pixels and more classes (e.g., National Land Cover Dataset, Landfire, Ecosystems of Northern Alaska), but the more general coverage was adequate and perhaps even an advantage for our work. The cross-walk between our categories and the Muller categories (and codes) was: wet = wet (5); moist = moist-grass prostrate (2) or moist-tussock grass shrub (9); uplands = tussock tundra (3) or moist-low shrub (4); unsuitable = dry prostrate barrens (1), water (6), or ice (7). Muller category eight, clouds, was rare in our study area but was classified as moist (because that was the most common type). Plots were assigned to habitat types based on whichever habitat was most common. For example, if the proportions of suitable habitat covered by wetlands, moist areas, and uplands were 0.3, 0.4, and 0.3, respectively, then the plot was assigned to the moist type. It is thus important to realize that the habitat in plots was heterogeneous, even though we refer to them using terms like "moist." This is perhaps especially important in interpreting numbers in plots classified as upland. Pure upland, in this area, usually had very few shorebirds. But plots classified as upland—because that was the most common type—often had small wetlands with shorebirds. It must also be recognized that different studies, using different definitions to classify plots, might produce different densities in, for example, "moist" plots. The assignments were thus of greatest use in letting us concentrate survey efforts in areas with the most birds rather than in elucidating habitat relationships (a task better done with the models we constructed).

We used the Arctic PRISM extension (see Bart et al., chapter 2, this volume) to partition the entire study area into 197,279 plots, most of which covered approximately 16 ha. We assigned plots to zones for logistic reasons. Plots were also assigned to strata based on region and habitat. Ideally, plots to be surveyed would have been selected using the comprehensive sampling frame. The process for partitioning the entire study area, however, was only worked out in the later years of the study. During the earlier years, plots were selected using a variety of random selection methods. On the Colville Delta and surrounding areas, plots were selected by partitioning that part of the survey area into plots manually (i.e., using GIS methods but without automatic creation and modification of plots) and plots to be surveyed were then selected randomly. Elsewhere, the most common method for choosing plots was to randomly select a point and partition the surrounding area of a predetermined size and shape into plots, each assigned to a habitat type. Plots to be surveyed were then randomly selected, with a restriction to yield the desired number of plots in each habitat type. Rapid surveys were conducted on 637 plots (Table 4.1). Sampling intensity varied substantially between regions (Fig. 4.3; Table 4.1).

We estimated detection ratios using six camps and 28 intensive plots (Table 4.2). Intensive plots were usually surveyed four times by rapid surveyors (range 1–8, Table 4.2). We calculated species-specific detection ratios for the 13 shorebird species encountered most often (Table 4.3). All detection ratios based on more than five birds were between 0.75 and 0.90. The only statistically significant $P < 0.05$) differences between species were for Buff-breasted and Pectoral Sandpipers (for scientific names, see Appendix C); however, the sample size for Buff-breasted Sandpipers was small, so the SE estimate is unreliable. Furthermore, with 13 species being compared, it is hardly surprising that one contrast was significant. The remaining estimates varied widely, as expected with small sample sizes. We therefore used all shorebirds combined to estimate an overall detection ratio. The estimate was 0.81,

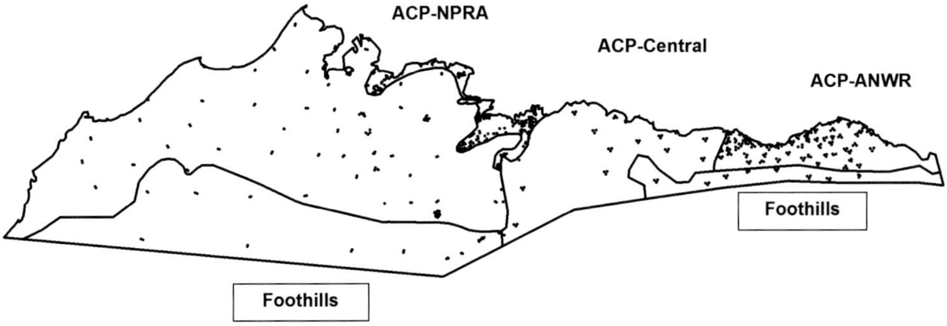

Figure 4.3. Plots in the ground survey (dots indicate groups of up to several plots).

TABLE 4.1
Areas and numbers of plots in each stratum.

	Area (km²)				Number of plots surveyed			
Region	Wetlands	Moist	Uplands	Total	Wetlands	Moist	Uplands	Total
ACP-Arctic NWR	1,207	2,233	1,353	4,793	85	70	24	179
ACP-Central	2,398	8,162	266	10,825	15	26	1	42
ACP-NPRA	8,804	15,324	10,362	34,489	205	142	10	357
Foothills	152	3,612	3,486	7,250	2	8	28	38
Colville River	10	10	10	30	0	2	19	21
Totals	12,570	29,342	15,476	57,388	307	248	82	637

TABLE 4.2
Number of intensive plots, indicated shorebird pairs, and rapid surveys of the intensive plots.

Camp				
Region	Year	No. of plots	No. of indicated pairs	No. of rapid surveys
ACP-ANWR	2002	4	27	16
ACP-ANWR	2004	4	14	13
ACP-NPRA	1998	8	143	30
ACP-NPRA	1999	4	65	28
ACP-NPRA	2000	4	84	16
ACP-NPRA	2001	4	59	32

with a SE of 0.08. We used these estimates in estimating densities and population sizes and their SEs for all shorebird species. Not using species-specific rates in our main analysis, however, certainly does not mean the rates are all identical. They undoubtedly do vary. For species with relatively large sample sizes, we therefore also present estimates with the species-specific detection ratio.

Although the PRISM surveys were designed for shorebirds, we recorded all species encountered. During intensive surveys, however, surveyors did not attempt to find all species; they only surveyed for shorebirds. As a result, we do

TABLE 4.3

Detection ratios for shorebirds, their standard errors (SE), and 95% confidence intervals (95% CI).

Species	Number on intensive plots	Detection ratio	SE	95% CI Lower bound	95% CI Upper bound
Semipalmated Sandpiper	111	0.75	0.13	0.50	1.00
Pectoral Sandpiper	96	0.83	0.06	0.71	0.95
Red-necked Phalarope	68	0.80	0.14	0.53	1.07
Red Phalarope	43	0.77	0.17	0.44	1.10
Dunlin	24	0.90	0.16	0.59	1.21
Black-bellied Plover	10	0.85	0.17	0.52	1.18
American Golden-Plover	9	0.88	0.26	0.37	1.39
Ruddy Turnstone	9	0.83	—	—	—
Long-billed Dowitcher	5	2.16	1.20	0.00	4.51
Buff-breasted Sandpiper	5	0.60	0.03	0.54	0.66
Stilt Sandpiper	4	0.61	0.11	0.39	0.83
Western Sandpiper	4	0.78	—	—	—
Bar-tailed Godwit	3	0.75	0.26	0.24	1.26

not have detection ratios for non-shorebirds. For non-shorebirds, we therefore report the total number of birds seen and the estimated number of pairs on rapid plot surveys. We assumed that a single indicated a pair (unless it was considered to be breeding off the plot). We calculated uncorrected, relative densities using these data.

Habitat Analyses

Habitat associations were investigated using regression analyses of the results from rapid plots. Most species occurred throughout our study area, but for a few (e.g., Western Sandpiper) that clearly occurred only in part of the study area, we deleted plots that were outside the range as depicted on the maps prepared by Ridgely et al. (2007). This section describes our modeling process.

We recorded spatial and habitat variables for each surveyed plot. The spatial variables were longitude (*ew*), mean (among pixels) elevation (*elev*), mean distance to the nearest wetland pixel (*diswet*), and smallest distance from the center of the plot to the coast (*discoa*). The habitat variables were recorded at three scales: micro (within the plot), meso (within 1 km of the center of the plot, and macro (within 10 km of the center of the plot). At each scale we recorded the proportion of the suitable area covered by pixels classified as wetlands, moist areas, and uplands. We thus had nine habitat variables: *micwet*, *micmoi*, and *micupl* at the micro scale; *meswet*, *mesmoi*, and *mesupl* at the meso scale; and *macwet*, *macmoi*, and *macupl* at the macro scale.

Habitat relationships were explored by building regression models that predicted density as a function of spatial and habitat variables. We used SAS (ver. 8.2, SAS Institute, Cary, NC) for the analysis. The response variable—number of birds recorded—was distinctly non-normal. Errors were correlated and random effects were present due to the selection of clusters. The recommended SAS procedure for such cases is GLIMMIX (Littell et al. 2006:16). GLIMMIX uses residual maximum likelihood to estimate parameters. With this approach, models that differ in their fixed effects cannot legitimately be compared using log likelihoods or AICs (Littell et al. 2006:754). We therefore compared models on the basis of the *P*-values of their regression coefficients.

Although we had relatively few variables for our sample size (637 plots), the number of possible models, with squared terms and interactions, far exceeds our sample size. It was thus necessary to reduce the number of models considered using biological knowledge. Below, we use a standard notation to describe the models in which only the names of the independent variables are provided. For example, consider the model

$$d_i = b_0 + b_1 ew_i + b_2 elev_i + b_3 ew_i{*}elev_i$$

where d_i, ew_i, and $elev_i$ are the observed density, longitude measured as the distance west from a north–south line at the east edge of the study area, and elevation, respectively, on the *i*th plot. This model would be described as *ew*, *elev*, and *ew**elev*. In listing models below, we generally start with the most complex model. Our model selection process depended in part on the number of positive counts. For abundant species we proceeded as follows:

1. Select the best model using the spatial variables *ew* and *elev*. The models were *ew ewsqd elev ewsqd**elev*, *ew ewsqd elev*, *ew elev ew**elev*, *ew elev*, *ew ewsqd*, *ew*, and *elev*. We included *ewsqd* because some species were most abundant in the middle (east to west) of the study area. *Elevsqd* was not included because the relationship between density and elevation appeared always to be monotonic (usually decreasing density with increasing elevation) and in most cases was not clearly different from linear. We followed the standard practices of calculating interactions with the highest-order terms of the main effects and of not including interactions unless the main effects were significant. Based on many years in the study area, we expected *ew*, *elev*, and *ew**elev* to be important correlates of density for many species.
2. Select the best model that adds habitat variables. We only considered the proportion of the plot covered by wetlands and moist areas since also including uplands would have been redundant. In these models, the intercept gives the predicted density if the plot was entirely upland (because then *x*-values for wetlands and moist areas are both zero) and the coefficients for wetlands and moist habitats express the change in density due to changing some of the upland to wetlands or moist areas. We did not consider higher-order terms or interactions in part because we doubted they would have much explanatory power and in part to keep the number of models relatively small. We therefore needed to consider three models at each spatial scale. Let *spatial* indicate the variables selected in Step 1. Then the models evaluated would be *spatial micwet micmoi*, *spatial micwet*, *spatial micmoi*, and so on for the meso and macro scales. This stage thus involved considering an additional nine models. We deleted any nonsignificant variables. Thus, *elev* might have been included at Step 1 but might be omitted from the model selected at Step 2.
3. Add distance to the coast. For a few species found only near the coast, we added *discoa*. The range for such species only included areas near the coast, so *discoa* was not necessarily significant for such species. This variable was added as the last step in the model-building process.

The process described above involves comparing approximately 20 models, though the model sets varied between species depending on which variables were selected in Step 1. This number seemed reasonable when we had hundreds of positive counts and all or nearly all of our 637 plots were within the species range. When we had only a few dozen positive counts, however, evaluating 20 models seemed likely to result in over-fitting (i.e., obtaining a model unlikely to have been chosen had we obtained a different random sample). Since we could not use AICs to compare models, it was difficult to know how serious this potential problem was. We elected only to consider habitat variables at the meso and macro scale when we had at least 100 positive counts. Thus for species with fewer than 100 positive counts, we compared roughly a dozen models (depending on the number compared in Step 3) rather than roughly 20 models.

Correlations between the independent variables were generally small (Table 4.4), as were correlations between density and independent variables (Table 4.5). These weak relationships meant that we could not expect habitat models to have high

TABLE 4.4

Correlation coefficients between independent variables.

Variable	Plot area	*ew*	*elev*	*micwet*	*micmoi*	*micupl*	*meswet*	*mesmoi*	*mesupl*	*macwet*	*macmoi*	*macupl*
ew	0.08											
elev	−0.08	−0.18										
micwet	0.1	0.06	−0.48									
micmoi	−0.04	−0.05	−0.08	−0.62								
micupl	−0.08	−0.02	0.66	−0.58	−0.28							
meswet	0.06	0.06	−0.61	0.76	−0.3	−0.62						
mesmoi	0.03	−0.03	−0.16	−0.27	0.66	−0.36	−0.4					
mesupl	−0.08	−0.03	0.74	−0.56	−0.22	0.91	−0.69	−0.38				
macwet	0.06	0.13	−0.67	0.63	−0.15	−0.61	0.87	−0.21	−0.71			
macmoi	0.08	−0.1	−0.22	−0.05	0.47	−0.43	−0.14	0.8	−0.48	−0.19		
macupl	−0.1	−0.05	0.74	−0.53	−0.16	0.82	−0.69	−0.32	0.95	−0.78	−0.47	
discoa	0.03	0.41	0.13	−0.1	0.06	0.07	−0.18	0.1	0.1	−0.22	0.1	0.13

TABLE 4.5

Correlation between density and independent variables for selected species (all species had >40 records).

Variable	AMGP	BARG	BBPL	BBSA	DUNL	LBDO	PESA	REPH	RNPH	RUTU	SESA	STSA	WESA
ew	−0.08	0.08	0.17	−0.04	0.28	0.17	0.04	0.12	0.02	0.02	0.13	−0.01	0.43
elev	0.06	−0.05	−0.11	−0.06	−0.20	−0.11	−0.18	−0.18	−0.12	−0.06	−0.19	−0.05	−0.08
micwet	0.05	0.05	0.17	−0.02	0.23	0.16	0.21	0.21	0.17	0.02	0.27	0.12	0.07
micmoi	−0.02	0	−0.10	0.06	−0.13	−0.07	−0.12	−0.11	−0.11	0.01	−0.15	−0.08	0.07
micupl	−0.04	−0.06	−0.11	−0.05	−0.16	−0.12	−0.14	−0.14	−0.09	−0.04	−0.18	−0.08	−0.11
meswet	−0.05	0.10	0.20	0.14	0.25	0.17	0.18	0.19	0.13	0.06	0.29	0.10	−0.07
mesmoi	0	−0.04	−0.13	−0.13	−0.12	−0.07	−0.07	−0.07	−0.07	−0.02	−0.13	−0.05	0.18
mesupl	0.05	−0.07	−0.12	−0.05	−0.17	−0.11	−0.13	−0.15	−0.08	−0.05	−0.19	−0.07	−0.07
macwet	−0.06	0.12	0.23	0.12	0.31	0.18	0.18	0.17	0.13	0.07	0.31	0.08	−0.13
macmoi	−0.01	−0.05	−0.14	−0.11	−0.14	−0.08	−0.03	−0.02	−0.07	−0.03	−0.11	−0.02	0.17
macupl	0.06	−0.07	−0.14	−0.05	−0.20	−0.12	−0.14	−0.15	−0.08	−0.05	−0.20	−0.07	−0.02
discoa	0.01	0.07	0.03	−0.04	−0.08	0.12	−0.12	−0.15	−0.04	−0.06	−0.02	0.06	−0.09

NOTE: Species are displayed with AOU 4-letter codes. See Appendix C.

explanatory power even if their coefficients were highly significant.

Aerial Surveys

Data from two aerial surveys were used in this analysis. The Arctic Coastal Plain Survey was flown in late June and early July. The survey area covered 61,645 km^2 including most of the area surveyed on the ground, but not extending into the foothills. A systematic sample of transects 18.8 km apart was flown with two observers. The plane flew 38 m above ground at 176 km/hr. Birds within 200 m of the airplane, on both sides of the transect, were recorded. The North Slope Eider Survey was similar, but the survey area was smaller, transects were half as far apart, and the survey was conducted during the middle part of June. We used data collected on both surveys during 1992–2005.

Survey results were summarized by determining, for each cell in the grid (Fig. 4.2), the total area surveyed (in all years) and the total number of birds recorded, defined as two times the number of singles plus the number in pairs and groups. We then calculated the density of observations (per km^2) of each species in each cell as "number recorded in all years/area surveyed in all years." Maps were then prepared for each species showing the numbers detected in each cell.

SPECIES ACCOUNTS

Shorebirds

Black-bellied Plover

Black-bellied Plovers were encountered on 142 plots and were judged to be breeding on 119 plots. Nearly all records were from the ACP-NPRA region (Fig. 4.4a). Sightings were split about equally between pairs (including nests and probable nests) and single birds. They were strongly associated with wetlands and almost never occurred in uplands (Table 4.6). The numbers of plots with four, five, and six sightings were seven, one, and one, and no plots had more than six sightings. Thus, they rarely occurred in high density. The estimated number of breeding pairs on surveyed plots was 175.

Ten Black-bellied Plovers were breeding in the intensive plots, all at one camp (Table 4.7). At that camp, the species was present on six of seven plots. Surveyors recorded a few birds on plots where the species was not breeding. The overall detection ratio was 0.85.

Estimated densities were highest in wetlands, lowest in uplands, and averaged a little less than 3 birds/km^2 with a CV of 0.23 (Table 4.6). Population size for the entire study area was estimated to be about 200,000 birds, the majority in the ACP-NPRA region.

The variables *ew*, *elev*, and *ewsqd* were each highly significant in at least one model and were therefore selected as the spatial variables. Both *wet* and *moi* were significant in the univariate models (models with one independent variable) and had about the same significance at the micro and meso scales but lower significance at the macro scale. When *wet* and *moi* were both in the model, *moi* was nonsignificant. We therefore used *micwet* as the only habitat variable. Adding *discoa* improved the fit, but adding *diswet* did not. All the variables and the intercept were highly significant in the final model (Table 4.8). According to the final model, density was higher in the west, at higher elevation, in wetlands, and farther from the coast.

The range map in the Birds of North America (BNA) species account is generally consistent with our results, though it shows the range extending farther east than we found birds (Paulson 1995). The BNA also states, "where tundra varies from low and wet to higher and drier, latter preferred." This would probably suggest to most readers that our upland habitat should have been preferred, whereas it was strongly avoided in our study area and even the density in moist areas was less than half the density in wetlands (Table 4.6). The BNA also makes it clear that habitat preferences vary across the range depending on the environment.

American Golden-Plover

American Golden-Plovers were encountered on 180 plots and were judged to be breeding on 122 plots. Records were widely distributed across the study area but were markedly less common in the western NPRA, although this may have been an artifact caused by small sample size (Fig. 4.4b). They were found in every region (Table 4.6). More than half were found in wetlands and most of the remaining records came

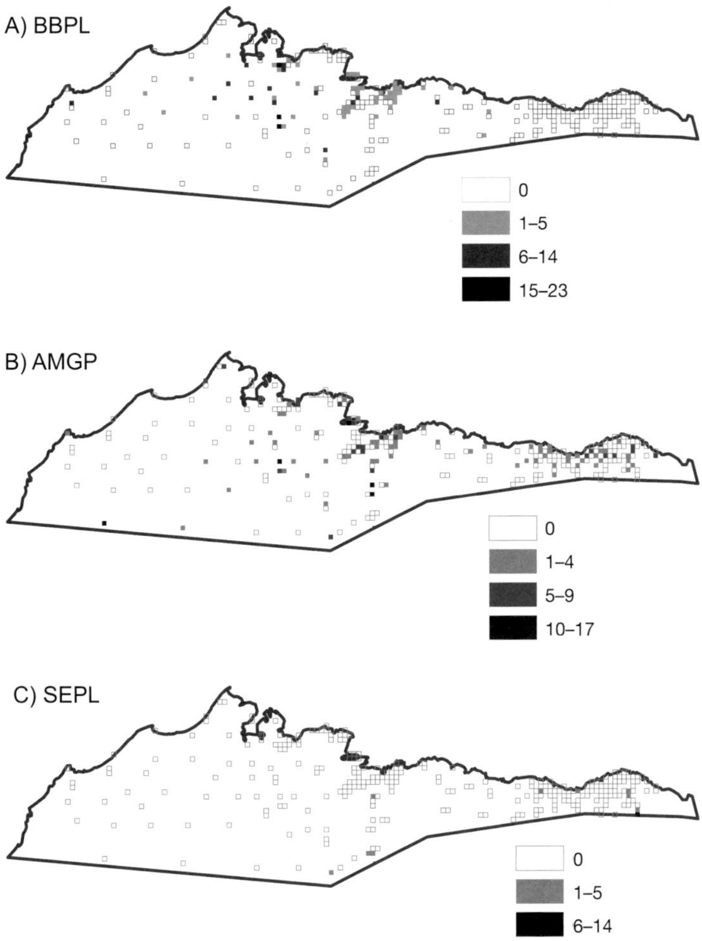

Figure 4.4. Densities (birds/km^2) of Black-bellied Plovers (a), American Golden-Plovers (b), and Semipalmated Plovers (c) recorded on the rapid surveys.

from moist habitat. The number of plots with three, four, and more than four indicated pairs were three, one, and zero. We did find one intensive plot with four pairs (all verified by nests); the other five plots with this species had one pair each. Thus, the species was generally distributed evenly rather than in a clumped fashion, but clusters of nesting pairs did exist at least rarely. The estimated number of breeding pairs on surveyed plots was 159.

Nine American Golden-Plovers were present on the intensive plots. The overall detection ratio was 0.88 (Table 4.7). As with most species, low numbers (including zero) tended to be overestimated and the high count of five actually present was underestimated.

The estimated densities, within regions, were highest in moist habitats in the ACP, though not in the Foothills region (Table 4.6). This is one of the few species that had substantial densities in uplands, though this was true only for the Foothills region. Densities in uplands in the ACP were low. The density across all areas was 3.75 birds/km^2. The estimated population size was about 275,000 for the entire study area. The Foothills region had a substantial estimated population but with a large CV.

None of the spatial parameters had clear predictive power. The only significant term in all of the models was *ew*elev* in the model *ew elev ew*elev*. We therefore did not include any spatial parameters in subsequent models. Among the habitat parameters, *wet* was not significant in any models but *moi* was significant at the micro and meso scales, and nearly significant at the macro scale, in both the univariate and bivariate models.

TABLE 4.6

Number of shorebirds recorded on rapid surveys, estimated densities and population size (with SEs).

Refer to Appendix C for common and scientific names of species 4-letter codes.

				Birds/km^2 (SE)				Population size	
Species	Region	Total recorded	Est. n pairs	Wetlands	Moist areas	Uplands	All habitats	Estimate (SE)	CV
AMGP	ACP-NPRA	113	114	2.52 (0.74)	8.63 (2.67)	0 (0)	3.82 (1.08)	131,827 (37,373)	0.28
	ACP-Central	19	8	1.05 (1.10)	3.34 (1.24)	0 (0)	2.22 (0.71)	24,061 (7,652)	0.32
	ACP-ANWR	59	29	2.44 (0.78)	4.27 (1.16)	0.41 (0.32)	2.46 (0.56)	11,805 (2,690)	0.23
	Foothills	10	6	0 (0)	0.7 (0.66)	5.1 (4.04)	4.93 (3.81)	114,564 (88,647)	0.77
	Colville River	2	2	0 (0)	0 (0)	1.77 (1.4)	0.99 (0.72)	20 (14)	0.73
	All	203	159	2.05 (1.05)	6.72 (0.95)	2.46 (2.92)	3.75 (1.15)	275,506 (84,298)	0.31
BBPL	ACP-NPRA	168	166	10.54 (2.18)	3.24 (1.28)	1.32 (1.43)	4.13 (1.03)	142,288 (35,599)	0.25
	ACP-Central	13	6	5.03 (2.97)	1.65 (1.15)	0 (0)	3.09 (1.41)	33,485 (15,285)	0.46
	ACP-ANWR	2	0	0 (0)	0 (0)	0 (0)	0 (0)	0 (0)	0
	Foothills	3	3	0 (0)	1.01 (1.00)	0 (0)	0.16 (0.16)	3,699 (3,684)	1.00
	Colville River	0	0	0 (0)	0 (0)	0 (0)	0 (0)	0 (0)	0
	All	186	175	8.13 (0.53)	2.59 (0.89)	0.61 (2.84)	2.74 (0.62)	201,162 (45,530)	0.23
BARG	ACP-NPRA	75	55	1.18 (0.65)	0.75 (0.39)	3.17 (3.23)	1.81 (1.30)	62,494 (44,797)	0.72
	ACP-Central	13	1	0 (0)	0.40 (0.40)	0 (0)	0.21 (0.21)	2,247 (2,272)	1.01
	ACP-ANWR	0	0	0 (0)	0 (0)	0 (0)	0 (0)	0 (0)	0
	Foothills	1	0	0 (0)	0 (0)	0 (0)	0 (0)	0 (0)	0.77
	Colville River	0	0	0 (0)	0 (0)	0 (0)	0 (0)	0 (0)	0
	All	89	56	0.72 (0.59)	0.57 (0.47)	1.47 (1.04)	1.03 (0.73)	75,602 (53,252)	0.7

WHIM	ACP-NPRA	13	12	0.01 (0.01)	1.04 (0.79)	0 (0)	0.39 (0.30)	13,520 (10,375)	0.77
	ACP-Central	0	0	0 (0)	0 (0)	0 (0)	0 (0)	0 (0)	0
	ACP-ANWR	5	2	0 (0)	0.54 (0.42)	0 (0)	0.22 (0.18)	1057 (847)	0.80
	Foothills	4	2	0 (0)	0.10 (0.11)	0.05 (0.05)	0.05 (0.04)	1232 (974)	0.79
	Colville River	1	1	0 (0)	0 (0)	0.40 (0.39)	0.22 (0.22)	4 (4)	0.98
	All	23	17	0.01 (0.15)	0.73 (0.17)	0.03 (0.23)	0.25 (0.17)	18,243 (12,529)	0.69
RUTU	ACP-NPRA	42	26	0.10 (0.06)	0.08 (0.06)	0 (0)	0.06 (0.04)	2,240 (1,216)	0.54
	ACP-Central	0	0	0 (0)	0 (0)	0 (0)	0 (0)	0 (0)	0
	ACP-ANWR	6	5	0.39 (0.23)	0.59 (0.49)	0 (0)	0.33 (0.21)	1,602 (1,019)	0.64
	Foothills	0	0	0 (0)	0 (0)	0 (0)	0 (0)	0 (0)	0.74
	Colville River	0	0	0 (0)	0 (0)	0 (0)	0 (0)	0 (0)	0
	All	48	31	0.08 (0.02)	0.10 (0.03)	0 (0)	0.05 (0.02)	3,419 (1,430)	0.42
DUNL	ACP-NPRA	303	350	30.37 (5.41)	5.6 (1.92)	4.83 (3.47)	10.91 (2.29)	376,370 (79,094)	0.21
	ACP-Central	19	14	8.21 (3.57)	3.89 (1.88)	0 (0)	5.68 (2.01)	61,451 (21,711)	0.35
	ACP-ANWR	22	14	2.87 (1.21)	0.04 (0.05)	0 (0)	0.67 (0.30)	3,232 (1,449)	0.45
	Foothills	0	0	0 (0)	0 (0)	0 (0)	0 (0)	0 (0)	0
	Colville River	0	0	0 (0)	0 (0)	0 (0)	0 (0)	0 (0)	0
	All	344	378	21.74 (1.01)	4.44 (1.92)	2.25 (4.66)	6.82 (1.34)	500,161 (98,089)	0.20
SESA	ACP-NPRA	1,127	1,174	66.24 (11.39)	24.72 (6.68)	4.77 (3.86)	26.18 (4.77)	902,819 (164,476)	0.18
	ACP-Central	87	59	28.86 (11.06)	18.66 (4.8)	0 (0)	22.57 (5.99)	244,309 (64,890)	0.27
	ACP-ANWR	143	100	15.22 (3.57)	8.09 (2.42)	0.18 (0.19)	6.87 (1.52)	32,941 (7,281)	0.22
	Foothills	4	2	0 (0)	2.38 (2.28)	0 (0)	0.37 (0.37)	8,708 (8,532)	0.98
	Colville River	0	0	0 (0)	0 (0)	0 (0)	0 (0)	0 (0)	0
	All	1,361	1,335	51.17 (2.54)	20.97 (3.55)	2.23 (13.58)	17.98 (2.98)	1,319,225 (218,761)	0.17

TABLE 4.6 (*continued*)

TABLE 4.6 (CONTINUED)

Species	Region	Total recorded	Est. *n* pairs	Birds/km² (SE)				Population size	
				Wetlands	Moist areas	Uplands	All habitats	Estimate (SE)	CV
WESA	ACP-NPRA	51	44	5.63 (2.7)	5.33 (2.82)	15.25 (6.57)	9.29 (2.99)	320,451 (103,233)	0.32
	ACP-Central	0	0	0 (0)	0 (0)	0 (0)	0 (0)	0 (0)	0
	ACP-ANWR	1	1	0.14 (0.14)	0 (0)	0 (0)	0.03 (0.03)	153 (153)	1
	Foothills	0	0	0 (0)	0 (0)	0 (0)	0 (0)	0 (0)	0.98
	Colville River	0	0	0 (0)	0 (0)	0 (0)	0 (0)	0 (0)	0
	All	52	45	3.58 (2.44)	3.44 (2.68)	7.08 (2.1)	5.21 (1.65)	382,443 (120,837)	0.32
PESA	ACP-NPRA	947	982	36.7 (6.57)	10.65 (2.14)	23.14 (10.08)	21.61 (4.81)	745,179 (165,849)	0.22
	ACP-Central	81	47	22.44 (6.21)	18.96 (5.32)	0 (0)	19.88 (4.42)	215,272 (47,842)	0.22
	ACP-ANWR	224	111	17.58 (4.02)	7.74 (2.11)	2.35 (1.5)	8.04 (1.65)	38,552 (7,922)	0.21
	Foothills	8	4	0 (0)	3.39 (2.27)	0.10 (0.10)	0.61 (0.38)	14,294 (8,780)	0.61
	Colville River	6	5	0 (0)	0 (0)	4.92 (5.87)	2.75 (3.05)	55 (61)	1.11
	All	1,266	1,149	30.82 (2.02)	12.07 (2.23)	10.93 (6.30)	15.23 (2.92)	1,117,937 (214,529)	0.19
BBSA	ACP-NPRA	21	14	1.36 (1.20)	0.87 (0.62)	0 (0)	0.63 (0.36)	21,788 (12,398)	0.57
	ACP-Central	12	3	2.28 (2.31)	0.91 (0.65)	0 (0)	1.49 (1.08)	16,081 (11,733)	0.73
	ACP-ANWR	27	8	1.09 (0.79)	0.04 (0.05)	0 (0)	0.27 (0.19)	1,284 (894)	0.70
	Foothills	0	0	0 (0)	0 (0)	0 (0)	0 (0)	0 (0)	1
	Colville River	0	0	0 (0)	0 (0)	0 (0)	0 (0)	0 (0)	0
	All	60	25	1.61 (0.36)	0.76 (0.31)	0 (0)	0.57 (0.25)	41,541 (18,538)	0.45
LBDO	ACP-NPRA	372	350	26.99 (5.24)	11.28 (2.60)	10.51 (5.17)	14.47 (2.95)	499,030 (101,783)	0.20
	ACP-Central	31	13	5.19 (2.30)	5.09 (2.28)	0 (0)	4.97 (1.61)	53,818 (17,429)	0.32
	ACP-ANWR	38	9	1.03 (0.74)	0.98 (0.57)	0.43 (0.45)	0.79 (0.34)	3,802 (1,634)	0.43
	Foothills	1	1	0 (0)	1.01 (1.00)	0 (0)	0.16 (0.16)	3,699 (3,684)	1
	Colville River	1	1	0 (0)	0 (0)	0.98 (1.17)	0.55 (0.61)	11 (12)	1.11
	All	443	374	18.66 (1.51)	8.61 (1.58)	4.90 (4.13)	8.83 (1.68)	647,695 (123,559)	0.19

STSA	ACP-NPRA	138	119	6.01 (1.65)	1.97 (0.99)	0 (0)	2.09 (0.55)	72,130 (18,891)	0.26
	ACP-Central	17	8	3.89 (2.51)	2.39 (1.32)	0 (0)	2.98 (1.30)	32,209 (14,033)	0.44
	ACP-ANWR	25	16	4.22 (1.56)	1.38 (0.79)	0 (0)	1.53 (0.53)	7,331 (2,559)	0.35
	Foothills	1	0	0 (0)	0 (0)	0 (0)	0 (0)	0 (0)	0.98
	Colville River	0	0	0 (0)	0 (0)	0 (0)	0 (0)	0 (0)	0
	All	181	143	5.24 (0.39)	1.90 (0.60)	0 (0)	1.65 (0.37)	121,213 (27,381)	0.23
RNPH	ACP-NPRA	544	458	17.75 (3.91)	9.62 (3.04)	10.21 (6.71)	11.67 (3.13)	402,491 (108,068)	0.27
	ACP-Central	30	14	8.09 (4.28)	3.70 (3.38)	0 (0)	5.52 (2.50)	59,775 (27,064)	0.45
	ACP-ANWR	167	78	14.94 (4.02)	2.88 (1.23)	2.48 (1.43)	5.49 (1.43)	26,295 (6,833)	0.26
	Foothills	2	2	0 (0)	1.19 (1.14)	0.10 (0.10)	0.27 (0.20)	6,241 (4,637)	0.74
	Colville River	3	3	0 (0)	0 (0)	2.95 (3.52)	1.65 (1.83)	33 (37)	1.11
	All	746	555	14.59 (1.41)	7.40 (2.04)	4.91 (4.77)	7.63 (1.82)	560,092 (133,782)	0.24
REPH	ACP-NPRA	632	672	38.22 (7.11)	9.91 (3.50)	0 (0)	12.41 (2.30)	427,952 (79,473)	0.19
	ACP-Central	42	17	10.57 (5.34)	3.81 (2.04)	0 (0)	6.67 (2.57)	72,246 (27,811)	0.38
	ACP-ANWR	75	46	6.38 (2.67)	1.44 (0.72)	0 (0)	2.05 (0.78)	9,825 (3,726)	0.38
	Foothills	0	0	0 (0)	0 (0)	0 (0)	0 (0)	0 (0)	0.61
	Colville River	0	0	0 (0)	0 (0)	0 (0)	0 (0)	0 (0)	0
	All	749	735	27.58 (1.23)	7.30 (2.37)	0 (0)	7.82 (1.39)	573,979 (102,199)	0.18

TABLE 4.7

The number of shorebirds recorded on rapid surveys of intensive plots ("Estimate"), the number determined to be present through intensive surveys ("Actual"), the detection ratio, and its standard error (SE).

Refer to Appendix C for common and scientific names of species 4-letter codes.

		Region and year							
Species	Count	ACP-NPRA 1998	ACP-NPRA 1999	ACP-NPRA 2000	ACP-NPRA 2001	ACP-ANWR 2002	ACP-ANWR 2004	Total	Detection ratio (SE)
BBPL	Estimate	7.25	0.14	0	0.88	0.25	0	8.52	0.85 (0.17)
	Actual	10	0	0	0	0	0	10	
AMGP	Estimate	2.75	0.14	3	0	1	1	7.89	0.88 (0.26)
	Actual	2	0	5	0	2	0	9	
BARG	Estimate	2	0	0.25	0	0	0	2.25	0.75 (0.26)
	Actual	2	0	1	0	0	0	3	
RUTU	Estimate	7.5	0	0	0	0	0	7.5	0.83 (—)
	Actual	9	0	0	0	0	0	9	
DUNL	Estimate	9.5	1.7	4	6.5	0	0	21.7	0.90 (0.16)
	Actual	11	1	2	10	0	0	24	
SESA	Estimate	37.5	16.3	19	5.1	3.8	2	83.7	0.75 (0.13)
	Actual	38	26	37	5	4	1	111	
WESA	Estimate	0	0	0	3.1	0	0	3.1	0.78 (—)
	Actual	0	0	0	4	0	0	4	
PESA	Estimate	37	8.4	13	15.1	3	2.8	79.3	0.83 (0.06)
	Actual	41	10	17	16	6	6	96	
BBSA	Estimate	2.5	0	0	0	0.5	0	3	0.60 (0.03)
	Actual	4	0	0	0	1	0	5	
LBDO	Estimate	1.5	0.3	1.5	5.3	2.3	0	10.8	2.16 (1.20)
	Actual	1	0	0	1	3	0	5	
STSA	Estimate	0.3	0.4	0.8	0	0	1	2.4	0.61 (0.11)
	Actual	0	1	1	0	0	2	4	
RNPH	Estimate	23.25	12.43	8.25	3.63	3.25	3.25	54.1	0.80 (0.14)
	Actual	21	16	12	4	11	4	68	
REPH	Estimate	7.75	4.57	4.75	16	0	0	33.1	0.77 (0.17)
	Actual	4	11	9	19	0	0	43	

Clear evidence thus existed that *moi* had explanatory power and that it was most powerful at the micro scale.

American Golden-Plovers usually breed in uplands, and usually close to wetlands. We thought that *diswet* therefore might strengthen the model. The *P*-value for *diswet*, with *micmoi* in the model, however, was 0.42, clearly indicating that this variable did not improve the model. In contrast, we did not expect *discoa* to contribute significantly but included it with *micmoi* and found that it was significant. The intercept was

TABLE 4.8

Parameter estimates and their significance from habitat models predicting counts of shorebirds on rapid surveys.

Refer to Appendix C for common and scientific names of species 4-letter codes.

		Variable[a]							
Species		Intercept	*ew*	*ew²*	*elev*	*ew*elev*	*micwet*	*micmoi*	*discoa*
BBPL	Value	1.221	0.148	−0.030	−0.016		−1.230		0.016
	Sig.	<0.001	0.002	<0.001	0.003		<0.001		0.004
AMGP	Value	−0.265						0.742	0.010
	Sig.	0.151						0.021	0.013
BARG	Value	−0.490	0.102		−0.013				
	Sig.	0.027	0.023		0.033				
RUTU	Value	−0.846							−0.106
	Sig.	0.040							0.013
DUNL	Value	2.316	0.182	−0.011	−0.043		−0.439		
	Sig.	<0.001	<0.001	0.026	<0.001		0.094		
SESA	Value	0.325	0.077		−0.015		2.854	2.067	
	Sig.	0.508	<0.001		<0.001		<0.001	<0.001	
WESA	Value	−0.943	0.368						
	Sig.	0.369	0.022						
PESA	Value	2.548	0.093		−0.024	−0.002	0.397		
	Sig.	<0.001	<0.001		<0.001	<0.001	0.006		
LBDO	Value	0.949	0.172	−0.009	−0.004		0.855		
	Sig.	<0.001	<0.001	0.021	0.081		<0.001		
STSA	Value	0.407			−0.009			−2.206	0.030
	Sig.	0.084			0.017			<0.001	<0.001
RNPH	Value	0.385			−0.011		1.390		0.019
	Sig.	0.060			<0.001		<0.001		<0.001
REPH	Value	1.768	0.203		−0.028		1.242		−0.024
	Sig.	<0.001	<0.001		<0.001		<0.001		0.002

[a] Variables not found to be significant in any model are excluded from table.

not significant; *micmoi* and *discoa* were significant, though not highly significant. In the final model, density was higher in moist areas and farther from the coast (Table 4.8).

The range map in the BNA shows its range covering our entire study area, which is consistent with our study except that we did not find it in the western NPRA (Johnson and Connors 1996a). The BNA summarized habitat preferences saying that "rocky, dry tundra" is generally preferred for nesting "but in some areas moist habitat with taller vegetation" is also used. In our study area, plots classified as moist were strongly preferred, which seems more consistent with the second habitat above than the first.

Semipalmated Plover

Semipalmated Plovers were encountered on 14 plots and were judged to be breeding on ten plots. The sightings were all on the Colville River or in the Arctic NWR (Fig. 4.4c). Throughout the

North Slope they occurred only in gravel stream beds, a relatively rare habitat that we did not focus on. For example, they occurred commonly on gravel bars along the Colville River upstream from the delta (where the substrate is silt). Our study does not provide a good basis for describing habitat associations quantitatively or for estimating population size.

The range map in the BNA shows this species occurring throughout our study area (Nol and Blanken 1999). That is probably accurate in the sense that the species probably does occur wherever gravel beds occur. Such areas, however, occur on much less than 1% of the North Slope.

Lesser Yellowlegs

One pair and one single Lesser Yellowlegs were encountered on one plot near the Colville River, about 70 km upstream from Umiat. They were judged to be breeding on the plot. The range map in the BNA is thus correct in showing that this species does not breed extensively on the North Slope (Tibbitts and Moskoff 1999).

Spotted Sandpiper

A single Spotted Sandpiper was recorded on a plot along the Colville River near Umiat. It was judged to be breeding off the plot. The range map in the BNA is thus correct in showing that the species does not breed extensively on the North Slope, except perhaps at its extreme southern edge (Oring et al. 1997).

Whimbrel

Whimbrels were encountered on 21 plots and were judged to be breeding on 15 plots. The records were widely distributed across the study area south of the coast (Fig. 4.5a). The species may have been less common in the western NPRA, or the lack of records there may have been due to the lower sampling intensity. Most detections were of single birds, and most were in moist areas (Table 4.6). The species did not occur on intensive plots. The estimated number of pairs breeding on surveyed plots was 17.

The estimated densities were much higher in moist areas than in wetlands or uplands, and this pattern was consistent across regions. Population size was estimated at about 18,000 (Table 4.6). Due to the small number of birds recorded, we did not attempt to characterize habitat relationships.

The BNA range map shows this species extending throughout the North Slope, which may be correct, or there may be a gap in its range in the western NPRA but we had too few samples there to be certain (Skeel and Mallory 1996). The BNA describes breeding habitat as "variable ranging from dry heath uplands to poorly drained hummocky, grass-sedge, dwarf shrub, and mossy lowlands." Breeding sites for this species elsewhere in Alaska are described as "wet, flat, dwarf-shrub tundra," "dry dwarf-shrub ridges and steep slopes," and "rolling, open, usually moist tundra." These descriptions seem consistent with the very strong association in our study area of Whimbrels with moist habitat. Many of the sites where we found breeding Whimbrels had low (and occasionally moderate height) shrubs.

Bar-tailed Godwit

Bar-tailed Godwits were encountered on 59 plots and were judged to be breeding on 43 plots, widely distributed across the study area (Fig. 4.5b). All but one of the birds judged to be breeding were in the ACP-NPRA region (Table 4.6). About half the records came from the Colville Delta and most were of single birds in either wetlands or moist areas. Five counts of four or more occurred, but all but three of the birds were judged to be non-territorial. Thus, the species was distributed fairly evenly across the surveyed plots. The estimated number of pairs breeding on the surveyed plots was 56.

Only three Bar-tailed Godwits were present on the intensive plots (Table 4.7). An average of 2.25 birds was recorded on rapid surveys for a detection ratio of 0.75. The species was not recorded on rapid surveys of any intensive plots where it was absent.

The estimated density was nearly 2 birds/km^2 in the ACP-NPRA region but low or zero in all other regions (Table 4.6). The point estimate was highest in uplands, but the estimate is not close to significantly different from the estimate for the wetlands and moist areas. Population size was estimated at about 75,000 with a large CV.

The variables *ew* and *elev* were significant alone and in combination. The only other significant spatial variable was *ewsqd*, and it was only significant when *ew* was not in the model. Neither

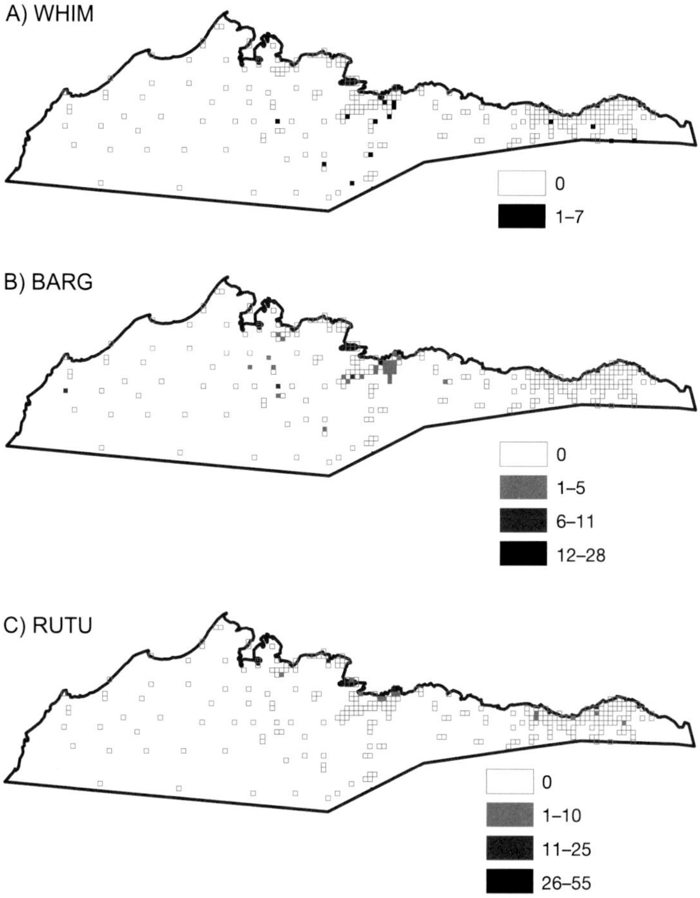

Figure 4.5. Densities (birds/km^2) of Whimbrels (a), Bar-tailed Godwits (b), and Ruddy Turnstones (c) recorded on the rapid surveys.

wet nor *moi* was significant at any spatial scale. Both of these variables and the intercept were significant (though not highly significant) predictors of the number obtained on counts. In the final model, density was higher in the west and at low elevations (Table 4.7).

The range map in the BNA shows the range ending before the Alaska–Canada border, which is consistent with our results (McCaffery and Gill 2001). This species was unusual in our study in not showing much association with habitat (no contrasts between habitats were close to significant). The BNA also refers to its use of many different habitats.

Ruddy Turnstone

Ruddy Turnstones were encountered on 22 plots and were judged to be breeding on 16 plots. Birds judged to be breeding were encountered largely in the ACP-NPRA region, though five indicated pairs were recorded in ANWR (Fig. 4.5c; Table 4.6). The species was notable for the high proportion (25%) of sightings that were nests or probable nests. Although nests were often on partially vegetated dunes, sightings were exclusively on plots classified as wetlands or moist areas. Most plots with the species present had one or two sightings, but 24 of the 48 sightings occurred on just two plots, one of which had 16 sightings. The species thus showed a highly clumped distribution on the surveyed plots. The estimated number of breeding pairs on surveyed plots was 31.

Nine Ruddy Turnstones were present on the intensive plots, all at one camp, and surveyors recorded an average of 7.5 birds for a detection ratio of 0.83 (Table 4.7). Since the species only occurred at one camp, we could not calculate a camp-to-camp variance and thus could not obtain a SE for the estimated detection ratio.

Estimated densities were less than 1 bird/km^2 in all regions and habitats (Table 4.6). All of the records were from wetlands or moist areas. The estimated population size was about 3,400. Despite a large CV, the results show that population size on the North Slope is fairly small (e.g., <10,000 birds).

None of the spatial or habitat variables was consistently significant. The factor *discoa* was nearly significant, and we thought it was justified on the basis of biological evidence (the species normally occurs close to the coast). In the final model, density was higher close to the coast (Table 4.8).

The BNA range map shows this species distributed in a narrow band throughout our study area (Nettleship 2000). Our finding of five to ten locations, nearly all right on the coast, is consistent with this description. We had too little data for a quantitative description of habitat associations, but our impression was that this species occurred in sites that were unusually barren (e.g., halophytic and sparsely vegetated) for our study area, consistent with their habitat use elsewhere.

Sanderling

A single Sanderling was recorded in the Arctic NWR. The BNA range map shows a small patch of range at Barrow (MacWhirter et al. 2002). We did not find Sanderlings there but could have missed such a small area.

Dunlin

Dunlin were encountered on 188 plots and were judged to be breeding on 175 plots. Most records were within 50 km of the coast (Fig. 4.6a). Nearly all the records were from the ACP-NPRA region and none were from the Foothills (Table 4.6). The range appeared to stop at the western edge of the Arctic NWR. Most records were from wetlands and hardly any were from uplands. The number of plots with seven, eight, and nine sightings were four, one, and one. No plot had more than nine sightings. They were thus one of the more abundant shorebird species. The estimated number of breeding pairs on the surveyed plots was 378.

Intensive plots had 24 territorial Dunlin, 21 of which were at two camps on the Colville Delta (Table 4.7). Surveyors recorded an average of nearly 22 birds/survey, for a detection ratio of 0.90.

Estimated densities were consistently highest in wetlands by a considerable margin (Table 4.6). Density in wetlands across all regions was significantly higher than in moist areas. Dunlin were rarely found in uplands. Overall density was nearly 7 birds/km^2. Population size was estimated at 500,000 birds with a CV of just 0.20. More than 70% of the birds were estimated to occur in the NPRA-coast region.

The variables *ew*, *elev*, and *ewsqd* were significant predictors of Dunlin counts in rapid plots. *Wet* was significant alone and in combination with *moi*. Its significance level was about the same for the micro and meso scales and lower at the macro scale. The factor *moi* was generally not significant either alone or in combination with *wet*. Neither *discoa* nor *diswet* improved the fit. All variables but *micwet* were significant. In the final model, density was higher in the west, at lower elevation, and in drier areas (Table 4.8).

The range map in the BNA is consistent with our findings except that we found few birds in the Arctic NWR, whereas the BNA map suggests they occur throughout the ACP (Warnock and Gill 1996). The BNA includes a detailed description of habitat for Dunlin near Prudhoe Bay, which is similar to our finding that density was highest in wetlands, intermediate in moist areas, and low in uplands.

Semipalmated Sandpiper

Semipalmated Sandpipers were encountered on 351 plots and were judged to be breeding on 325 plots. They occurred throughout the ACP but were rare in the Foothills (Fig. 4.6b). Among the 1,361 records (Table 4.6) were 174 nests and probable nests. The largest number of birds was found on the Colville Delta. Most records were in wetlands and few were in uplands. The species was often abundant on plots; 29 plots had ten or more sightings, and four had 20 or more sightings. The estimated number of pairs breeding on surveyed plots was 1,335.

More than 100 Semipalmated Sandpipers occurred on the intensive plots, and the species was present at every camp (Table 4.7). The overall detection ratio was 0.75 with a small CV of 0.13.

The estimated density in wetlands was >50 birds/km^2, more than twice as many as in moist areas ($P < 0.001$; Table 4.6). Even fewer birds were seen in uplands. The overall density was

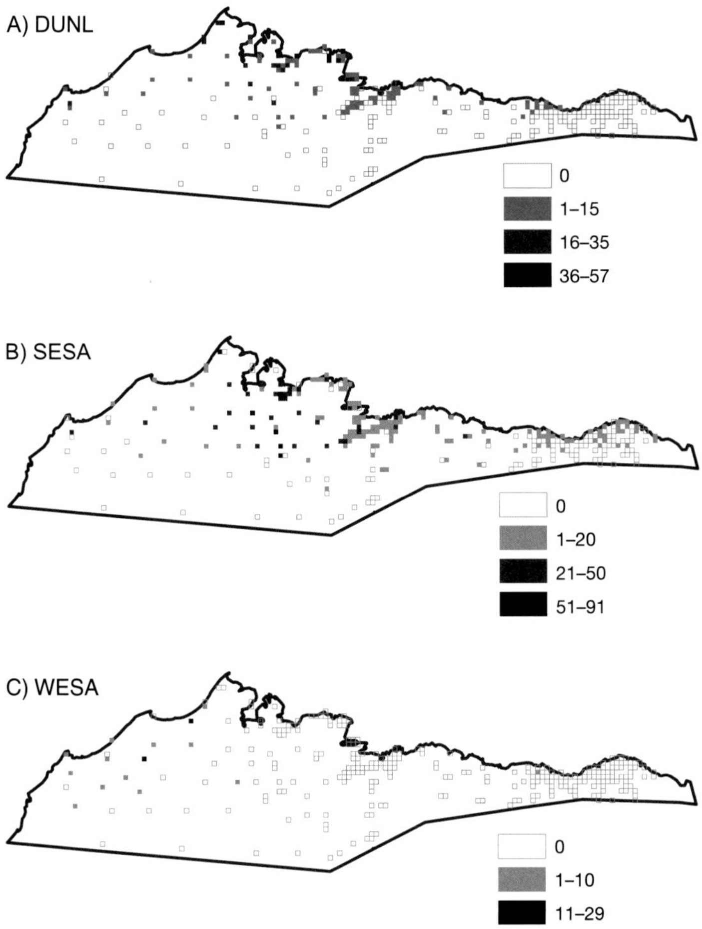

Figure 4.6. Densities (birds/km²) of Dunlin (a), Semipalmated Sandpipers (b), and Western Sandpipers (c) recorded on the rapid surveys.

nearly 20 birds/km². The estimated population size was about 1.3 million birds. Most of the birds were estimated to be in the ACP-NPRA. Unlike many species, significant numbers (>240,000) were estimated to occur in the Central region.

The spatial variables *ew* and *elev* were highly significant alone and in combination. When we added *ewsqd*, it was significant but *ew* became nonsignificant. We thus used *ew* and *elev* as the spatial variables. When we added *wet*, *moi*, or *wet moi* to *ew elev*, *wet* and *moi* were both significant. The factors *wet* and *moi* were highly significant at the micro and meso scale. Convergence did not occur when *discoa* was added to *ew elev micoi micwet*. Adding *diswet* to *ew elev micmoi micwet* was not significant. All variables were highly significant. According to the final model, density was higher in the west, at lower elevations, and in wetlands or moist areas (Table 4.8).

The range map in the BNA is consistent with our findings as are the descriptions of this species' habitat as "moist or wet sedge-grass or heath tundra" (Gratto-Trevor 1992).

Western Sandpiper

Western Sandpipers were encountered on 23 plots and were judged to be breeding on 20 plots. They were encountered almost exclusively in the western part of the ACP-NPRA region, with a single record in the Arctic NWR (Fig. 4.6c). They were unusual among shorebirds in occurring regularly in uplands, although they were not found in the Foothills region (Table 4.6). More sightings occurred in wetlands than in uplands or moist areas. The numbers per plot were almost always four or fewer, although one plot had eight sightings. The number of estimated pairs breeding on surveyed plots was 45.

Four Western Sandpipers were present on the intensive plots, all at one camp (Table 4.7). The mean number recorded/survey was 3.13 for a detection ratio of 0.78. Since the species occurred only at one camp, we could not estimate the SE of the point estimate.

The estimated density in the ACP-NPRA region, 9.29, is not very meaningful because the species occurred only west of Barrow (excluding the single bird recorded in the Arctic NWR). The high estimated density in uplands is unusual for shorebirds. The estimated population size was nearly 400,000.

The only significant spatial or habitat variable, including *discoa* and *diswet*, was *ew*. According to the final model, density was higher in the west (Table 4.8).

The BNA range map shows three small, discrete patches of range for Western Sandpipers, whereas we found it regularly west of Barrow (especially close to the coast, Wilson 1994). A review using all available data now would probably be worthwhile. The BNA says this species breeds in areas dominated by prostrate woody plants and dwarf shrubs. This is consistent with our finding that they were most common in moist and upland areas.

White-rumped Sandpiper

White-rumped Sandpipers were encountered on six plots and were judged to be breeding on four plots mainly between Barrow and Cape Halkett (Fig. 4.7a). The other two records were in the Arctic NWR and were probably migrants. The estimated number of breeding pairs in surveyed plots was five.

The BNA range map shows this species' range extending from the Canadian border to Barrow, whereas we found only five pairs of White-rumped Sandpipers (all along the coast near Barrow, Parmelee 1992b).

Baird's Sandpiper

Baird's Sandpipers were encountered on 12 plots and were judged to be breeding on six plots. Two pairs and ten single Baird's Sandpipers were encountered, but only seven of the sightings were considered to be of birds nesting on the surveyed plots. Seven of the 12 sightings were in the Arctic NWR and two were in the Central region. Only three sightings were west of the Colville River. The records were obtained mainly in wetlands; four were from moist areas and none were in uplands. The BNA range map shows the range for this species extending along the coast throughout our study area (Moskoff and Montgomerie 2002). Our results suggest the range may be less extensive.

Pectoral Sandpiper

Pectoral Sandpipers were encountered on 376 plots and were judged to be breeding on 345 plots. They were widely distributed north of the Foothills (Fig. 4.7b). Most sightings were of single birds in wetlands (Table 4.6). Only about 1% of the sightings were in uplands. This species was often quite abundant on plots; 23 plots had ten or more sightings and five plots had 20 or more sightings. The estimated number of breeding pairs on surveyed plots was 1,149.

Nearly 100 Pectoral Sandpipers were present on the intensive plots and the species occurred at all six camps (Table 4.7). Surveyors recorded an average of 79 birds/survey for a detection ratio of 0.83. The mean numbers recorded/survey were less than the actual numbers present at each of the six camps.

Estimated densities were about 30 birds/km^2 in wetlands and about 12 birds/km^2 in moist areas (Table 4.6). The difference between the two habitats was highly significant ($P < 0.001$). Estimated densities in uplands were more variable. The overall estimate for uplands (10.93) was not significantly different from 0.0. The estimated population size was greater than 1 million. Nearly all birds were in the ACP, with nearly 75% in the NPRA.

The variables *ew* and *elev* were significant in several models, as was their interaction. The interaction *ewsqd*elev* was significant, but *ew* was not significant. We therefore used *ew*, *elev*, and *ew*elev* as the spatial parameters. With these variables included, *wet* was generally significant, sometimes highly so, whereas *moi* was either not significant or at least not highly significant. The factor *wet* was equally significant at the micro and meso scale and not significant at the macro scale. When *discoa* and *diswet* were added, neither was significant. The model selected was thus *ew elev*

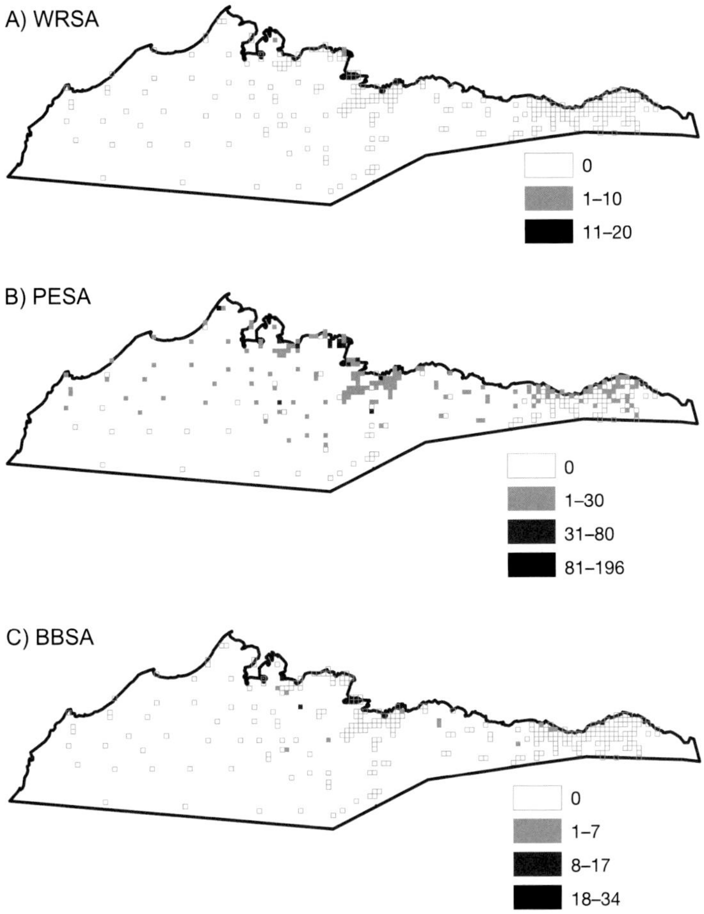

Figure 4.7. Densities (birds/km²) of White-rumped Sandpipers (a), Pectoral Sandpipers (b), and Buff-breasted Sandpipers (c) recorded on the rapid surveys.

*ew*elev micwet*. According to the final model, density was higher in the west, at lower elevations, and in wetlands (Table 4.8).

The BNA range map and descriptions of breeding habitats are consistent with our results (Holmes and Pitelka 1998).

Buff-breasted Sandpiper

Buff-breasted Sandpipers were encountered on 28 plots and were only judged to be breeding on 16 plots. They were recorded widely across the ACP (Fig. 4.7c) east of Barrow. Most records were of single birds in wetlands, but a substantial number of records occurred in moist areas as well (Table 4.6).

Few birds were recorded on the intensive plots (Table 4.7). No birds were recorded when the species was actually absent. The observed detection ratio was 0.6. The low SE results from only two intensive plots having birds and the detection ratio happening to be similar (0.5 and 0.63). We do not consider this a reliable estimate of the true SE.

Estimated densities were highest in wetlands, intermediate in moist areas, and zero in uplands (Table 4.6). All CVs were large. Population size was estimated at a little over 40,000 but with a large CV of 0.45. The BNA range map for our study area is consistent with our observations (Lanctot and Laredo 1994).

Long-billed Dowitcher

Long-billed Dowitchers were encountered on 229 plots and were judged to be breeding on 198 plots. They were recorded throughout the ACP,

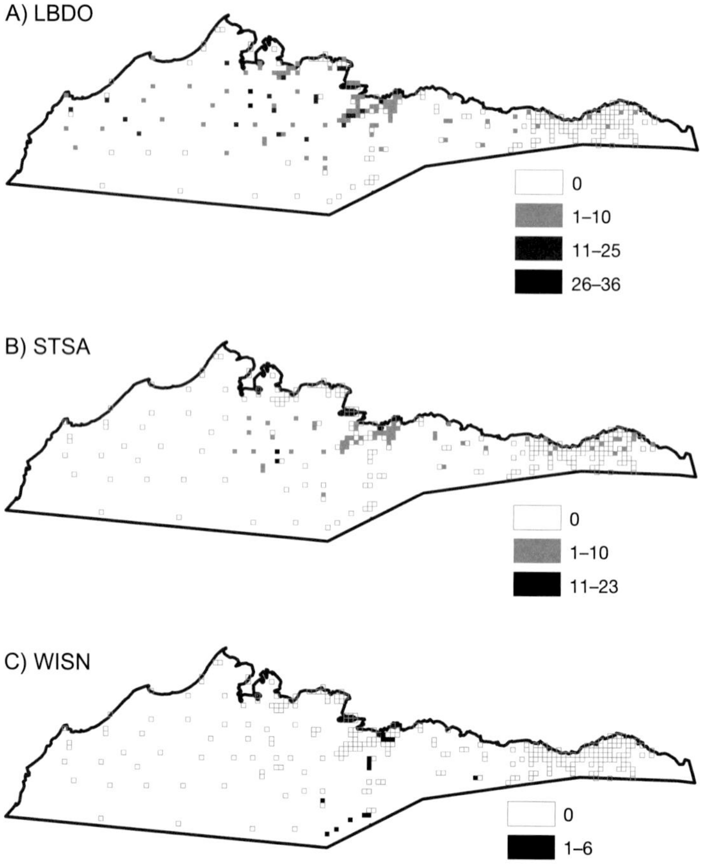

Figure 4.8. Densities (birds/km^2) of Long-billed Dowitchers (a), Stilt Sandpipers (b), and Wilson's Snipe (c) recorded on the rapid surveys.

but relatively few records came from the Arctic NWR (Fig. 4.8a; Table 4.6). About a third of the sightings were of nests, probable nests, or pairs; the rest were of single birds. About two-thirds of the sightings were in wetlands, and nearly all the remaining sightings were in moist areas. Single plots had 10, 11, 12, and 13 sightings each and 18 plots had four or more sightings. Thus, detecting multiple birds per plot was not uncommon. The estimated number of pairs breeding on surveyed plots was 374.

Only five Long-billed Dowitchers were present on the intensive plots (Table 4.7). Surveyors consistently recorded this species in larger numbers than were present. The overall detection ratio was 2.16, but due to the small sample size the confidence interval for the estimate extended almost to zero. Our subjective impression is that this species was secretive at nests and moved around a lot. It may often have been counted on plots where it was not breeding. If this is true, then our estimated densities and population sizes (which used the combined detection rate, not 2.16) could have substantial positive bias.

Estimated region-wide densities were highest in wetlands and lowest in uplands (Table 4.6). The difference between densities in wetlands and moist areas was highly significant. The estimated population size was about 650,000, making this one of the most abundant shorebirds. The estimate was quite precise (CV = 0.19) but, as noted above, we are concerned that the detection ratio might have been substantially underestimated. Most of the birds were estimated to occur in the ACP-NPRA region. A small population was estimated to occur in the Foothills and on the Colville River.

The variables *ew*, *elev*, and *ewsqd* were all significant when included together in the model. No other variables were significant. With these variables in the model, *wet* was significant, especially at the micro scale. The factor *moi* was far from significant alone but was significant when *wet* was included. Because *moi* was not significant alone, we did not include it. When *discoa* was added to this model, the estimator did not converge. When *diswet* was added instead, *diswet* was significant but the significance level for *elev* dropped to 0.21, which, overall, seemed like a less satisfactory model. The final model was thus *ew elev ewsqd micwet* (Table 4.8). In this model, all variables except *elev* were significant. According to the final model, density was higher in the west, at lower elevations, and in wetlands (Table 4.8).

The BNA range map is consistent with our results (Takekawa and Warnock 2000). The description of habitat as "wet, grassy meadows" (assuming grass includes sedge) is consistent with our finding that density was highest in wetlands and moist areas.

Stilt Sandpiper

Stilt Sandpipers were encountered on 103 plots and were judged to be breeding on 86 plots. They were widely distributed, though never abundant, across the ACP but were not encountered in the southern or western parts of the study area (Fig. 4.8B; Table 4.6). They were strongly associated with wetlands, though about a third of the sightings were in moist areas. Only one sighting was in uplands. Most plots on which the species occurred had relatively few birds. Seven plots had five or more birds. The largest numbers of sightings were individual plots with 10, 10, and 14 sightings. The estimated number of pairs breeding on surveyed plots was 143.

Four pairs were present at three intensive camps (Table 4.7). The mean number recorded per survey was 2.43 for an overall detection ratio of 0.61. The SE was fairly small, but we suspect this was just due to the camp-specific rates being similar by chance. We do not place high confidence in an estimate of the detection ratio based on only four birds.

Densities were much higher ($P = 0.001$) in wetlands than in moist areas, and the species was never recorded in uplands (Table 4.6). Densities were highest in wetlands in all three ACP regions. Estimated population size was about 120,000 with a moderate CV of 0.23. More than half the birds were in the NPRA.

The only significant spatial variable was *elev* with no other spatial variables included. The habitat variables *wet* and *moi* were both significant when added to *elev*, but only *wet* was significant in *elev wet moi*. Significance values were slightly higher at the micro scale than at the meso scale. In *elev micmoi discoa*, *discoa* was significant. In *elev micmoi diswet*, *diswet* was significant but *elev* was then not significant ($P = 0.83$). The final model was thus *elev micmoi discoa* (Table 4.8). All variables were significant; *micmoi* and *discoa* were highly significant. According to the final model, density was higher at lower elevations, lower in moist areas, and higher farther from the coast (Table 4.8).

The BNA range map for our study is fairly accurate except that we found the species farther south than shown in the BNA (see Fig. 4.8b, Klima and Jehl 1998). The BNA's description of this species' habitat, especially near Prudhoe Bay, is consistent with our results.

Wilson's Snipe

Wilson's Snipe were encountered on 24 plots and were judged to be breeding on 18 plots. Most records occurred along the Colville River (Fig. 4.8c). About equal numbers were recorded in wetlands and moist areas. The estimated number of pairs breeding on surveyed plots was 21.

The BNA range map shows our entire study area as within the range of Wilson's Snipe (Mueller 1999). This may be accurate, depending on how low density can be within the range, but our results suggest that only the central part of our study area, and especially the Colville River, was within the range. The tundra is an unusual breeding habitat for this species and is not explicitly acknowledged in the BNA.

Red-necked Phalarope

Red-necked Phalaropes were encountered on 247 plots and were judged to be breeding on 220 plots. They were widely distributed across the ACP but were nearly absent from the Foothills (Fig. 4.9a; Table 4.6). Nests and probable nests were uncommon, but about a third of the sightings were pairs. They were encountered mainly in wetlands, but a substantial number were found in moist areas. Large groups were uncommon, but 12 plots

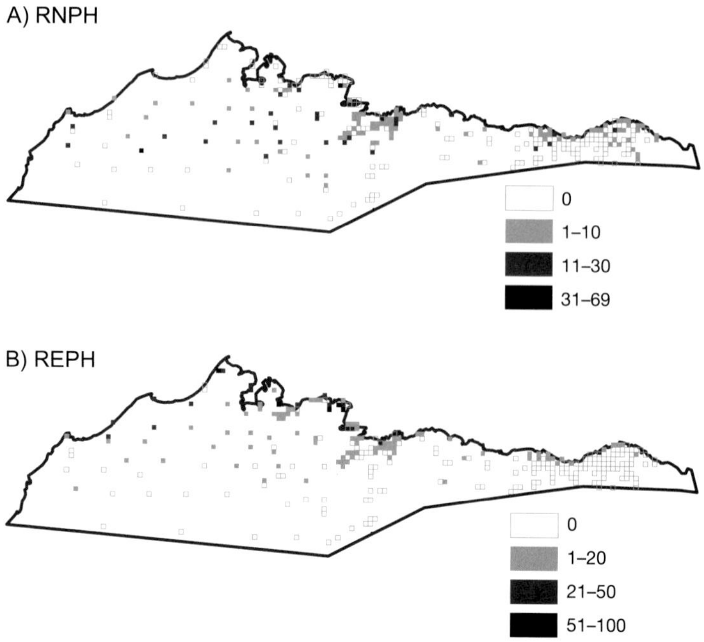

Figure 4.9. Densities (birds/km^2) of Red-necked Phalaropes (a) and Red Phalaropes (b) recorded on the rapid surveys.

had ten or more sightings and individual plots had 21, 22, and 27 sightings. The estimated number of pairs breeding on surveyed plots was 555.

Sixty-eight Red-necked Phalaropes were present on the intensive plots, and the species was present at all of the six camps (Table 4.7). Surveyors recorded an average of 54 birds/survey for a detection ratio of 0.80.

Densities were much higher in wetlands than in moist areas and were lowest in uplands (Table 4.6). The difference in density between wetlands and moist areas was highly significant. The overall density was 7.6 birds/km^2. The estimated population size was more than half a million birds, with the great majority in the ACP-NPRA region.

The spatial variable *ew* was significant alone but was not significant when *elev* was in the model. The factor *elev* was highly significant in all models. No other spatial variables were significant. With *elev* in the model, *wet* and *moi* were both significant alone, but only *wet* was significant when both were in the model. The factor *wet* was only significant at the micro scale. This led us to the model *elev micmoi*. Adding *discoa* to this model, was significant but *diswet* was not. The final model was thus *elev micwet discoa* (Table 4.8). All variables were significant but the intercept was just nonsignificant. According to the final model, density was higher at lower elevations, in wetlands, and farther from the coast (Table 4.8)

The BNA range map includes all of our study area, which is consistent with our results (Rubega et al. 2000). The habitat description, especially for populations in the Mackenzie Delta, is consistent with our results.

Red Phalarope

Red Phalaropes were encountered on 219 plots and were judged to be breeding on 193 plots. They were widely recorded across the ACP, though they were noticeably rare in the Arctic NWR (Fig. 4.9b; Table 4.6). More than half of the sightings were of nests, probable nests, or (especially) pairs. They were more restricted to wetlands (77% of the sightings) than most other shorebirds. None were recorded in uplands. They were sometimes abundant on plots; 14 plots had ten or more sightings, and two plots had 35 and 51 sightings. The estimated number of pairs breeding on surveyed plots was 735.

Forty-three Red Phalaropes occurred on the intensive plots, with four of six camps having the species present (Table 4.7). Surveyors recorded an average of 33 birds/survey for a detection ratio of 0.77.

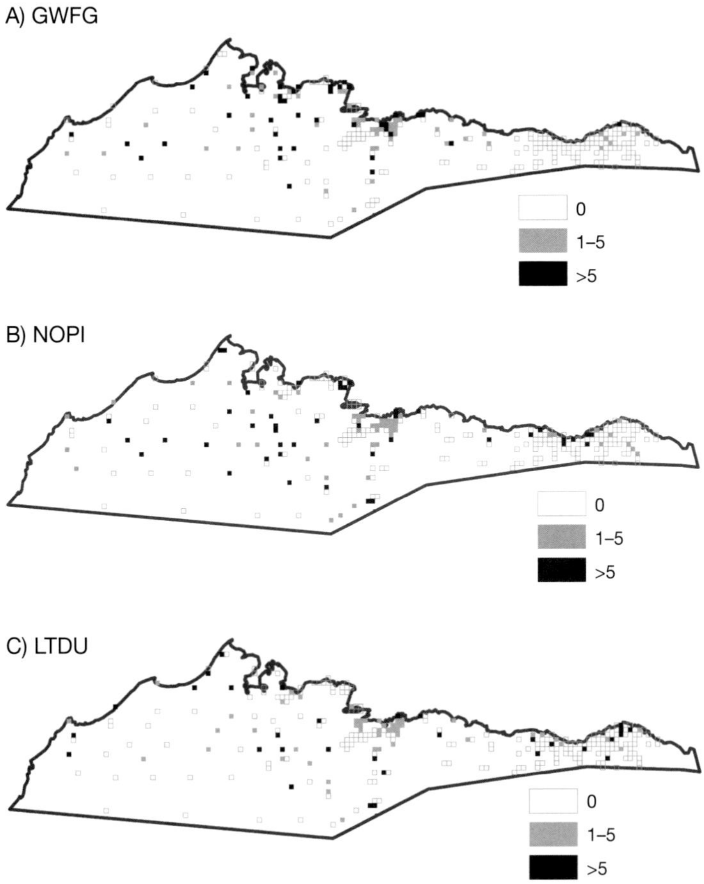

Figure 4.10. Densities (birds/km^2) of Greater White-fronted Geese (a), Northern Pintail (b), and Long-tailed Ducks (c) recorded on the rapid surveys.

Estimated densities were highest in wetlands in the ACP and were 0.0 in uplands in the other regions (Table 4.6). The region-wide density in wetlands was significantly higher than in moist areas. Overall density was nearly 8 birds/km^2. Estimated population size for all regions was about 570,000. Most of the estimated population was in the NPRA.

The only significant spatial variables were *ew* and *elev*, and they were highly significant in all combinations. The habitat variables *wet* and *moi* were both significant alone but *moi* was not significant with *wet*. The factor *wet* was equally significant at the micro and meso scale but not at the macro scale. The factor *discoa* was significant when added to *ew elev wet* but *diswet* was not. The final model was thus *ew elev micwet discoa* (Table 4.8). All variables and the intercept were highly significant. According to the final model, density was higher in the west, at lower elevations, in wetlands and closer to the coast (Table 4.8).

The BNA range map includes all of our study area, which is consistent with our results (Tracy et al. 2002). The habitat description is consistent with our results.

Other Species

Greater White-fronted Goose

Greater White-fronted Geese were encountered during ground surveys on 206 plots and were judged to be breeding on 135 plots. They occurred mainly west of Prudhoe Bay and north of the Foothills (Fig. 4.10a). This species was recorded on 1,583 cells during the aerial surveys. Results were similar to the ground surveys, though the patterns were much easier to discern (Fig. 4.11a). Few birds were recorded east of Prudhoe Bay. The aerial surveys suggest a band of high density extending west and a little north from Prudhoe Bay.

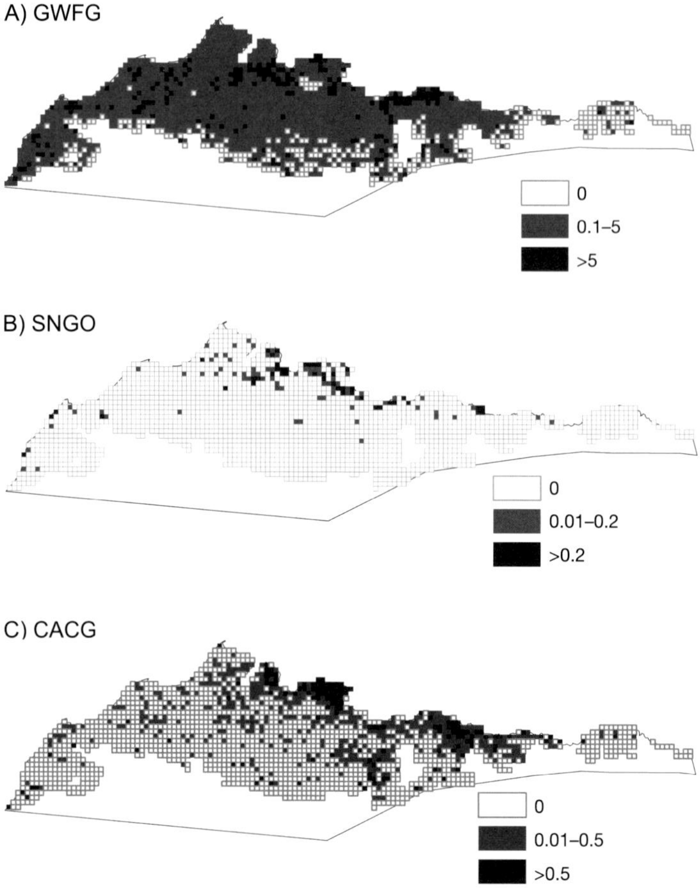

Figure 4.11. Densities (birds/km^2) of Greater White-fronted Geese (a), Snow Geese (b), and Cackling Geese (c) recorded on the aerial surveys.

On ground surveys, most records were of single birds in wetlands (Table 4.9). About one-third of the birds were in moist areas, and only about 1% were in uplands. The estimated number of pairs breeding on surveyed plots was 347.

Density estimates from the ground surveys were approximately 8, 3, and 3 birds/km^2 in wetlands, moist areas, and uplands, respectively, with an overall density of 4.1. The uncorrected estimate of population size was about 300,000 with a moderate CV of 0.29 (Table 4.9). Nesting birds of this species often leave a survey plot when the surveyor is still far away and it is difficult to know whether departing birds were on the survey plot. Incubating birds usually do not flush unless the surveyor comes within 10–20 m. For both reasons, it seems likely that the detection ratio achieved on ground surveys was less than 1, which would suggest the population was probably even larger than the estimate based on the ground surveys.

The range and habitat relationships as described in the BNA are consistent with our results (Ely and Dzubin 1994).

Snow Goose

Aerial surveys recorded Snow Geese in 124 cells (Fig. 4.11b). During ground surveys, Snow Geese were encountered on ten plots and were judged to be breeding on three plots. Flocks of non-breeders were observed on four plots. Most records were close to the coast between just east of the Colville Delta and Dease Inlet. Other records were widely scattered.

Detection ratios for this species are probably close to 1.0. The estimated population size was 485 (0.70). This estimate, however, does not include birds judged to be non-breeders.

The BNA range map has a dashed line along the southern border of our study site (Mowbray

TABLE 4.9

Number of waterfowl, waterbirds, and landbirds recorded on rapid surveys, estimated densities, and population size (with SEs).

Refer to Appendix C for common and scientific names of species 4-letter codes.

				Birds/km^2 (SE)				Population size	
Species	Region	Total recorded	Est. *n* pairs	Wetlands	Moist areas	Uplands	All habitats	Estimate (SE)	CV
GWFG	ACP-NPRA	636	326	11.07 (1.69)	3.68 (1.18)	6.87 (5.14)	6.61 (2.11)	228,112 (72,914)	0.32
	ACP-Central	190	10	4.18 (2.08)	2.7 (1.37)	0 (0)	3.27 (1.14)	35,362 (12,301)	0.35
	ACP-ANWR	140	9	1.42 (0.70)	0.91 (0.63)	0 (0)	0.70 (0.31)	3,344 (1,485)	0.44
	Foothills	2	0	0 (0)	0 (0)	0 (0)	0 (0)	0 (0)	1.02
	Colville River	2	2	0 (0)	0 (0)	1.12 (0.99)	0.63 (0.51)	13 (10)	0.81
	All	970	347	8.19 (0.64)	3.03 (1.12)	3.19 (3.14)	4.13 (1.2)	303,028 (87,860)	0.29
CACG	ACP-NPRA	97	26	0.10 (0.06)	0.32 (0.32)	0 (0)	0.14 (0.12)	4,945 (4,127)	0.83
	ACP-Central	14	3	1.31 (0.99)	0.45 (0.45)	0 (0)	0.81 (0.46)	8,791 (5,014)	0.57
	ACP-ANWR	39	21	2.00 (0.76)	1.65 (0.89)	0 (0)	1.13 (0.42)	5,439 (2,019)	0.37
	Foothills	2	0	0 (0)	0 (0)	0 (0)	0 (0)	0 (0)	1.00
	Colville River	8	8	0 (0)	0 (0)	5.45 (5.72)	3.04 (2.95)	61 (59)	0.97
	All	160	58	0.59 (0.14)	0.43 (0.15)	0.02 (0.23)	0.27 (0.09)	19,539 (6,937)	0.36
TUSW	ACP-NPRA	71	21	0.20 (0.12)	0.01 (0.01)	0 (0)	0.05 (0.03)	1,750 (943)	0.54
	ACP-Central	3	1	0 (0)	0.45 (0.45)	0 (0)	0.23 (0.24)	2,542 (2,592)	1.02
	ACP-ANWR	12	1	0 (0)	0.19 (0.18)	0 (0)	0.08 (0.08)	366 (374)	1.02
	Foothills	0	0	0 (0)	0 (0)	0 (0)	0 (0)	0 (0)	0.28
	Colville River	0	0	0 (0)	0 (0)	0 (0)	0 (0)	0 (0)	0
	All	86	23	0.12 (0.04)	0.12 (0.05)	0 (0)	0.06 (0.04)	4,483 (2,616)	0.58
NOPI	ACP-NPRA	442	224	10.41 (2.17)	4.87 (1.21)	3.89 (2.05)	5.71 (1.10)	196,961 (38,044)	0.19
	ACP-Central	24	12	13.02 (11.45)	1.40 (1.08)	0 (0)	6.49 (5.17)	70,307 (55,973)	0.80

TABLE 4.9 *(continued)*

TABLE 4.9 (CONTINUED)

Species	Region	Total recorded	Est. *n* pairs	Birds/km² (SE)				Population size	
				Wetlands	Moist areas	Uplands	All habitats	Estimate (SE)	CV
	ACP-ANWR	147	51	2.67 (2.07)	3.37 (1.91)	4.92 (1.79)	3.77 (1.17)	18,071 (5,619)	0.31
	Foothills	3	0	0 (0)	0 (0)	0 (0)	0 (0)	0 (0)	0.16
	Colville River	10	14	0 (0)	0 (0)	11.69 (4.49)	6.53 (1.99)	131 (40)	0.31
	All	626	301	10.67 (1.51)	3.71 (0.94)	2.07 (1.93)	4.30 (0.93)	315,559 (68,429)	0.22
GRSC	ACP-NPRA	67	37	1.75 (1.03)	1.17 (0.65)	0 (0)	0.83 (0.40)	28,692 (13,892)	0.48
	ACP-Central	5	2	0 (0)	0.62 (0.44)	0 (0)	0.33 (0.23)	3,547 (2,510)	0.71
	ACP-ANWR	9	5	0.31 (0.23)	0.07 (0.07)	1.42 (1.37)	0.61 (0.51)	2,930 (2,422)	0.83
	Foothills	0	0	0 (0)	0 (0)	0 (0)	0 (0)	0 (0)	1.02
	Colville River	3	4	0 (0)	0 (0)	4.36 (1.85)	2.44 (1.05)	49 (21)	0.43
	All	84	48	1.13 (0.32)	0.90 (0.26)	0.08 (0.44)	0.55 (0.23)	40,534 (16,844)	0.42
SPEI	ACP-NPRA	27	21	0.80 (0.49)	0.01 (0.01)	0 (0)	0.18 (0.11)	6,369 (3,802)	0.6
	ACP-Central	1	1	1.85 (1.87)	0 (0)	0 (0)	0.82 (0.82)	8,868 (8,891)	1
	ACP-ANWR	0	0	0 (0)	0 (0)	0 (0)	0 (0)	0 (0)	0
	Foothills	0	0	0 (0)	0 (0)	0 (0)	0 (0)	0 (0)	0.28
	Colville River	0	0	0 (0)	0 (0)	0 (0)	0 (0)	0 (0)	0
	All	28	22	1.06 (0.13)	0.01 (0.15)	0 (0)	0.21 (0.13)	15,584 (9,333)	0.60
LTDU	ACP-NPRA	195	131	4.77 (1.27)	3.93 (1.17)	1.07 (1.15)	2.97 (0.72)	102,345 (24,791)	0.24
	ACP-Central	20	14	16.78 (7.52)	1.03 (1.02)	0 (0)	7.96 (3.42)	86,173 (36,983)	0.43
	ACP-ANWR	61	30	4.09 (1.33)	0.75 (0.52)	1.42 (1.37)	1.76 (0.60)	8,415 (2,897)	0.34
	Foothills	1	0	0 (0)	0 (0)	0 (0)	0 (0)	0 (0)	0.16
	Colville River	6	7	0 (0)	0 (0)	8.13 (4.17)	4.54 (2.23)	91 (45)	0.49
	All	283	182	8.35 (0.84)	2.82 (0.79)	0.59 (2.52)	2.85 (0.62)	208,928 (45,327)	0.22

ROPT	ACP-NPRA	41	34	1.81 (0.53)	3.56 (1.09)	0 (0)	1.73 (0.44)	59,828 (15,259)	0.26
	ACP-Central	4	3	0.52 (0.50)	1.28 (0.97)	0 (0)	0.90 (0.56)	9,727 (6,032)	0.62
	ACP-ANWR	19	14	0.55 (0.31)	1.59 (0.74)	1.92 (1.73)	1.47 (0.70)	7,050 (3,346)	0.47
	Foothills	9	9	1.3 (1.97)	1.78 (1.59)	1.7 (1.64)	1.89 (1.57)	44,032 (36,527)	0.83
	Colville River	1	1	0 (0)	0 (0)	1.38 (1.34)	0.77 (0.76)	15 (15)	0.98
	All	74	61	1.35 (0.46)	2.96 (0.44)	0.90 (1.44)	1.66 (0.46)	121,592 (34,046)	0.28
WIPT	ACP-NPRA	329	303	9.27 (1.88)	20.11 (2.92)	18.69 (6.64)	17.01 (2.96)	586,801 (102,249)	0.17
	ACP-Central	16	11	2.13 (1.53)	3.81 (1.36)	48.31 (0)	4.51 (0.96)	48,771 (10,356)	0.21
	ACP-ANWR	30	22	0.67 (0.41)	1.22 (0.53)	7.01 (2.92)	3.18 (1.13)	15,224 (5,428)	0.36
	Foothills	10	11	0 (0)	0 (0)	7.46 (3.80)	7.03 (3.59)	163,472 (83,458)	0.51
	Colville River	30	30	0 (0)	0 (0)	18.77 (5.45)	10.48 (2.82)	210 (56)	0.27
	All	415	377	6.51 (2.41)	13.9 (1.82)	13.07 (3.35)	12.04 (1.90)	883,500 (139,683)	0.16
PALO	ACP-NPRA	102	67	1.79 (0.62)	0.96 (0.60)	0 (0)	0.76 (0.26)	26,341 (9,128)	0.35
	ACP-Central	7	0	0 (0)	0 (0)	0 (0)	0 (0)	0 (0)	0
	ACP-ANWR	28	16	2.27 (0.89)	0.15 (0.15)	0.35 (0.36)	0.70 (0.28)	3,372 (1,319)	0.39
	Foothills	0	0	0 (0)	0 (0)	0 (0)	0 (0)	0 (0)	0.97
	Colville River	5	5	0 (0)	0 (0)	3.52 (3.81)	1.97 (1.96)	39 (39)	1.00
	All	142	88	1.27 (0.15)	0.63 (0.29)	0.03 (0.32)	0.47 (0.15)	34,426 (10,932)	0.32
GLGU	ACP-NPRA	102	67	1.79 (0.62)	0.96 (0.60)	0 (0)	0.76 (0.26)	26,341 (9,128)	0.35
	ACP-Central	7	0	0 (0)	0 (0)	0 (0)	0 (0)	0 (0)	0
	ACP-ANWR	28	16	2.27 (0.89)	0.15 (0.15)	0.35 (0.36)	0.70 (0.28)	3,372 (1,319)	0.39
	Foothills	0	0	0 (0)	0 (0)	0 (0)	0 (0)	0 (0)	0.97
	Colville River	5	5	0 (0)	0 (0)	3.52 (3.81)	1.97 (1.96)	39 (39)	1.00
	All	142	88	1.27 (0.15)	0.63 (0.29)	0.03 (0.32)	0.47 (0.15)	34,426 (10,932)	0.32

TABLE 4.9 (*continued*)

TABLE 4.9 (CONTINUED)

Species	Region	Total recorded	Est. *n* pairs	Birds/km² (SE)				Population size	
				Wetlands	Moist areas	Uplands	All habitats	Estimate (SE)	CV
SAGU	ACP-NPRA	20	18	1.38 (0.53)	0.88 (0.67)	0 (0)	0.64 (0.28)	21,978 (9,559)	0.43
	ACP-Central	0	0	0 (0)	0 (0)	0 (0)	0 (0)	0 (0)	0
	ACP-ANWR	17	9	0.27 (0.27)	0.29 (0.29)	0 (0)	0.18 (0.14)	870 (654)	0.75
	Foothills	0	0	0 (0)	0 (0)	0 (0)	0 (0)	0 (0)	0.83
	Colville River	0	0	0 (0)	0 (0)	0 (0)	0 (0)	0 (0)	0
	All	37	27	0.89 (0.14)	0.59 (0.27)	0 (0)	0.37 (0.15)	26,949 (11,349)	0.42
ARTE	ACP-NPRA	112	76	4.59 (1.85)	3.00 (1.95)	0 (0)	2.15 (0.83)	74,275 (28,733)	0.39
	ACP-Central	13	1	1.49 (1.44)	0 (0)	0 (0)	0.66 (0.65)	7,122 (7,009)	0.98
	ACP-ANWR	59	9	0.77 (0.42)	0.18 (0.18)	0 (0)	0.25 (0.12)	1,202 (588)	0.49
	Foothills	0	0	0 (0)	0 (0)	0 (0)	0 (0)	0 (0)	0
	Colville River	0	0	0 (0)	0 (0)	0 (0)	0 (0)	0 (0)	0
	All	184	86	3.40 (0.48)	1.95 (0.85)	0 (0)	1.30 (0.47)	95,594 (34,824)	0.36
PAJA	ACP-NPRA	46	27	1.14 (0.55)	0.90 (0.66)	0 (0)	0.59 (0.28)	20,344 (9,612)	0.47
	ACP-Central	20	7	2.44 (1.26)	1.19 (0.66)	0 (0)	1.70 (0.61)	18,413 (6,613)	0.36
	ACP-ANWR	61	9	0.58 (0.27)	1.07 (0.57)	0 (0)	0.57 (0.25)	2,737 (1,194)	0.44
	Foothills	3	1	0 (0)	0.97 (0.92)	0 (0)	0.15 (0.15)	3,538 (3,447)	0.97
	Colville River	0	0	0 (0)	0 (0)	0 (0)	0 (0)	0 (0)	0
	All	130	44	1.49 (0.22)	1.06 (0.28)	0 (0)	0.64 (0.18)	46,761 (13,379)	0.29
LTJA	ACP-NPRA	44	22	0.43 (0.32)	1.12 (0.71)	1.4 (1.40)	1.07 (0.62)	36,830 (21,242)	0.58
	ACP-Central	11	3	0.99 (1.00)	1.57 (1.10)	0 (0)	1.26 (0.73)	13,636 (7,905)	0.58
	ACP-ANWR	40	6	0 (0)	0.79 (0.48)	0.58 (0.42)	0.53 (0.25)	2561 (1,209)	0.47
	Foothills	2	0	0 (0)	0 (0)	0 (0)	0 (0)	0 (0)	0.16

	Colville River	0	0	0 (0)	0 (0)	0 (0)	0 (0)	0 (0)	0
	All	97	31	0.56 (0.51)	1.13 (0.37)	0.68 (0.77)	0.80 (0.36)	58,673 (26,377)	0.45
YEWA	ACP-NPRA	51	48	1.47 (0.60)	2.62 (1.27)	2.12 (1.53)	2.15 (0.78)	74,104 (26,935)	0.36
	ACP-Central	8	3	0 (0)	1.45 (1.07)	48.31 (0.00)	2.32 (0.57)	25,153 (6,146)	0.24
	ACP-ANWR	18	12	0.11 (0.12)	1.39 (0.88)	3.83 (2.53)	1.98 (1.02)	9,473 (4,873)	0.51
	Foothills	5	4	7.39 (3.57)	1.70 (1.59)	0 (0)	0.34 (0.27)	7,988 (6,177)	0.77
	Colville River	16	16	0 (0)	0 (0)	14.99 (6.66)	8.37 (3.31)	167 (66)	0.40
	All	98	83	1.05 (0.67)	2.36 (0.67)	1.65 (0.75)	1.76 (0.45)	129,399 (33,330)	0.26
SAVS	ACP-NPRA	285	258	3.73 (1.08)	7.28 (1.78)	27.55 (7.72)	14.55 (3.37)	501,727 (116,155)	0.23
	ACP-Central	6	3	0 (0)	2.08 (1.37)	48.31 (0)	2.65 (0.72)	28,737 (7,827)	0.27
	ACP-ANWR	49	45	1.39 (0.84)	5.54 (2.13)	11.95 (3.98)	6.90 (1.75)	33,073 (8,404)	0.25
	Foothills	43	41	23.48 (8.74)	4.16 (1.88)	14.8 (4.32)	14.80 (4.09)	344,276 (95,204)	0.28
	Colville River	28	28	0 (0)	0 (0)	16.05 (3.14)	8.96 (2.24)	179 (45)	0.25
	All	411	375	2.68 (3.34)	6.16 (2.52)	20.90 (2.52)	12.52 (2.15)	919,060 (157,607)	0.17
LALO	ACP-NPRA	450	405	11.88 (4.62)	15.62 (5.20)	6.64 (2.31)	11.24 (2.53)	387,694 (87,123)	0.22
	ACP-Central	62	56	23.27 (8.26)	17.13 (5.62)	0 (0)	19.29 (4.50)	208,860 (48,769)	0.23
	ACP-ANWR	538	484	57.52 (17.62)	47.37 (5.13)	43.07 (9.49)	48.14 (5.60)	230,734 (26,848)	0.12
	Foothills	38	44	14.79 (7.14)	8.52 (2.67)	9.89 (1.72)	10.72 (1.68)	249,322 (39,173)	0.16
	Colville River	1	1	0 (0)	0 (0)	0.93 (0.87)	0.52 (0.50)	10 (10)	0.96
	All	1,089	990	18 (2.91)	18.67 (2.74)	9.75 (2.04)	14.27 (1.53)	1,047,043 (112,027)	0.11

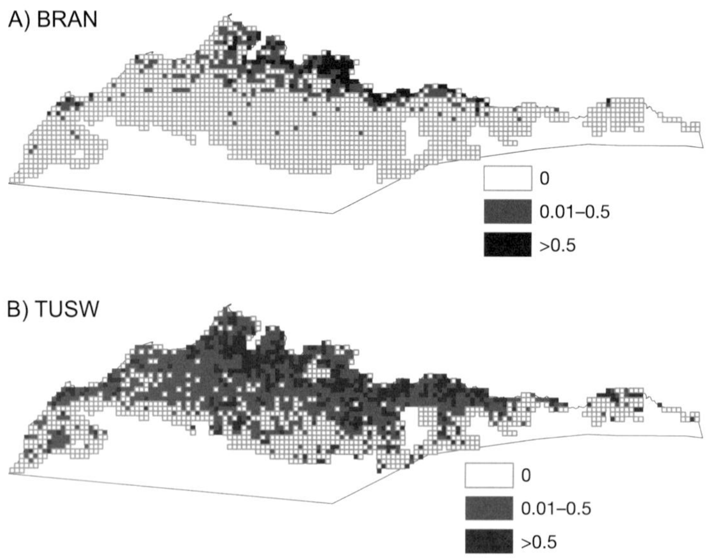

Figure 4.12. Density (birds/km^2) of Brant (a) and Tundra Swans (b) recorded on the aerial surveys.

et al. 2000). Results from the aerial surveys suggest that most of this area should be included in the range.

Cackling Goose

For this description, we assume all small geese resembling the Canada Goose were the Cackling Goose, though a few may have been *Branta hutchinsii hutchinsii* or *B. canadensis parvipes* (Mlodinow et al. 2008). During aerial surveys, Cackling Geese were recorded in 578 cells (Fig. 4.11c). They were particularly dense along the coast from Prudhoe Bay to Barrow, but they were also recorded widely across the ACP. They were uncommon in the Arctic NWR.

During ground surveys, Cackling Geese were encountered on 55 plots and were judged to be breeding on 32 plots. Most records were of single birds on wetlands. The species was rarely recorded in uplands (Table 4.9). The estimated number of pairs breeding on surveyed plots was 58.

Estimated densities from the rapid surveys were similar in wetlands and moist areas and much lower on uplands (Table 4.9). It seems likely that detection ratios were less than 1.0. The uncorrected, estimated population size was about 20,000 with a medium CV of 0.36.

The range map in the BNA for this species is consistent with our results (Mowbray et al. 2002).

Brant

During ground surveys, Brant were encountered on 22 plots and were judged to be breeding on 15 plots. All of the records were within several kilometers of the coast. Although 55 indicated pairs were recorded, 36 were on just two plots. The ground surveys thus provided relatively little information about this species' distribution. In contrast, during the aerial surveys Brant were recorded in 368 cells (Fig. 4.12a). Most records were between Prudhoe Bay and Barrow and within 50 km of the coast. Density was highest right along the coast.

The range map in the BNA is consistent with our results, except that we found them farther south than depicted in the BNA map, especially in the central NPRA (Reed et al. 1998).

Tundra Swan

During aerial surveys, Tundra Swans were recorded in 1,183 cells (Fig. 4.12b). Records occurred widely across the study area, though they were much less common in the Arctic NWR and near the southern edge of the coastal plain.

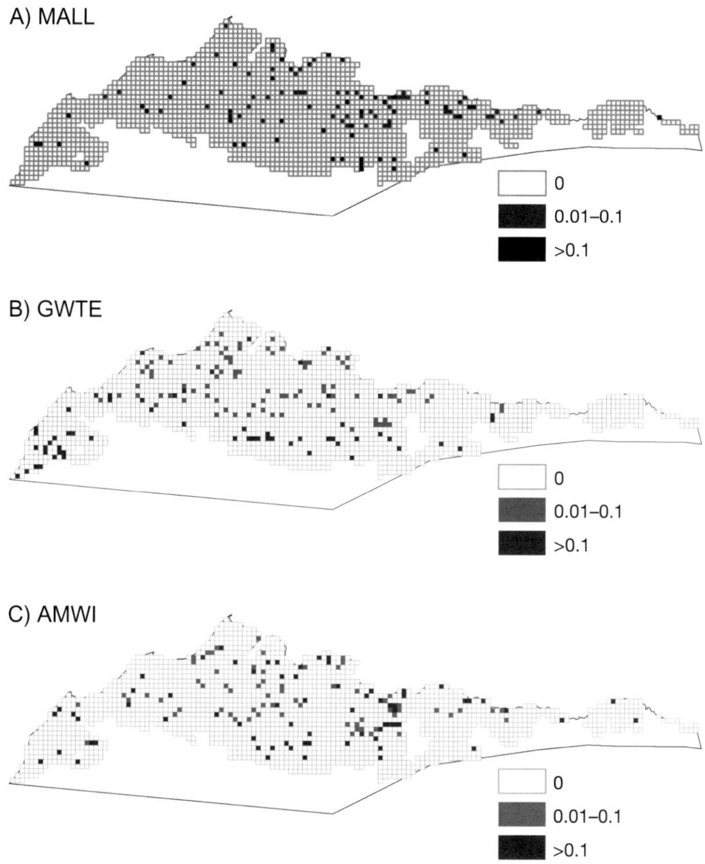

Figure 4.13. Densities (birds/km^2) of Mallards (a), Green-winged Teal (b), and American Wigeon (c) recorded on the aerial surveys.

On ground surveys, Tundra Swans were encountered on 45 plots and were judged to be breeding on 21 plots. Too few records were obtained for meaningful comparisons with the aerial survey data.

We suspect detection ratios for swans were close to 1.0. The uncorrected population estimate was 4,483 (Table 4.9). This estimate did not include non-breeding flocks, which are common on the North Slope.

The range map in the BNA is consistent with our results, except that we found them farther south than depicted in the BNA in the central NPRA (Limpert and Earnst 1994).

Mallard

During aerial surveys, Mallards were recorded in 107 cells (Fig. 4.13a). Records were widely distributed across the study area west of Prudhoe Bay. Density was perhaps slightly higher well south of the coast.

During ground surveys, Mallards were encountered on 11 plots and were judged to be breeding on eight plots. Only nine pairs were judged to be breeding.

The range map in the BNA for this species shows our entire study area as being within the range (Drilling et al. 2002). This seems reasonable, despite the low density with less than 1,500 pairs even if detection ratios are only 10%.

Green-winged Teal

During aerial surveys, Green-winged Teal were recorded in 149 cells (Fig. 4.13b). The records were almost entirely from west of the Colville River. Density was highest in the southern part of the coastal plain.

During ground surveys, Green-winged Teal were encountered on nine plots in the NPRA and were judged to be breeding on all of them. Most sightings were of pairs or single birds in wetlands. The estimated number of birds breeding in surveyed plots was 12.

The region- or habitat-specific SEs were too large to make useful comparisons between regions or habitats. The estimated population size was 5,950 with a large CV of 0.73. Results indicate that population size was probably less than 15,000.

The BNA range map is consistent with our results, except that we found them in the western part of the ACP-NPRA region, whereas the BNA map does not include this area (Johnson 1995).

American Wigeon

During aerial surveys, American Wigeon were encountered on 135 cells (Fig. 4.13c). Nearly all records were obtained from west of the Colville River. Records were widely distributed across the ACP-NPRA region, but density appeared to be slightly higher in the southern part of the coastal plain.

During ground surveys, American Wigeon were encountered on six plots and were judged to be breeding on all of them. All records were in the ACP-NPRA region. The estimated number of pairs breeding on surveyed plots was seven. The estimated population size was 407 with a large CV of 0.71.

The BNA range map does not include any of our study area (Mowbray 1999). The number of records for this species on aerial surveys was similar to results for Mallard and Green-winged Teal, both of which are depicted in the BNA as occurring across most or all of the North Slope. It may therefore be more appropriate that the range map for American Wigeon include all of our study area, or at least the ACP-NPRA region portion.

Northern Pintail

During aerial surveys, Northern Pintails were recorded in 1,613 cells (Fig. 4.14a). Records were obtained from throughout the study area including the Arctic NWR. Density appeared to be highest west of the Colville River and in the northern portion of the ACP.

During ground surveys, Northern Pintails were encountered on 219 plots and were judged to be breeding on 139 plots (Fig. 4.10b). Results were similar to results on the aerial survey. Records were widely distributed across the study area, including in the Arctic NWR (Table 4.9). Ground surveys revealed only a few birds in the Foothills, but 14 indicated pairs were recorded in the Colville River region. The estimated number of breeding birds was often much less than the observed number because non-breeding birds, often in flocks, were common. The estimated number of pairs breeding on surveyed plots was 301 (Table 4.9).

Data from the ground surveys suggested that densities were highest in wetlands but were substantial even in uplands (Table 4.9). Variation in density between ACP regions appeared to be small. The uncorrected estimate of population size was about 315,000 with a moderate CV of 0.22 (Table 4.9). We suspect the detection ratio was less than 1.0. The estimated population size excludes birds judged to be non-breeders.

The BNA range map includes all of our study area, which is consistent with our results (Austin and Miller 1995).

Northern Shoveler

During aerial surveys, Northern Shovelers were recorded in 96 cells (Fig. 4.14b). Records occurred almost exclusively west of the Colville River and were widely distributed across the NPRA.

Northern Shovelers were encountered on 15 plots and were judged to be breeding on 13 plots. They were observed almost exclusively on the Colville Delta, though a probable nest was recorded in the Arctic NWR and two pairs were recorded on one plot in the western part of the study area. They were recorded mainly as pairs or single birds in wetlands. The estimated number of pairs breeding on the surveyed plots was 22. The uncorrected estimate of population size was 1,110 with a large CV of 0.42.

The BNA range map excludes most of the NPRA, whereas we found this species sparsely distributed throughout the NPRA (Dubowy 1996).

Canvasback

Canvasbacks were not recorded on the ground surveys and were seen in only three cells, widely distributed across the study area, on the aerial survey. The BNA range map, which excludes the North Slope, is thus consistent with our results (Mowbray 2002).

Redhead

Four pairs of Redheads were encountered on one plot west of Fish Creek but were judged to

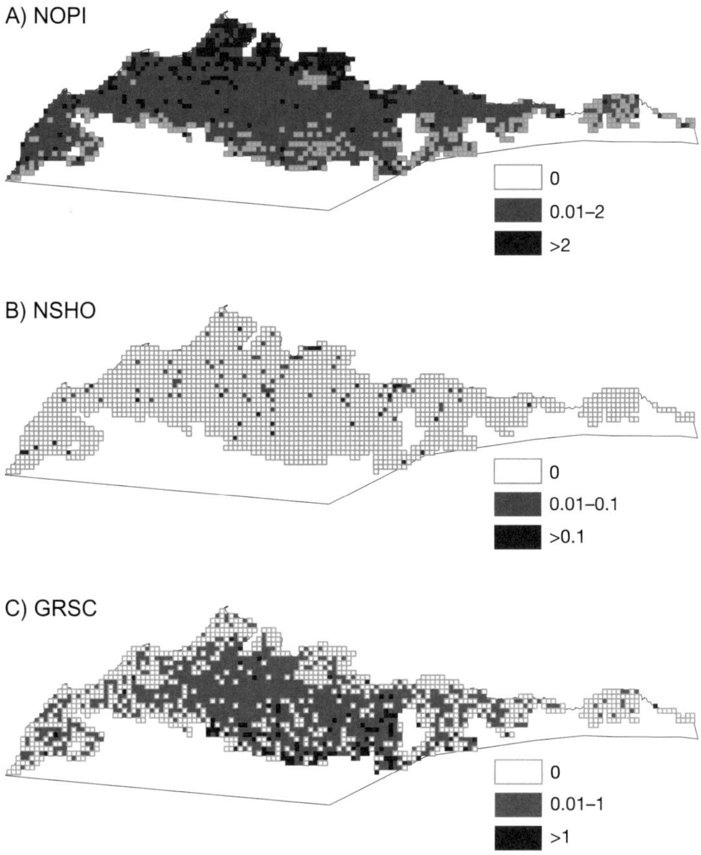

Figure 4.14. Densities (birds/km^2) of Northern Pintails (a), Northern Shovelers (b), and Greater Scaup (c) recorded on the aerial surveys.

be breeding off the plot. They were not recorded on the aerial survey. The BNA range map, which excludes the North Slope, is thus consistent with our results (Woodin and Michot 2002).

Greater Scaup

During the aerial surveys, Greater Scaup were recorded in 1,090 cells (Fig. 4.14c). Records were obtained from throughout the study area but were less common east of Prudhoe Bay and more common at the southern edge of the coastal plain in the eastern half of the NPRA.

During ground surveys, Greater Scaup were encountered on 46 plots and were judged to be breeding on 35 plots. Results were similar to the aerial survey, with birds being recorded widely across the study area, including in the Arctic NWR (Table 4.9). No birds were recorded in the Foothills. Most records were of pairs in wetlands or moist areas. Nearly half the detections were considered to be birds not breeding on the plot. The estimated number of pairs breeding on the surveyed plots was 48.

The uncorrected, estimated population size was about 40,000 with a large CV of 0.42 (Table 4.9). We suspect the detection ratio was below 1.0, though perhaps not far below it given the large sample size.

The BNA range map portrays the study area as being occupied by "isolated nest site(s) or small, isolated breeding populations(s) of uncertain persistence" (Kessel et al. 2002). Our results suggest the entire study area should be considered within the species' range.

Common Eider

None of the surveys reported in this chapter covered the primary breeding grounds for this

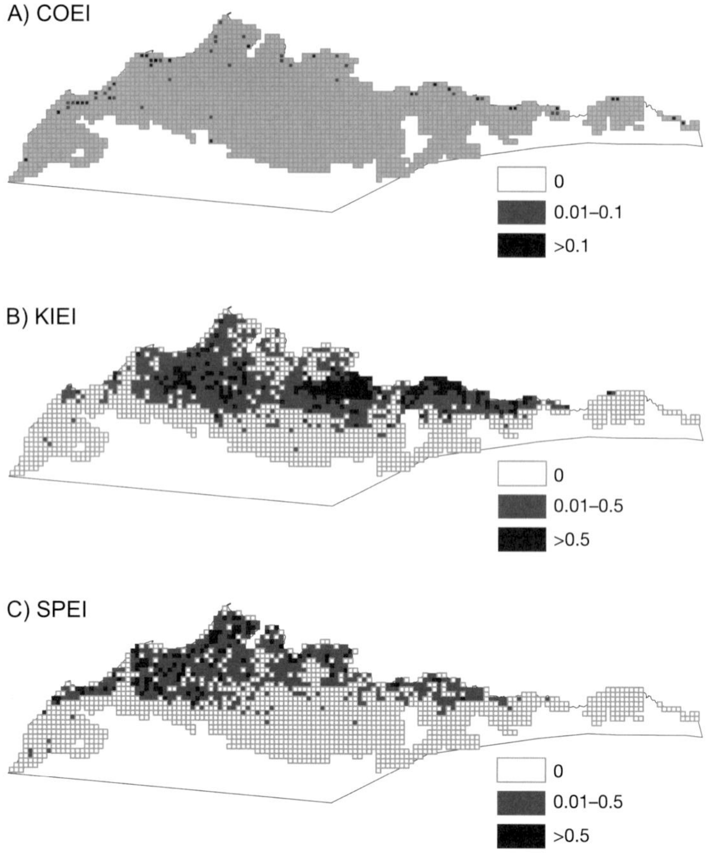

Figure 4.15. Densities (birds/km^2) of Common Eiders (a), King Eiders (b), and Spectacled Eiders (c) recorded on the aerial surveys.

species (offshore islands, covered by a separate USFWS survey). During aerial surveys, Common Eiders were recorded in 57 cells (Fig. 4.15a). They were observed along the entire coast including in the Arctic NWR but appeared to be most common southwest of Wainwright. A few records were obtained from inland areas. Only one Common Eider was observed during ground surveys.

The BNA range map shows the species occurring in a narrow band along the entire coast in our study area (Goudie et al. 2000). This is probably the only way to depict its range, but it should be realized that all or nearly all nests are on the barrier islands, not on the mainland.

King Eider

During aerial surveys, King Eiders were recorded in 853 cells (Fig. 4.15b). Nearly all records came from west of Prudhoe Bay. Density was highest in a band extending west from Prudhoe Bay into the NPRA. Few birds were seen in the southern third of the coastal plain.

During ground surveys, King Eiders were encountered on 37 plots and were judged to be breeding on 22 plots. The pattern was similar, with nearly all records being in the band of high density evident from the aerial surveys. Most records were of pairs in wetlands. The estimated number of pairs breeding on surveyed plots was 40 (Table 4.9).

Males were present for most of the survey period, so detection ratios were probably quite high. The uncorrected, estimated population size was 38,531 with a moderate CV of 0.42 (Table 4.9).

The range map in the BNA is consistent with our results, except that we found the species

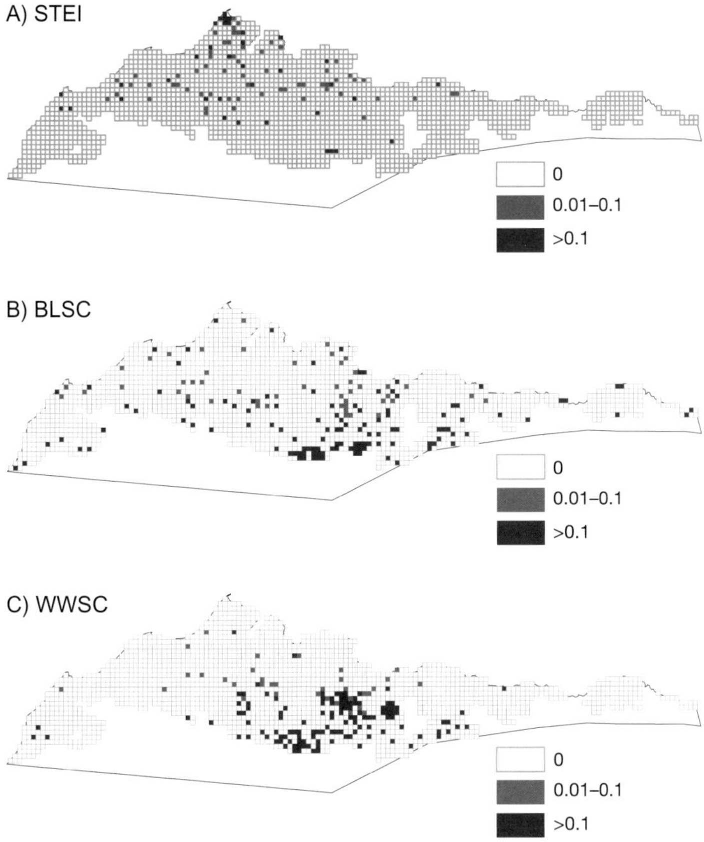

Figure 4.16. Densities (birds/km^2) of Steller's Eiders (a), Black Scoters (b), and White-winged Scoters (c) recorded on the aerial surveys.

farther south in the NPRA than indicated on the BNA range map (Suydam 2000).

Spectacled Eider

During aerial surveys, Spectacled Eiders were recorded in 629 cells (Fig. 4.15c). Records occurred almost exclusively west of Prudhoe Bay and were most common in the northwestern quarter of the NPRA. Almost no birds were recorded in the southern portion of the coastal plain.

During ground surveys, Spectacled Eiders were encountered on 20 plots and were judged to be breeding on 15 plots. Most of the indicated breeding birds were on the Colville Delta (Table 4.9).

Males were present during most of the surveys, so we suspect the detection ratio was close to 1.0. The uncorrected population estimate was 15,584 birds but the CV (0.60) was so large that this estimate is of little value (Table 4.9). Furthermore, more than half the estimate is due to a single bird observed in the ACP-Central region. If this bird had not been seen, the estimate would have been 6,369. This species provides a good example of why measures of precision should not be ignored.

The range map in the BNA corresponds closely to our results, except that we found it farther south in the NPRA than depicted on the BNA map (Petersen et al. 2000).

Steller's Eider

During aerial surveys, Steller's Eiders were recorded in 97 cells (Fig. 4.16a). Records occurred almost exclusively in the NPRA. They were widely distributed across the coastal plain. This species was not recorded on ground surveys.

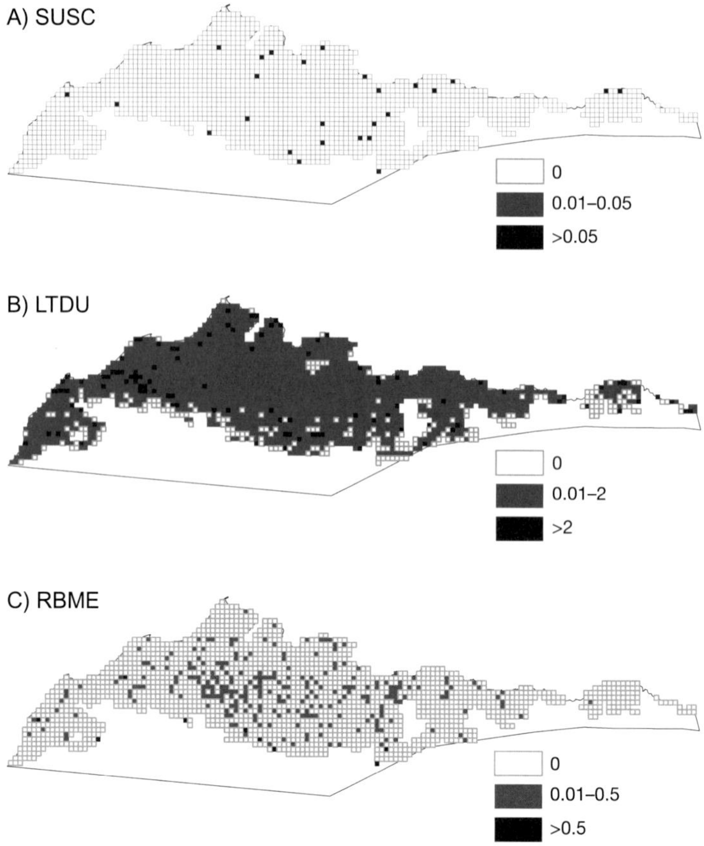

Figure 4.17. Densities (birds/km^2) of Surf Scoters (a), Long-tailed Ducks (b), and Red-breasted Mergansers (c) recorded on the aerial surveys.

The BNA range map is consistent with our results (Frederickson 2001).

Black Scoter

During aerial surveys, Black Scoters were recorded in 171 cells (Fig. 4.16b). They were recorded throughout the study area but were uncommon east of Prudhoe Bay. They were most common in the southern part of the coastal plain.

The BNA range map classifies the entire study area as "breeding probable" (Bordage and Savard 1995). We found the species throughout the study area, especially in the southern half of the NPRA and ACP-Central region, but we do not have evidence for or against breeding.

White-winged Scoter

During aerial surveys, White-winged Scoters were recorded in 158 cells (Fig. 4.16c). Most records were from the southern part of the coastal plain in a broad band along the Colville River. During ground surveys, White-winged Scoters were encountered on only four plots.

The BNA range map includes only the ACP-Arctic NWR region in our study area (Brown and Frederickson 1997). In contrast, we did not record in that region but did find it commonly in the south-central part of the study area and occasionally elsewhere in the western two-thirds of the study area.

Surf Scoter

During aerial surveys, Surf Scoters were recorded in 28 cells (Fig. 4.17a). Most records were from the eastern half of the NPRA, though a few came from the western NPRA and the Arctic NWR. The species was not recorded on the ground surveys.

The BNA stated that breeding on the North Slope was not confirmed except for one nest close to the Colville River (Savard et al. 1998). We found the species widely, but sparsely distributed across the entire study area (though we did not confirm breeding). Density may have been slightly higher in the south-central region, similar to the White-winged Scoter.

Long-tailed Duck

During aerial surveys, Long-tailed Ducks were recorded in 1,689 cells (Fig. 4.17b). They occurred throughout the study area. They were less common in the Arctic NWR and perhaps more common in the southern part of the coastal plain.

During ground surveys, Long-tailed Ducks were encountered on 156 plots and were judged to be breeding on 106 plots. Results were similar to results from the aerial surveys. The species was not recorded breeding in the Foothills (Table 4.9). Most records were pairs or single birds on wetlands. The estimated number of birds breeding in surveyed plots was 182 (Table 4.9).

Long-tailed Ducks were conspicuous for most of the survey period, so we suspect that detection ratios were close to 1.0. Densities were much higher on wetlands, but some birds were encountered in uplands (Table 4.9). The overall density was nearly 3 birds/km^2. The uncorrected, estimated population size was about 200,000 (Table 4.9).

The BNA range map includes the entire study area, which is consistent with our results (Robertson and Savard 2002).

Common Goldeneye

During aerial surveys, Common Goldeneyes were recorded in 11 cells, mainly in the central NPRA. They were not recorded on ground surveys. The range map in the BNA excludes our study area, which is consistent with our results (Eadie et al. 1995).

Red-breasted Merganser

During aerial surveys, Red-breasted Mergansers were recorded in 262 cells (Fig. 4.17c). Nearly all records were in the NPRA. Density appeared to be somewhat higher in the southern part of the coastal plain. Red-breasted Mergansers were encountered on only 11 plots and were judged to be breeding on six plots.

The BNA range map includes the entire study area, which is consistent with our results except perhaps in the Arctic NWR (Titman 1999).

Rock Ptarmigan

Rock Ptarmigan and Willow Ptarmigan were neither consistently recorded nor distinguished during the aerial surveys. On ground surveys, Rock Ptarmigan were encountered on 62 plots and were judged to be breeding on 53 plots (Fig. 4.18a). Nearly all records were from the ACP regions (Table 4.9). Most detections were of single birds. Female ptarmigan rarely flush (surveyors occasionally stepped on incubating birds), so nests, probable nests, and pairs were rarely encountered. The species occurred in all habitats. The estimated number of pairs breeding on surveyed plots was 61 (Table 4.9).

Male Rock Ptarmigan kept their winter plumage throughout the survey period and were nearly always conspicuous. We therefore suspect the detection ratio during ground surveys was close to 1.0. On this assumption, it appears that density is approximately the same across habitats and regions (1–3 birds/km^2) and that the population size is probably around 100,000–150,000 birds (Table 4.9).

The BNA range map includes the entire study area, which is consistent with our results (Montgomerie and Holder 2008).

Willow Ptarmigan

During ground surveys, Willow Ptarmigan were recorded on 211 plots and were judged to be breeding on 193 plots (Fig. 4.18b). Most records were single birds in wetlands or moist areas, but many records were also obtained in uplands. The estimated number of pairs breeding on surveyed plots was 377 (Table 4.9).

Male Willow Ptarmigan retained their winter plumage throughout the survey period and were conspicuous on their territories. We therefore suspect that detection ratios were close to 1.0. The estimated density across all habitats was 12 birds/km^2, making it one of the most abundant species (Table 4.9). Density was higher in moist areas and uplands than in wetlands. Population size was estimated at nearly 900,000 birds with a low CV of 0.16.

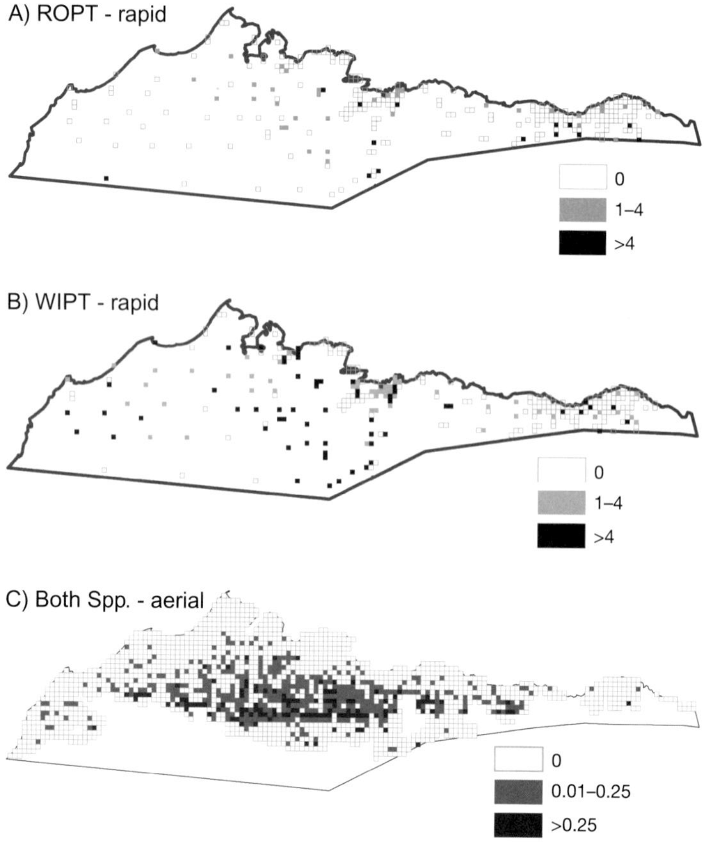

Figure 4.18. Densities (birds/km^2) of Rock Ptarmigan (a) and Willow Ptarmigan (b) recorded on the rapid surveys, and of both species recorded on the aerial surveys (c).

Although ptarmigan were not generally classified to species or recorded consistently on the aerial surveys, records of ptarmigan were obtained from 520 cells (Fig. 4.18c). Most records were from west of Prudhoe Bay. They were uncommon or absent in the northern part of the NPRA and appeared to be less common at the southern edge of the coastal plain.

The BNA range map includes the entire study area, which is consistent with our results (Hannon et al. 1998).

Red-throated Loon

During aerial surveys, Red-throated Loons were recorded in 850 cells (Fig. 4.19a). Sightings were distributed fairly evenly across the study area, except that they were less common in the Arctic NWR and along the southern edge of the coastal plain.

During ground surveys, Red-throated Loons were encountered on 35 plots and were judged to be breeding on 21 plots. Most records were from the Colville Delta or immediately west of there along the coast, though a few records were from the Arctic NWR. The estimated number of pairs breeding on surveyed plots was 24. Because most of our records came from the Colville Delta, and yet the species obviously occurred widely across the study area, we did not estimate densities from the ground surveys.

The BNA range map includes our entire study area, which is consistent with our results (Barr et al. 2000).

Pacific Loon

During aerial surveys, Pacific Loons were recorded in 1,575 cells (Fig. 4.19b). Sightings came from throughout the surveyed area but were least common in the Arctic NWR and may have been slightly more common in the west-central portion of the NPRA.

During ground surveys, Pacific Loons were encountered on 94 plots and were judged to be

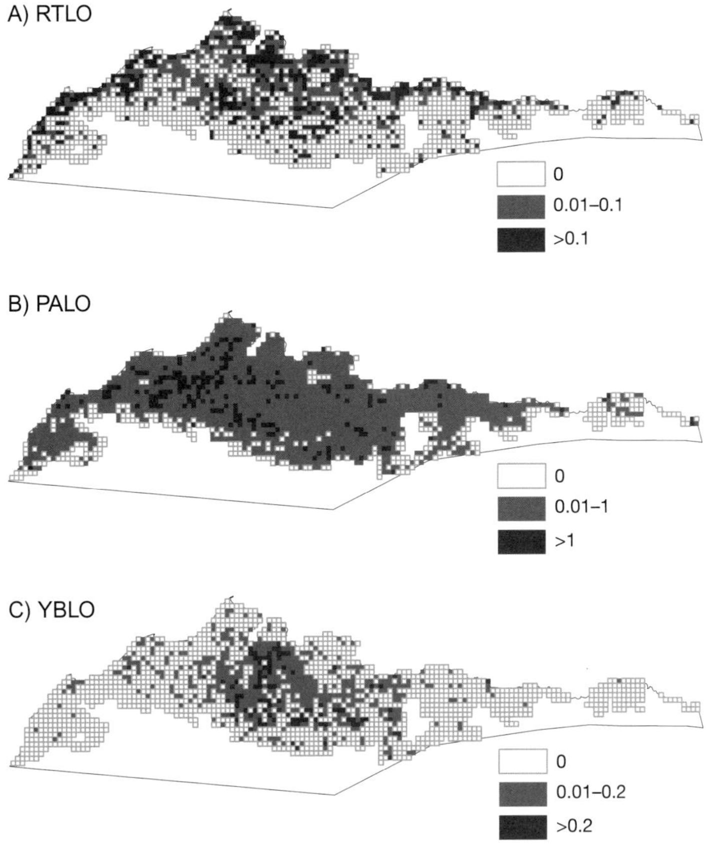

Figure 4.19. Densities (birds/km^2) of Red-throated Loons (a), Pacific Loons (b), and Yellow-billed Loons (c) recorded on the aerial surveys.

breeding on 66 plots. Nearly all sightings were in the ACP (Table 4.9). Most observations were pairs and almost all were in wetlands. The estimated number of pairs breeding on surveyed plots was 88.

We suspect detection ratios were close to 1.0, though birds at nests often hid from surveyors and could be hard to detect. Densities were highest in wetlands and almost zero in uplands (Table 4.9). Most of the population was estimated to occur in the ACP-NPRA. Population size was estimated at about 35,000 with a moderate CV of 0.32 (Table 4.9).

The BNA range map includes the entire study area, which is consistent with our results (Russell 2002).

Common Loon

During aerial surveys, Common Loons were recorded in nine cells, mainly in the NPRA, and widely distributed across it, but including three cells near the coast close to Prudhoe Bay. The species was not recorded on the ground surveys. The BNA does not include any of our study area, which is consistent with our results (Mcintyre and Barr 1997).

Yellow-billed Loon

During aerial surveys, Yellow-billed Loons were recorded in 575 cells (Fig. 4.19c). Sightings were concentrated in the central and eastern NPRA and appeared to be less common both close to the coast and close to the Foothills. Few sightings occurred east of Prudhoe Bay.

This species was only encountered during ground surveys on 14 plots and was only judged to be breeding on five plots. All but one of the records was from the Colville Delta.

The BNA range map includes our entire study area, which is consistent with our results (North 1994).

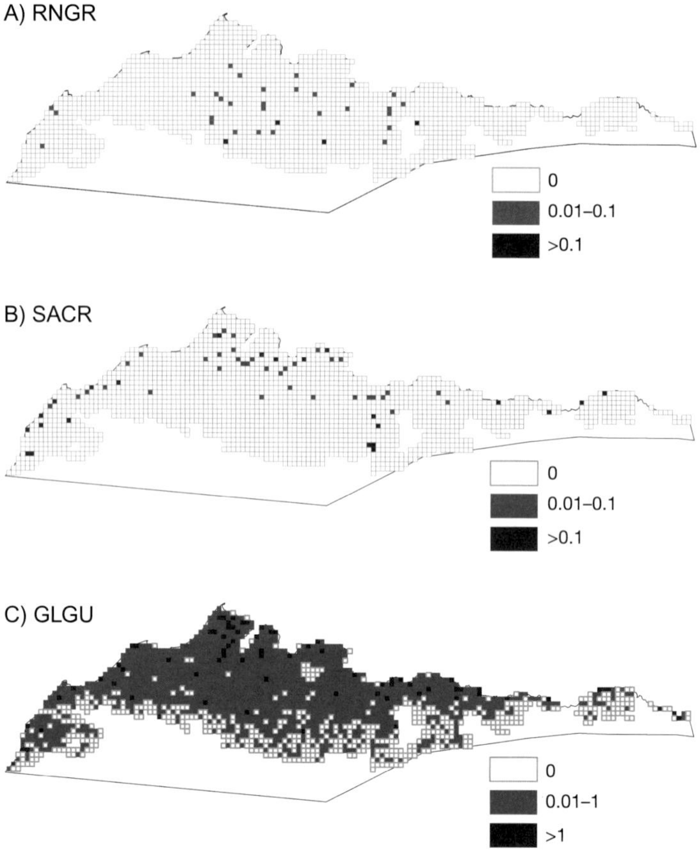

Figure 4.20. Densities (birds/km^2) of Red-necked Grebes (a), Sandhill Cranes (b), and Glaucous Gulls (c) recorded on the aerial surveys.

Red-necked Grebe

During aerial surveys, Red-necked Grebes were recorded in 33 cells (Fig. 4.20a). The sightings were mainly in the central and eastern part of the NPRA. Only a few sightings were east of the Colville. Density appeared to be lower close to the coast and in the southern part of the coastal plan. The species was not encountered on ground surveys.

The BNA range map does not include any of our study area (Stout and Nuechterlein 1999). Even though the species was recorded in the central part of the study area, it could be argued that density was too low to include this region in the range.

Northern Harrier

On the ground surveys, surveyors recorded six single Northern Harriers, three in the Arctic NWR and one each in each of the three NPRA regions. None of the birds was judged to be breeding. They were not recorded on aerial surveys. The BNA range map excludes the entire study area, which is consistent with our results (MacWhirter and Bildstein 1996).

Golden Eagle

During aerial surveys, Golden Eagles were recorded in 140 cells (Fig. 4.21a). The sightings were widely distributed across the survey area but were a little more common in southern parts. The species was only encountered on three plots during the ground surveys.

The BNA range map excludes our study area (Kochert et al. 2002). We suspect many of the birds recorded on aerial surveys were not breeding, so whether to include our study area in the species' range depends on whether birds summering (or visiting) north of their usual breeding range are to be included.

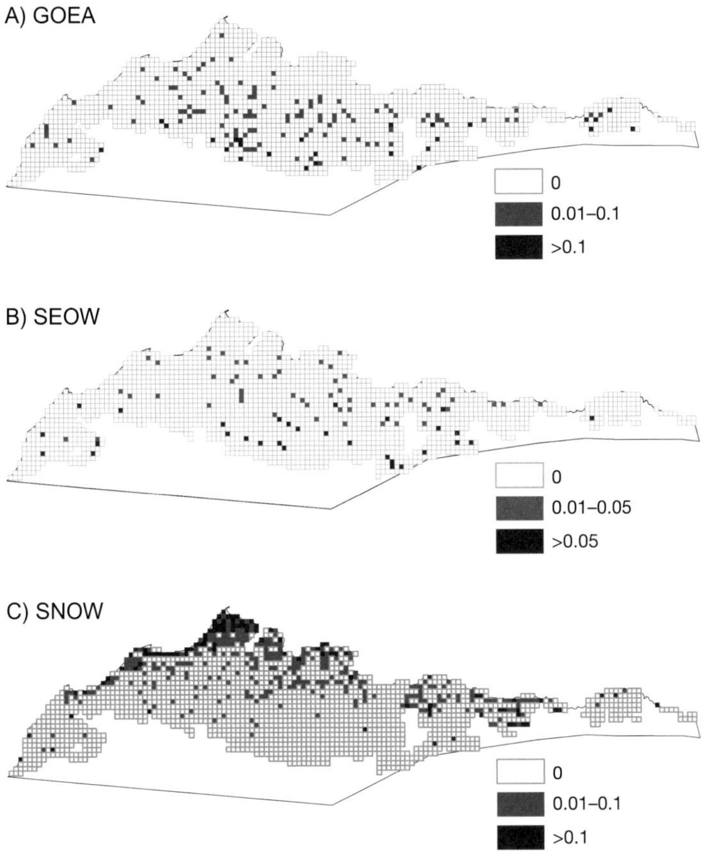

Figure 4.21. Densities (birds/km^2) of Golden Eagles (a), Short-eared Owls (b), and Snowy Owls (c) recorded on the aerial surveys.

Rough-legged Hawk

Rough-legged Hawks were observed breeding outside our plots on the Colville River upstream from Umiat. They were also recorded on seven plots: two in the Arctic NWR and five in the NPRA. They were not judged to be breeding on any of these plots. They were not recorded on the aerial surveys. We suspect that few pairs breed within our survey area except along the Colville River. The BNA includes our entire study area as within the range (Bechard and Swem 2002). Based on our results, excluding most of our study area except the upper Colville River may be more appropriate.

Peregrine Falcon

Peregrine Falcons were encountered on ten plots and were judged to be breeding on one plot. They were observed along the Colville River, where they breed, and once in the Arctic NWR. They were not recorded on the aerial survey, which seems odd given the number of flights that crossed the Colville River. The BNA shows most of our study area in the range (White et al. 2002). Based on our results, excluding most of our study area, other than the Colville River upstream from Ocean Point, might be more appropriate.

Gyrfalcon

Two Gyrfalcons were recorded, one in the Arctic NWR at the southern edge of the study area and one along the coast at the extreme western edge of the study area. None were recorded on the aerial survey. The BNA excludes most our study area from the range, which is consistent with our results (Booms et al. 2008).

Sandhill Crane

During aerial surveys, Sandhill Cranes were recorded in 59 cells (Fig. 4.20b). Sightings occurred mainly within 50 km of the coast and up the Colville River. During ground surveys, Sandhill Cranes were encountered on 13 plots and were judged to be breeding on three plots. We assume the detection ratio was close to 1.0. The estimated population size was 212 with a large CV of 0.74. The results indicate that the breeding population is almost certainly less than 500 birds.

The BNA range map does not include any of the North Slope, but our results suggest that the species occurs regularly in the northern portion of the study area and along the Colville River (Tacha et al. 1992).

Mew Gull

Aerial surveyors recorded Mew Gulls in seven cells in the southern part of the NPRA. The species was not encountered during ground surveys. The BNA excludes our entire study area, which is consistent with our results (Moskoff and Bevier 2002).

Herring Gull

A single Herring Gull was recorded in the NPRA-Coastal region on the ground surveys but was not thought to be breeding on the plot. The species was not recorded on aerial surveys. The BNA excludes our entire study area (Pierotti and Good 1994).

Glaucous Gull

During aerial surveys, Glaucous Gulls were recorded in 1,401 cells (Fig. 4.20c). Sightings were recorded throughout the surveyed area, though they were less common in the Arctic NWR and along the southern border of the coastal plain.

Glaucous Gulls were encountered on 138 plots and were judged to be breeding on 46 plots. They occurred widely across the study area during ground surveys (Table 4.9). They were most frequently recorded in the ACP-NPRA region. They often came to surveyors from great distances and so were encountered far more often (232 records) than they were recorded as breeding pairs (56 records).

Glaucous Gulls are one of the most conspicuous and easy to count species on the tundra. We suspect detection ratios were close to 1.0. Densities were similar in wetlands and moist areas (Table 4.9). Region-specific estimates were of low precision. Population size was estimated at about 35,000 with a moderate CV of 0.27 (Table 4.9).

The BNA range map includes most of our study area, but we found the species farther south in the NPRA than depicted on the BNA map (Gilchrist 2001).

Sabine's Gull

During aerial surveys, Sabine's Gulls were recorded in 831 cells (Fig. 4.22a). Most sightings were west of Prudhoe Bay and well north of the Foothills region. There was a slight suggestion that density was lower close to the coast, west of Barrow.

During the ground surveys, Sabine's Gulls were encountered on 22 plots and were judged to be breeding on 16 plots (Table 4.9). Most records were of single breeding pairs, but small colonies occurred occasionally and two were on surveyed plots (eight and five pairs). The estimated number of pairs breeding on surveyed plots was 27.

Sabine's Gulls were generally conspicuous. Where groups occurred, especially if some were on and some were off the plot, estimating the number of pairs could be difficult, but this happened rarely and probably had little impact on estimated densities. We therefore suspect the detection ratio was close to 1.0. The estimated population size was about 27,000 with a moderate CV of 0.42 (Table 4.9).

The BNA range map is consistent with our results (Day et al. 2001).

Arctic Tern

During the aerial surveys, Arctic Terns were encountered in 1,226 cells (Fig. 4.22b). They were widespread across the ACP-NPRA region but were most dense in the south-central part. They occurred in only a few cells in the Arctic NWR.

During the ground surveys, Arctic Terns were encountered on 100 plots and were judged to be breeding on 54 plots (Table 4.9). Most records were of single birds in wetlands. Foraging groups

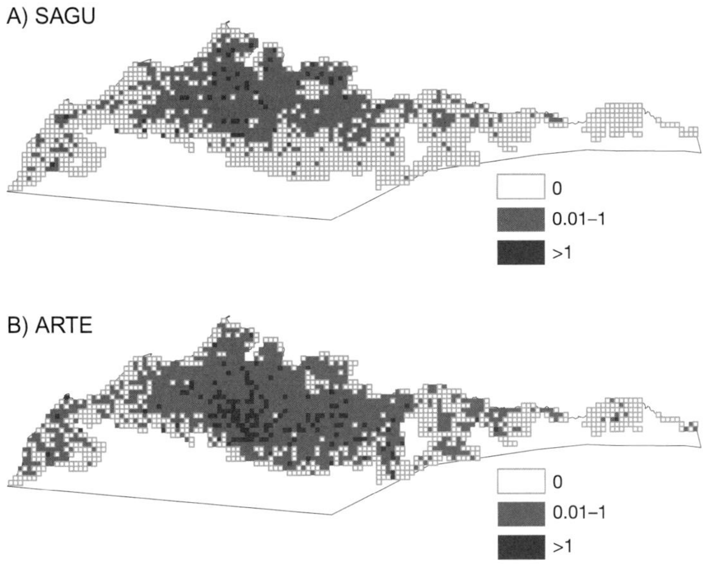

Figure 4.22. Densities (birds/km^2) of Sabine's gulls (a) and Arctic Terns (b) recorded on the aerial surveys.

were often encountered, so the total number observed (184) on all plots was much higher than the number (86) estimated to be breeding on these plots.

Arctic Terns are usually obvious foraging above wetlands. We therefore suspect the detection ratio is close to 1.0. Relative densities were highest in wetlands (Table 4.9). Most of the estimated population was in the ACP-NPRA region. The estimated, uncorrected population size was nearly 100,000 with a moderate CV of 0.36 (Table 4.9).

The BNA range map includes the entire study area, which is consistent with our results (Hatch 2002).

Pomarine Jaeger

Pomarine Jaegers were encountered on 25 plots and were judged to be breeding on six plots. The other records were of foraging birds. The species migrates eastward across the study area throughout most of June, but birds that were clearly migrating were not recorded. Jaegers were counted but not identified to species on the aerial surveys. The BNA range map shows the species breeding in a narrow band along the coast throughout our study area (Wiley and Lee 2000). This may be a reasonable depiction, but the species certainly does not breed throughout this area each year. For example, we never found it breeding on the Colville Delta during ten years of intensive study there.

Parasitic Jaeger

Parasitic Jaegers were encountered on 129 plots and were judged to be breeding on 42 plots (Fig. 4.23a). Records were widely distributed across the study area except possibly in the westernmost part of the NPRA (Table 4.9). The most common records were single birds in wetlands; however, they occurred frequently in moist areas. They often approached surveyors from considerable distances, so the estimated number of breeding birds (44) was much smaller than the number encountered (130).

The estimates of relative densities show that the species achieved highest densities in wetlands but was also common in moist areas (Table 4.9). Parasitic Jaegers are usually conspicuous when surveyors are near to the nest. We therefore suspect the detection ratio is fairly close to 1.0. The estimated population size was nearly 50,000 birds with a CV of 0.29 (Table 4.9).

The BNA range map includes the entire study area, which is consistent with our results (Wiley and Lee 1999).

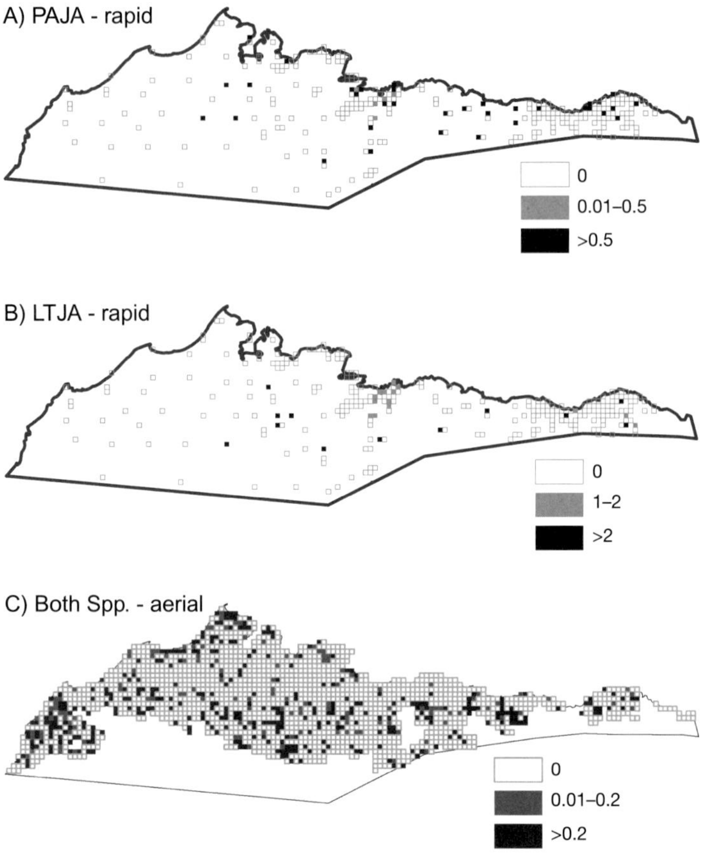

Figure 4.23. Densities (birds/km^2) of Parasitic Jaegers (a) and Long-tailed Jaegers (b) recorded on the rapid surveys, and of both species recorded on the aerial surveys (c).

Long-tailed Jaeger

During ground surveys, Long-tailed Jaegers were encountered on 90 plots and were judged to be breeding on 27 plots (Fig. 4.23b). The records were widely distributed across the study area but were concentrated in the Colville River Delta and were absent from the western two-thirds of the NPRA (Table 4.9). As with other jaegers, birds flying over the plot, often clearly coming to investigate the surveyor, were common, with the result that many more birds were recorded than were judged to be breeding within the plot. The estimated number of birds breeding on surveyed plots was 31.

We suspect detection ratios for jaegers are close to 1.0. Uncorrected densities were similar across habitats and regions (Table 4.9). Population size was estimated at about 60,000 birds.

As noted above, jaegers were seldom identified to species on the aerial survey. Combined results for jaegers (Fig. 4.23c) showed that they were widely distributed across the study area and were somewhat more common near the south border of the study area. Jaegers were recorded in 1,424 cells.

The BNA range map includes nearly all of our study area, which is consistent with our results (Wiley and Lee 1998). The map shows a small, unoccupied area in the vicinity of Barrow. We do not know whether that is accurate.

Short-eared Owl

During the aerial surveys, Short-eared Owls were recorded in 86 cells (Fig. 4.21b). Sightings were widely distributed across the surveyed areas but were least common in the Arctic NWR and appeared to be more common in the southern part of the ACP. During ground surveys, Short-eared Owls were encountered on 21 plots and were only judged to be breeding on three plots.

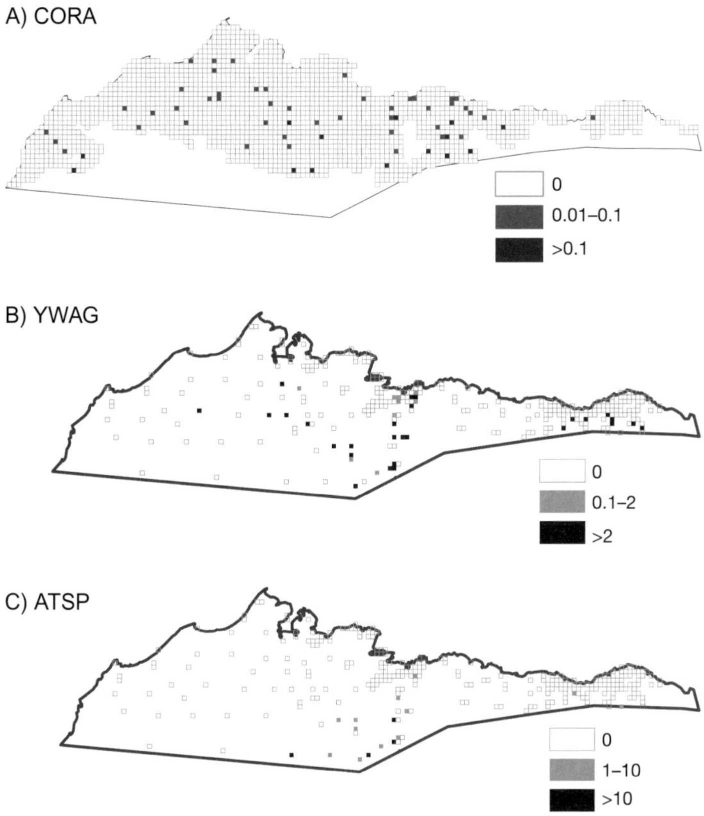

Figure 4.24. Densities (birds/km^2) of Common Ravens recorded on the aerial surveys (a), and Eastern Yellow Wagtails (b) and American Tree Sparrows (c) recorded on the rapid surveys.

The BNA range map includes the entire study area, which is consistent with our results (Wiggins et al. 2006).

Snowy Owl

During the aerial surveys, Snowy Owls were recorded in 384 cells (Fig. 4.21c). They were recorded widely across the surveyed area but were most abundant along the coast from Barrow to Wainwright. During the ground surveys, Snowy Owls were encountered on four plots and were not judged to be breeding on any of them.

The BNA range map includes the entire study area, which is consistent with our results (Parmelee 1992a).

Common Raven

During the aerial surveys, Common Ravens were recorded in 59 cells (Fig. 4.24a). Sightings were widely distributed across the surveyed area but were most common in the southern part of the coastal plain and least common close to the coast. During the ground surveys, the species was encountered on 17 plots but was not judged to be breeding on any of them.

The BNA range map includes the entire study area, which is consistent with our results (Boarman and Heinrich 1999).

Tree Swallow

Tree Swallows were encountered on nine plots and were judged to be breeding on seven plots. All but one record was from the Colville River upstream from Umiat. The BNA shows the species occurring only south of the Brooks Range (Robertson et al. 1992). Our results suggest that they also regularly occur in the upper Colville River near the southern border of our study area.

Arctic Warbler

Arctic Warblers were encountered on only one plot (three birds estimated to indicate three breeding pairs) on the Colville River near Umiat. The BNA shows the range extending into our study area mainly in the upper Colville River, which is consistent with our results (Lowther and Sharbaugh 2008).

Bluethroat

Bluethroats were encountered on ten plots and were judged to be breeding on all ten of them. They were encountered frequently in low shrub along the Colville river and occasionally elsewhere. Thirteen indicated pairs were recorded. The BNA includes our entire study area within the range (Guzy and McCaffery 2002). Our results suggest that the species occurs rarely away from the Colville River upstream from Ocean Point.

Gray-cheeked Thrush

Gray-cheeked Thrushes were encountered on four plots and were judged to be breeding on all of them. They were recorded only along the Colville River in tall shrubs. They were abundant on two of these plots (17 indicated pairs on each plot). The BNA excludes most of our study area except for a small portion of the upper Colville River, which is consistent with our results (Lowther et al. 2001).

American Robin

American Robins were encountered on three plots in the southern NPRA and were judged to be breeding on all of them. The BNA excludes all our study area, which is consistent with our results (Sallabanks and James 1999).

Eastern Yellow Wagtail

Yellow Wagtails were encountered on 57 plots and were judged to be breeding on 51 plots (Fig. 4.24b). The records were widely distributed in areas with shrubs at least 1 m tall. Single birds were the most common record. About equal numbers of records were obtained in wetlands and moist areas (Table 4.9). The estimated number of pairs breeding in surveyed plots was 83.

Yellow Wagtails sang actively throughout the survey period and have a distinctive note, given in flight and detectable from considerable distances. They remain largely on territory and are easy to count. We suspect the detection ratio was close to 1.0. Overall density was an impressive 1.8 birds/km^2 (Table 4.9). Densities were highest in moist areas. Population size was estimated at about 130,000 birds with a moderate CV of 0.26 (Table 4.9).

The BNA includes our entire study area within the range (Badyaev et al. 1998). This may be correct, though our results suggest that densities are low in the northwestern part of the NPRA and close to the coast throughout our study area.

American Pipit

American Pipits were encountered on two plots in the Arctic NWR and were not judged to be breeding on either of them. The BNA excludes most of our study area, which is consistent with our results (Verbeek and Hendricks 1994).

Yellow Warbler

Yellow Warblers were common on three plots on the Colville River upstream from Umiat. The estimated number of pairs breeding in these plots was 24. The BNA range map excludes our entire study area, whereas our results suggest that the species may breed commonly, if locally, on the upper Colville River (Lowther et al. 1999).

American Tree Sparrow

Tree Sparrows were encountered on 23 plots and were judged to be breeding on 22 plots (Fig. 4.24c). They were encountered commonly along the Colville River in the shrubs and occasionally elsewhere in the southern part of the study area. They were nearly always single birds, though a few nests and probable nests were found. They occurred exclusively in moist areas and uplands. The estimated number of pairs breeding in the surveyed plots was 80.

The BNA range map excludes our study area, whereas we found the species breeding commonly along the Colville at least as far north as

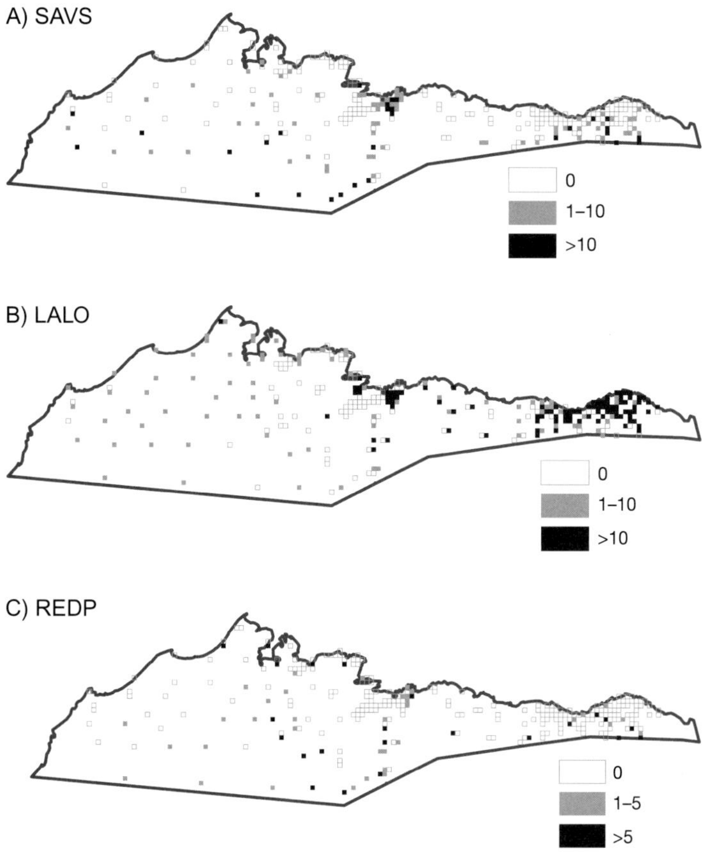

Figure 4.25. Densities (birds/km^2) of Savannah Sparrows (a), Lapland Longspurs (b), and Redpolls (c) recorded on the rapid surveys.

Ocean Point and sporadically almost to the coast (Naugler 1993).

Fox Sparrow

Fox Sparrows were encountered on eight plots (23 indicated pairs) along the Colville River about 100 km upstream from Umiat. They occurred in medium and tall shrubs. The range map in the BNA is consistent with our results (Weckstein et al. 2002).

Savannah Sparrow

Savannah Sparrows were encountered on 153 plots and were judged to be breeding on all but two of them (Fig. 4.25a). Most plots had zero to three pairs, but 28 plots had four or more pairs. They were recorded commonly in all habitats and in all regions (Table 4.9). The estimated number of pairs breeding on surveyed plots was 375.

Savannah Sparrows sang frequently throughout the survey period and unlike the other two common landbirds, Lapland Longspurs and Redpolls, were easy to count because their density was low and they generally sang from within their territories rather than covering large areas as both Longspurs and Redpolls frequently do. We suspect the detection ratio was slightly less than 1.0. Density was much higher in uplands than in other habitats (Table 4.9). The estimated density for all habitats was an extremely high 12.5 birds/km^2. The estimated population size was about 920,000 with a small CV of 0.17 (Table 4.9).

The BNA includes all of the study area, which is consistent with our results (Wheelwright and Rising 2008).

White-crowned Sparrow

White-crowned Sparrows were encountered on nine plots and were judged to be breeding on all of them. All records were from the Colville River, mainly above Umiat. The BNA shows the species breeding throughout the southern quarter of our study area, whereas we found it only along the Colville River upstream from Ocean Point (Chilton et al. 1995).

Lapland Longspurs

Lapland Longspurs were encountered on 318 plots and were judged to be breeding on 306 plots (Fig. 4.25b). Density appeared to be highest in the Arctic NWR. Single birds in wetlands or moist areas were the most common record. The estimated number of birds breeding on surveyed plots was 990 (Table 4.9).

Intensive work with banded birds on the Colville Delta showed that Longspur density was often 2–3 pairs/ha. At these densities, it was difficult to count them when all other species were being counted. As a result, surveyors often just indicated "present" rather than trying to estimate numbers present. This occurred on 108 plots out of 299 plots on which positive values were recorded. We converted these values to "1" for estimates of relative density.

Even with these sources of negative bias, the estimated densities in the Arctic NWR exceeded 40/km^2 and the estimated population size was over 1 million (Table 4.9). Given the substantial degree to which actual numbers were underestimated, it is plausible that the study area might hold 5 million Lapland Longspurs.

The BNA range map includes the entire study area, which is consistent with our results (Hussell and Montgomerie 2002).

Snow Bunting

Snow Buntings occurred on only three plots, all in the Colville Delta near buildings. The BNA shows the species breeding throughout all but the southernmost part of our study area, whereas we found it almost exclusively near settlements and other man-made features (Lyon and Montgomerie 1995).

Redpolls

We did not attempt to distinguish between Common and Hoary Redpolls, though the full range of color variations was encountered. Redpolls were encountered on 94 plots and were judged to be breeding on 74 plots (Fig. 4.25c). They were hard to count because they were mainly detected from flight calls and they traveled widely. As a result, we did not try to estimate densities. They were found widely across the study area wherever brush occurred, especially when it was one or more meters tall. They were particularly common along the Colville River in the taller brush there.

The BNAs for the redpolls show both species occurring throughout our study area, which is consistent with our results, though they may be uncommon in the northwestern part of the NPRA (Knox and Lowther 2000a, 2000b).

DISCUSSION

Geographic Patterns

Results in the species accounts may be used to compare regions and species. We do this below by examining patterns in density and abundance for waterfowl, waterbirds, shorebirds, and landbirds, using the most reliable survey data (i.e., ground or aerial) available. To estimate the proportions of the populations in each region from the aerial surveys, we multiplied the density of observations in each cell by 36 to obtain an estimate of the number of birds that would have been recorded in a single survey covering the entire cell. We referred to the sum, across all cells, of these estimates as the "aerial population index." We used the proportion of the index in each region as an estimate of the proportion of the population in the region and the density of observations as an estimate of the relative density among regions.

Waterfowl and Waterbirds

By far the most abundant species (based on the aerial surveys) were Northern Pintails, Greater White-fronted Geese, and Long-tailed Ducks, which together comprised 81% of the common waterfowl (Table 4.10). The estimates for Spectacled and Steller's Eiders and for Yellow-billed Loons have high CVs, which is unfortunate given the conservation concerns for these species.

For all waterfowl combined, 80% of the population was in the NPRA and only 5% was in the Arctic NWR (Table 4.11). Most of this difference is due to the much larger size of the NPRA, but most densities were also lower in the Arctic

TABLE 4.10

Ranked estimates of abundance and density for selected waterfowl and waterbirds, shorebirds, and landbirds based on the ground surveys

Taxa	Species	*n* birds recorded	Estimated population size	Estimated density (birds/km^2)	CV
Waterfowl and waterbirds	Northern Pintail	301	315,559	4.3	0.22
	Greater White-fronted Goose	347	303,028	4.13	0.29
	Long-tailed Duck	182	208,928	2.85	0.22
	Arctic Tern	86	95,594	1.3	0.36
	Long-tailed Jaeger	31	58,673	0.8	0.45
	Parasitic Jaeger	44	46,761	0.64	0.29
	Greater Scaup	48	40,534	0.55	0.42
	King Eider	40	38,531	0.53	0.42
	Glaucous Gull	56	35,132	0.48	0.27
	Pacific Loon	88	34,426	0.47	0.32
	Sabine's Gull	27	26,949	0.37	0.42
	Cackling Goose	58	19,539	0.27	0.36
	Spectacled Eider	22	15,584	0.21	0.60
	Brant	55	11,045	0.15	0.51
	White-winged Scoter	2	8,338	0.11	1.00
	Green-winged Teal	12	5,950	0.08	0.73
	Tundra Swan	23	4,483	0.06	0.58
	Red-breasted Merganser	12	3,722	0.05	0.87
	Yellow-billed Loon	5	2,570	0.04	0.85
	Red-throated Loon	24	1,737	0.02	0.41
	Mallard	9	1,465	0.02	0.75
	Northern Shoveler	22	1,110	0.02	0.42
Shorebirds	Semipalmated Sandpiper	1,335	1,319,225	17.98	0.17
	Pectoral Sandpiper	1,149	1,117,937	15.23	0.19
	Long-billed Dowitcher	374	647,695	8.83	0.19
	Red Phalarope	735	573,979	7.82	0.18
	Red-necked Phalarope	555	560,092	7.63	0.24
	Dunlin	378	500,161	6.82	0.20
	Western Sandpiper	45	382,443	5.21	0.32
	American Golden-Plover	159	275,506	3.75	0.31
	Black-bellied Plover	175	201,162	2.74	0.23
	Stilt Sandpiper	143	121,213	1.65	0.23
	Bar-tailed Godwit	56	75,602	1.03	0.70
	Buff-breasted Sandpiper	25	41,541	0.57	0.45

TABLE 4.10 (*continued*)

TABLE 4.10 (CONTINUED)

Taxa	Species	n birds recorded	Estimated population size	Estimated density (birds/km²)	CV
	Baird's Sandpiper	7	22,252	0.3	0.69
	Whimbrel	17	18,243	0.25	0.69
	Wilson's Snipe	21	14,599	0.2	0.58
Landbirds	Lapland Longspur	990	1,047,043	14.27	0.11
	Savannah Sparrow	375	919,060	12.52	0.17
	Willow Ptarmigan	377	883,500	12.04	0.16
	Redpolls	89	225,699	3.08	0.14
	American Tree Sparrow	80	194,539	2.65	0.39
	Eastern Yellow Wagtail	83	129,399	1.76	0.26
	Rock Ptarmigan	61	121,592	1.66	0.28
	Bluethroat	13	39,163	0.53	0.51

NWR. For all waterfowl, density was more than twice as high in the NPRA as in the Arctic NWR. The Central region was intermediate with respect to both population size and density.

Among gulls, terns, and jaegers, five species were common and widespread, with CVs ≤ 0.45 from the ground surveys (Table 4.9). Other members in this group, including Pomarine Jaeger, were much rarer. Regional estimates of distribution and relative density, based on the aerial surveys, were similar to results for waterfowl (Table 4.11). Most of the population of each species occurred in the NPRA, and densities were lower on the Arctic NWR. The difference was small, however, for Glaucous Gulls and the jaegers. Results from the ground surveys were generally similar (Table 4.9); however, the small samples within regions made comparisons difficult.

Shorebirds

Fifteen species of shorebirds had estimated population sizes of 14,000 or more based on the ground surveys (Table 4.10). The two most abundant of these comprised about half of all shorebirds.

The aerial surveys also recorded shorebirds, though not to species level. Results were fairly similar to the results for waterfowl. The NPRA contained 87% of all shorebirds, with 10% in the Central region and the remainder in the ANWR. Densities were more than twice as high in the NPRA as in the ANWR (0.56 sightings/km² vs. 0.23 sightings/km²). Results from the ground surveys showed similar patterns (Table 4.12). Most species were highly concentrated in the NPRA. Density was markedly higher in the NPRA for nearly all species.

Landbirds

Ground surveys demonstrated that eight species of landbirds were common and widespread (Table 4.9). Most other landbirds were restricted to shrubs, especially along the Colville River. Regional patterns were generally similar to other groups, with the majority of the ACP population in the NPRA, and densities being highest there (Table 4.12). Lapland Longspurs were the only exception. Density for this species was much higher in the Arctic NWR, though it is possible that this pattern was due to surveyors recording estimated densities (rather than a "1" indicating "present") in the Arctic NWR.

Upper Colville River

The upper Colville River, above Ocean Point, contains the tallest shrubs in the study area and supports a diverse bird fauna quite unlike the rest of the study area (Table 4.13). During our brief surveys of this area, we recorded 34 species on plots plus several additional species off plots. Tree Sparrows were by far the most abundant species.

TABLE 4.11

Regional estimates of relative population size and relative density for selected waterfowl and waterbirds in the Arctic Coastal Plain based on the aerial surveys.

Species	*n* cells with positive counts	Proportion of population[a]			Relative density[b]		
		ACP-NPRA	ACP-Central	ACP-ANWR	ACP-NPRA	ACP-Central	ACP-ANWR
All waterfowl	1,858	0.80	0.15	0.05	5.43	3.97	2.50
Greater WF Goose	1,583	0.80	0.19	0.01	1.72	1.57	0.17
Long-tailed Duck	1,689	0.75	0.14	0.11	0.80	0.56	0.90
Northern Pintail	1,613	0.88	0.09	0.03	1.17	0.47	0.29
Pacific Loon	1,575	0.83	0.15	0.02	0.44	0.29	0.08
Glaucous Gull	1,401	0.80	0.13	0.08	0.28	0.17	0.21
Greater Scaup	1,090	0.76	0.22	0.02	0.23	0.26	0.04
Arctic Tern	1,226	0.89	0.09	0.02	0.32	0.12	0.06
Brant	368	0.74	0.08	0.17	0.15	0.06	0.27
Cackling Goose	578	0.81	0.17	0.02	0.20	0.16	0.04
Tundra Swan	1,183	0.73	0.18	0.10	0.13	0.12	0.13
Black Scoter	171	0.41	0.13	0.46	0.03	0.04	0.30
King Eider	853	0.74	0.25	0.01	0.15	0.19	0.01
Jaegers	1,424	0.77	0.19	0.04	0.11	0.10	0.05
Sabine's Gull	831	0.97	0.03	0	0.15	0.02	0.01
Spectacled Eider	629	0.91	0.08	0	0.12	0.04	0
Red-throated Loon	850	0.82	0.12	0.05	0.06	0.03	0.03
Snow Goose	124	0.83	0.17	0	0.06	0.05	0
White-wgd. Scoter	158	0.83	0.17	0	0.05	0.04	0
Common Eider	57	0.38	0.06	0.56	0.01	0	0.07
Yellow-billed Loon	575	0.91	0.08	0	0.04	0.01	0
Red-br. Merganser	262	0.83	0.16	0.01	0.02	0.02	0
Mallard	107	0.66	0.24	0.10	0.01	0.01	0.01
American Wigeon	135	0.76	0.18	0.07	0.01	0.01	0.01
Steller's Eider	97	0.96	0.04	0	0.01	0	0
Northern Shoveler	96	0.85	0.15	0	0.01	0	0
Green-winged Teal	149	0.91	0.09	0	0.01	0	0
Red-necked Grebe	33	0.80	0.20	0	0	0	0

[a] Proportion of the aerial population index in each region.

[b] Sightings/km^2.

TABLE 4.12

Regional estimates of relative population size and relative density for selected shorebirds and landbirds in the Arctic Coastal Plain based on the ground surveys.

	Proportion of population[a]			Relative density[b]		
Species	ACP-NPRA	ACP-Central	ACP-ANWR	ACP-NPRA	ACP-Central	ACP-ANWR
Lapland Longspur	0.47	0.25	0.28	11.26	19.29	48.14
Semipalmated Sandpiper	0.77	0.21	0.03	26.19	22.57	6.87
Pectoral Sandpiper	0.75	0.22	0.04	21.62	19.88	8.04
Willow Ptarmigan	0.90	0.07	0.02	17	4.51	3.18
Savannah Sparrow	0.89	0.05	0.06	14.54	2.65	6.90
Red-necked Phalarope	0.82	0.12	0.05	11.67	5.52	5.49
Red Phalarope	0.84	0.14	0.02	12.42	6.67	2.05
Long-billed Dowitcher	0.90	0.10	0.01	14.46	4.97	0.79
Dunlin	0.85	0.14	0.01	10.92	5.68	0.67
Western Sandpiper	1.00	0	0	9.28	0	0.03
American Golden-Plover	0.79	0.14	0.07	3.82	2.22	2.46
Black-bellied Plover	0.81	0.19	0	4.13	3.09	0
Stilt Sandpiper	0.65	0.29	0.07	2.09	2.98	1.53
Eastern Yellow Wagtail	0.68	0.23	0.09	2.15	2.32	1.98
Redpoll	0.72	0.22	0.05	2.25	2.22	1.21
Rock Ptarmigan	0.78	0.13	0.09	1.73	0.90	1.47
Baird's Sandpiper	0.17	0.14	0.70	0.11	0.29	3.36
American Tree Sparrow	0.81	0.19	0.01	1.47	1.09	0.08
Buff-breasted Sandpiper	0.56	0.41	0.03	0.63	1.49	0.27
Bar-tailed Godwit	0.97	0.03	0	1.81	0.21	0
Bluethroat	0.99	0	0.01	0.59	0	0.03
Whimbrel	0.93	0	0.07	0.39	0	0.22
Wilson's Snipe	1	0	0	0.04	0	0

[a]Proportion of the aerial population index in each region.
[b]Sightings/km^2.

Several other species were found only on the Colville River.

Summary

The main geographic pattern demonstrated in this study was that birds on the North Slope occurred mainly in the ACP. Few species were present in the Foothills. Within the ACP a distinct east–west gradient existed, with higher densities, for most species, in the west. This trend, in combination with the much larger areas in the west, resulted in the great majority of the birds of most species being in the NPRA, with the fewest in the Arctic NWR.

Habitat Relationships

The analysis allowed us to determine which spatial scale (plot, area within 1 km, and area within 10 km) was most useful for predicting abundance. Detailed results are presented in the shorebird accounts. Combining results (Table 4.14) shows

TABLE 4.13

Species recorded on the Colville River upstream from Ocean Point.

Species	*n* birds recorded	Estimated population size	Estimated density (birds/km^2)	CV
American Tree Sparrow	50	3134	156.69	0.82
Tree Swallow	45	482	24.10	0.46
Gray-cheeked Thrush	40	519	25.94	0.57
Willow Ptarmigan	30	210	10.48	0.27
Savannah Sparrow	28	179	8.96	0.25
Yellow Warbler	24	339	16.95	0.65
Fox Sparrow	22	250	12.48	0.43
Eastern Yellow Wagtail	16	167	8.37	0.40
Northern Pintail	14	131	6.53	0.31
Redpoll	14	134	6.71	0.35
White-crowned Sparrow	11	668	33.41	0.72
Wilson's Snipe	9	774	38.68	0.77
Bluethroat	8	47	2.37	0.43
Cackling Goose	8	61	3.04	0.97
Long-tailed Duck	7	91	4.54	0.49
American Robin	6	70	3.51	0.48
Pectoral Sandpiper	5	55	2.75	1.11
Arctic Warbler	5	71	3.57	0.67
Pacific Loon	5	39	1.97	1.00
Greater Scaup	4	49	2.44	0.43
Red-necked Phalarope	3	33	1.65	1.11
Semipalmated Plover	3	43	2.16	0.49
American Golden-Plover	2	20	0.99	0.73
Lesser Yellowlegs	2	9	0.44	0.98
Glaucous Gull	2	18	0.89	1.11
Greater White-fronted Goose	2	13	0.63	0.81
King Eider	2	18	0.89	1.11
Mallard	2	28	1.40	0.64
Long-billed Dowitcher	1	11	0.55	1.11
Whimbrel	1	4	0.22	0.98
Lapland Longspur	1	10	0.52	0.96
Red-breasted Merganser	1	10	0.52	0.96
Rock Ptarmigan	1	15	0.77	0.98
Sandhill Crane	1	4	0.18	0.98

TABLE 4.14

Significance levels of habitat variables, measured at three spatial scales, in a multiple regression to predict bird density.

Species	Variable	Scale			Best scale
		Micro	Meso	Macro	
Black-bellied Plover	Wetlands	0.0001	0.0001	0.0008	???
	Moist areas	0.0017	0.01	0.0052	micro
American Golden-Plover	Wetlands	0.568	0.719	0.837	micro
	Moist areas	0.01	0.031	0.063	micro
Bar-tailed Godwit	Wetlands	0.0935	0.1753	0.3607	micro
	Moist areas	0.1194	0.2546	0.7601	micro
Ruddy Turnstone	Wetlands	0.6496	0.1323	0.2512	meso
	Moist areas	0.7673	0.2849	0.2512	macro
Dunlin	Wetlands	0.0153	0.0154	0.0485	micro
	Moist areas	0.0939	0.3682	0.8067	micro
Semipalmated Sandpiper	Wetlands	0	0	0	???
	Moist areas	0	0.0001	0.3715	micro
Western Sandpiper	Wetlands	0.3412	0.1786	0.1052	macro
	Moist areas	0.5816	0.18	0.3631	meso
Pectoral Sandpiper	Wetlands	0.0059	0.0058	0.8266	meso
	Moist areas	0.048	0.0814	0.1842	micro
Long-billed Dowitcher	Wetlands	0.0005	0.0173	0.0551	micro
	Moist areas	0.1986	0.9504	0.7053	micro
Stilt Sandpiper	Wetlands	0	0.0039	0.8896	micro
	Moist areas	0	0.0876	0.3149	micro
Red-necked Phalarope	Wetlands	0	0.0508	0.9989	micro
	Moist areas	0	0.0205	0.3125	micro
Red Phalarope	Wetlands	0	0	0.4788	micro
	Moist areas	0	0.0001	0.0307	micro

that in 17 of 24 cases the significance level was highest when habitat was measured at the micro scale. In three cases the meso scale had the highest significance level, in two analyses the macro scale was best, and in two the issue was uncertain because significance was extremely high at all scales. In our analysis, there was thus little advantage to be gained by measuring habitat at a scale larger than the plot. When the analyses can all be completed using GIS methods, as was the case for us, there is little harm in carrying out analyses at the other scales. If fieldwork were required, our analysis suggests that confining the work to the plot would be prudent.

The analysis also permits a crude comparison of whether spatial or habitat variables were more useful in predicting density (Table 4.8). Among the 12 models, spatial variables were used in all models, and a total of 27 times. In contrast, habitat variables were used in only nine of the models, and a total of ten times. Furthermore, spatial variables usually dominated the model so that predicted numbers varied in a smooth manner across the landscape, showing little influence of habitat. This does not mean, of course, that habitat is unimportant to these species, as any field biologist knows. Instead, it shows that our level of resolution (100-m pixels) and small number of

TABLE 4.15

Estimated detection ratios on aerial surveys.[a]

Species	Ground surveys Estimate	Ground surveys CV	Aerial surveys Estimate	Aerial surveys CV	Detection ratio	SE
Jaegers	105,435	0.28	5,647	0.14	0.05	0.02
Short-eared Owl	11,084	0.90	204	0.45	0.02	0.02
Long-tailed Duck	208,928	0.22	46,328	0.10	0.22	0.05
Northern Pintail	315,559	0.22	68,180	0.12	0.22	0.05
Green-winged Teal	5,950	0.73	397	0.55	0.07	0.06
Arctic Tern	95,594	0.36	23,036	0.14	0.24	0.09
Gr. Wh.-fronted Goose	303,028	0.29	129,879	0.11	0.43	0.13
Glaucous Gull	35,132	0.27	16,216	0.16	0.46	0.15
King Eider	38,531	0.42	12,216	0.24	0.32	0.15
Northern Shoveler	1,110	0.42	238	0.61	0.21	0.16
Greater Scaup	40,534	0.42	16,626	0.21	0.40	0.19
Sabine's Gull	26,949	0.42	11,758	0.25	0.44	0.21
Pacific Loon	34,426	0.32	25,331	0.08	0.74	0.24
Red-necked Grebe	264	1.00	50	1.18	0.19	0.29
Spectacled Eider	15,584	0.60	6,706	0.70	0.43	0.40
Mallard	1,465	0.75	633	0.57	0.43	0.41
White-winged Scoter	8,338	1.00	3,806	0.43	0.46	0.50
Red-breasted Merganser	3,722	0.87	2,018	0.56	0.54	0.56
Brant	11,045	0.51	9,861	0.63	0.89	0.73
Cackling Goose	19,539	0.36	21,693	0.62	1.11	0.80
Yellow-billed Loon	2,570	0.85	2,440	0.22	0.95	0.83
Sandhill Crane	212	0.74	221	0.58	1.04	0.98
Red-throated Loon	1,737	0.41	3,955	0.16	2.28	1.00
Tundra Swan	4,483	0.58	11,717	0.17	2.61	1.58
American Wigeon	407	0.71	845	0.51	2.08	1.81
Snow Goose	485	0.70	5,296	1.43	10.92	17.35

[a]Estimated from aerial surveys provided by Dr. R. S. Stehn and were based on the Arctic Coastal Plain surveys (Mallek et al. 2004) except for Spectacled and King Eiders, which were based on the North Slope eider survey (R. S. Stehn, pers. comm.).

categories (wetland, moist, upland) largely failed to capture the important variation, another indication that small- rather than large-scale variation was important to our species.

Range Map Adjustments

This study provides a great deal of new information that may be useful in adjusting range maps. These adjustments should be made by the biologists most familiar with the species in our study area. Our results suggest that consideration be given to expanding the ranges for American Wigeon, Red-necked Grebe, Golden Eagle, and Sandhill Crane and that consideration be given to contracting the ranges for Least Sandpiper, Baird's Sandpiper, Wilson's Snipe, Rough-legged Hawk, and Bluethroat. Smaller adjustments might be made for about two dozen other species, as described in the species accounts.

Detection Ratios on Aerial Surveys

We compared our estimates from ground surveys of population size with the indices of population size published by the USFWS (Mallek 2006). A summary of 26 species, sorted by CV of the detection ratio (Table 4.15), suggests that detection ratios vary widely between species. Large, white, monomorphic species such as Tundra Swans and Snow Geese, and obvious species such as loons, had detection ratios above 0.5. Many other species had lower rates, often <0.3. CVs for the detection ratios, however, were often large, indicating that these data do not lend themselves to estimating species-specific detection ratios, nor do the aerial surveys provide a good basis for estimating population size except for the most obvious species. As stated in Bart et al. (chapter 2, this volume), bias in population size estimates, in this case based on the aerial survey index, does not necessarily cause any bias in trend estimates; it only makes such bias possible.

CONCLUSIONS

This is one of the first avian studies in the arctic to combine results from aerial and ground surveys and to have large samples collected over a large area in both programs. The fact that well-defined sampling plans were used in both surveys meant the results could be compared and combined when appropriate. Although the aerial surveys did not include estimation of detection ratios, they still revealed spatial patterns and proportions of the population in different areas, assuming only that detection ratios were fairly uniform across the study area. This chapter has presented hundreds of estimated densities and population sizes, most of them for the first time ever for this region. These estimates, all of which are accompanied by measures of precision, will help managers identify the most important species in a given region and the most important region for a given species. Major findings applicable to groups of species are that (1) density was generally higher, and population size much higher, in the western ACP than east of the Colville River; (2) in predicting density, habitat variables measured at the plot level were consistently better than, or as least as good as, measurements at a larger scale; (3) spatial variables were consistently more useful than habitat variables in predicting density; (4) many distribution maps can now be revised using our findings; and (5) detection ratios on the aerial survey probably varied widely, approaching 1.0 for large, white birds, but being substantially lower for most other species.

ACKNOWLEDGMENTS

The surveys were carried out by D. Battaglia, W. Boyd, D. Brann, B. Clock, P. Cotter, S. Dieni, S. Earnst, C. Eldermire, S. Fellows, B. Harrington, R. Hunnewell, A. Johnson, H. Johnson, P. Lemons, M. McGarvey, P. Mullen, R. Pagen, N. Parker, L. Payne, B. Peterjohn, D. Poinsette, E. Rasmussen, A. Schmidt, S. Schulte, N. Senner, E. Urban, E. Wells, and B. Winn. M. McGarvey, S. Earnst, and P. Cotter assisted with numerous logistical details. We appreciate the insightful comments provided by numerous reviewers on prior drafts of the manuscript. We especially thank K. Wohl for supporting this project from its inception. Funding for this study was provided by the U.S. Fish and Wildlife Service, the U.S. Geological Survey, and Manomet Center for Conservation Sciences.

ONLINE CONTENT

Abstracts are available in French and Spanish from www.ucpress.edu/go/sab.
Une traduction du résumé est disponible en français.
Una traducción del resumen está disponible en español.

CHAPTER FIVE

Yukon North Slope and Mackenzie Delta

Jennie Rausch and Victoria Johnston

Abstract. We conducted surveys on the Yukon North Slope and in the Mackenzie Delta eastward past the tip of the Tuktoyaktuk Peninsula to the Anderson River Delta, NWT, in June and July of 2005 through 2008. We surveyed 171 12-ha "rapid" plots and recorded 417 breeding pairs of 13 species of shorebirds. On an additional 20 plots, we conducted intensive nest searching and found 113 nests of seven species of shorebirds. By correcting our density estimates with a detection ratio and extrapolating these corrected densities across the region, we estimated that there are half a million shorebirds in this region. The Yukon North Slope and the Mackenzie Delta is the most diverse region in the Canadian arctic surveyed to date. Arctic PRISM data from this region have had value-added uses for local managers of species at risk (e.g., Short-eared Owl and Eskimo Curlew) and environmental assessments such as the proposed Mackenzie Gas Project. Comparison of our results with studies in the area done in the late 1960s to early 1990s revealed that shorebird populations in this region have remained stable. The data collected through Arctic PRISM may provide a focus for research into determining the causes of continental shorebird population declines.

Key Words: arctic, Kendall Island Migratory Bird Sanctuary, Mackenzie Delta, Mackenzie Gas Project, monitoring, Northwest Territories, Pectoral Sandpiper, population size, PRISM, Red-necked Phalarope, Semipalmated Sandpiper, shorebirds, species at risk, Whimbrel, Yukon North Slope.

Region 12 of the Arctic PRISM program covers 20,756 km^2 and stretches from the Alaska–Yukon border through Yukon and into the Northwest Territories to the tip of the Tuktoyaktuk Peninsula (Fig. 5.1). It is the smallest Arctic PRISM region in Canada, but is located in an area of high shorebird diversity and density relative to other parts of the Canadian arctic (Johnston and Pepper 2009: appendix 3, Northwest Territories/Nunavut Bird Checklist Survey 2009). It covers one of only two breeding areas in Canada for Whimbrel, Hudsonian Godwit, and Long-billed Dowitcher (Northwest Territories/Nunavut Bird Checklist Survey 2009).

We divided Region 12 into four subregions: Yukon North Slope (Yukon NS; 9,156 km^2),

Rausch, J., and V. Johnston. 2012. Yukon North Slope and Mackenzie Delta. Pp. 97–112 *in* J. Bart and V. Johnston (editors). Arctic shorebirds in North America: a decade of monitoring. Studies in Avian Biology (no. 44). University of California Press, Berkeley, CA.

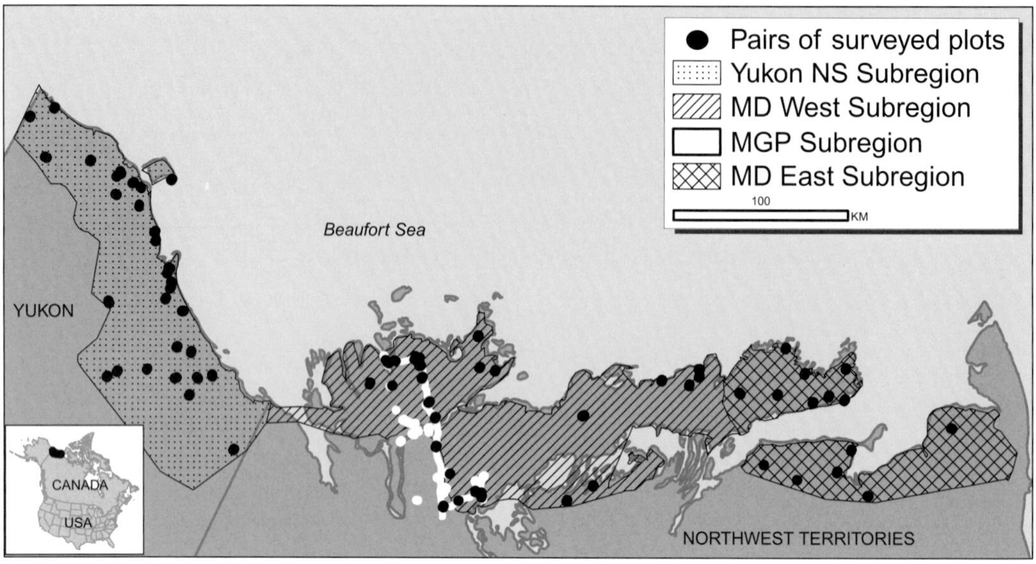

Figure 5.1. Region 12 subregions and locations of pairs of rapid survey plots.

Mackenzie Delta West (MD West; 6,913 km^2), Mackenzie Delta East (MD East; 4,385 km^2), and Mackenzie Gas Project (MGP; 302 km^2). The four subregions are more than we would normally have for a region as small as Region 12; however, delineation of the extra subregion (MGP) for management purposes (i.e., environmental assessment) has allowed us to conduct extra surveys in this area of interest while still contributing to the overall Region 12 results.

Region 12 is located in the Taiga Cordillera and Southern arctic ecozones (Marshall and Schut 1999). The geology of the eastern part of the region is dominated by the British Mountains, which influenced the creation of gentle slopes and V-shaped river valleys. Deltaic lowlands predominate in the western part of the region. The landscape is dotted with thousands of lakes, ponds, and wetlands. The soil is composed of cryosols. Permafrost features such as patterned ground (e.g., low- and high-centered polygons), pingos, and tussocks are present throughout the region. Sinuous eskers and dry heath uplands also occur widely across the region (Marshall and Schut 1999).

Part of Region 12 overlaps with the proposed Mackenzie Gas Project (MGP). The MGP involves the largest environmental assessment in Canadian history. It began in 2002 when a consortium of gas producers proposed building a 1,220-km pipeline from the Mackenzie Delta to southern Canada (Mackenzie Gas Project 2004). The data collected during the Region 12 surveys were used to help prepare the environmental impact assessment for the Mackenzie Gas Project (Environment Canada 2006).

The MGP report focuses on the estimated densities, population sizes, and habitat relationships of shorebirds in Region 12. It also describes the valued-added uses of the Arctic PRISM data for local conservation managers.

Previous Work Done in the Region

Numerous studies of birds have been conducted in the region since the early 1970s, primarily of waterfowl (Barry 1967, Salter and Davis 1974, Schweinsberg 1974, Koski 1975, Richardson et al. 1975, Vermeer and Anweiler 1975, Richardson and Johnson 1981, Alexander et al. 1988, Armstrong 1998, Hines and Wiebe Robertson 2006, J. Hines, unpubl. data). Surveys specifically for shorebirds were conducted by Gratto-Trevor (1994, 1996) and Dickson et al. (1989), but the surveys covered only a small portion of Region 12 (Kendall Island Migratory Bird Sanctuary and adjacent areas). Arctic PRISM is the first large-scale, ground-based shorebird survey of the entire region.

METHODS

Methods for plot selection, surveys, and calculation of estimated densities and population sizes

TABLE 5.1

Location of Region 12 base camps and dates utilized.

Camp (survey type)	Location	Dates
Yukon (rapid)	69°34′00″N, 138°55′00″W	7–10 June 2005 6–11 June 2006
Fish Island (intensive)	69°22′23″N, 134°53′36″W	7 June–19 July 2007 6 June–13 July 2008
Niglintgak 1 (intensive)	69°19′43″N, 135°15′56″W	7 June–9 July 2005
Niglintgak 2 (intensive)	69°18′52″N, 135°19′33″W	9 June–12 July 2006
Taglu (intensive)	69°22′05″N, W134°57′18″W	9 June–17 July 2006
Swimming point 1 (rapid)	69°09′00″N, W134°34′41″W	8–18 June 2006
Swimming point 2 (rapid)	69°05′56″N, W134°23′00″W	8–18 June 2007

are described in Bart et al. (chapter 2, this volume). This section presents additional information on methodology that is specific to the work in Region 12.

Over the four study years, seven base camps were established (one in Yukon and six in the Northwest Territories; Table 5.1). Reconnaissance surveys were undertaken in 2004 to allow surveyors to become familiar with the area, and to facilitate proper stratification of the region. Rapid surveys were completed in the Yukon NS and MD east subregions in 2005–2006, and in the MD west and MGP subregions in 2005–2007.

Weather data (temperature, average and maximum wind speed, and wind direction), were recorded twice daily at all camps using a handheld weather meter. Descriptive comments on cloud cover and general weather conditions were also recorded. Precipitation was not recorded at the camp sites, so the data were obtained from the Tuktoyaktuk and Inuvik airports, located approximately 75 km to the east and 120 km to the south of the camp locations, respectively.

Habitat Classification

The final four-class (wetland, vegetated upland, dry, water) Region 12 habitat classification was created by stitching together the best available classifications in the region (Table 5.2) from:

1. Government of the Northwest Territories, Environment and Natural Resources: supervised classification from 2000 imagery with ground-truthing;
2. MGP Proponents: supervised classification from 1991, 1992, and 2000 imagery with ground-truthing;
3. Canadian Wildlife Service, Yukon: supervised classification from late 1980s imagery with ground-truthing; and
4. U.S. Geological Survey, Jonathan Bart: unsupervised classification from 2004 imagery.

All of the input habitat classifications were originally created from Landsat5 and/or Landsat7 imagery with 30-m ground resolution. The proportion of coverage each of the above contributed to the Region 12 habitat classification was 3%, 39%, 37%, and 21%, respectively.

Rapid and Intensive Plots

Rapid plots covered 12 ha (400 m N–S, 300 m E–W). They were assigned to habitat classes as follows:

If more than 50% of the plot was assigned to the wetland habitat type, then the plot was classified as wetland.

Then, if more than 40% of the plot was assigned to the dry habitat type, the plot was classified as dry.

All remaining plots were assigned to the vegetated upland type.

Intensive plots were selected non-randomly to ensure the plots would contain breeding shorebirds. Experienced camp leaders spent one to

TABLE 5.2

Classes combined from several habitat classifications to create the Arctic PRISM Region 12 habitat classification.

Four classes based on suitability of habitat for nesting shorebirds.

PRISM class	Classification source: Government of the Northwest Territories (3% of Region 12 classification)[a]	MGP Proponent (39% of Region 12 classification)[a]	Canadian Wildlife Service (37% of Region 12 classification)[a]
Wetland	Mesic dry meadow Dwarf shrub wet Wet graminoid	Delta low-centred polygons Delta sedge–cottongrass High-centered polygons Low-centered polygons Riparian sedge–cottongrass Water emergents	Aquatic tundra Wet tundra Moist wet tundra complex Moist graminoid non-tussock
Vegetated upland	Tussock tundra Low shrub tussock tundra Aquatic vegetation Emergent	Black spruce/shrub heath Dwarf shrub heath Recent burn <2 years since burn Riparian black spruce/shrub Sedge–cottongrass tussock	Wet graminoid, low shrub Cottongrass tussock tundra Dwarf shrub tundra
Dry	Non-vegetated soil Sparse Dwarf shrub other Dwarf shrub lichen Lichen Low shrub willow alder Open tall shrub Closed tall shrub Closed birch Closed poplar Open deciduous Open spruce Closed mixed needleleaf Woodland needleleaf	Airstrips (general) All-weather paved roads Barge landings Buildings Clearings along pipeline route Delta shrub Ikhil pipeline Medium shrub (all) Primary industrial Riparian shrub Saxifrage heath Secondary industrial Summer roads Urban Winter road	Wet partially vegetated/wet barren Low shrub tundra Shrub thicket Dry partially vegetated/dry barren
Water	Turbid water Clear water	Dark water Light water	Silty water Clear water
No data	Mesic dry forb[b] No data	Non-vegetated, includes clouds	Ice and snow Cloud and shadow

[a]The remaining 21% was the unsupervised, non–ground-truthed MD east portion.

[b]There was a classification problem with this class so it was excluded.

two days surveying within walking or boating distance of the intensive camp to choose suitable plot locations. Some of the intensive plots were rope dragged twice with approximately one week between the surveys.

Regional Population Estimates

Complete details of the population estimation procedure are found Bart et al. (chapter 2, this volume). In summary, we obtain estimated densities

TABLE 5.3

Weather conditions in the Mackenzie Delta during the survey periods for Niglintgak Island (NIG), Taglu Island (TAG), and Fish Island (FISH).

	Mean temperature (C)				Mean wind speed (km/hr)		
	June		July				
Location[a]	0800 h	2000 h	0800 h	2000 h	June	July	Precipitation June–July (mm)[b]
2005 (NIG)	7.3	8.5	7.5	7.9	9.8	7.4	43.4
2006 (NIG)	9.9	13.6	8.5	9.5	11.9	14.4	83.8
2006 (TAG)	10.1	12.6	10.6	13	11.9	13.6	
2007 (FISH)	5.7	9.1	13.4	16.5	12.6	9.4	18.5
2008 (FISH)	9.3	14.5	15.3	18.5	11.2	11.5	68.0

[a] Weather data for Yukon North Slope (YNS) were not collected.

[b] Data from Environment Canada (2009) weather stations at the airports in Tuktoyaktuk and Inuvik, NT. Total precipitation for June–July was averaged between stations except for 2008, for which only Inuvik data were available.

that are corrected using a detection ratio. The detection ratio is calculated from the intensive plots as the number of birds estimated to be on a plot (using the rapid surveys) compared to the number actually present (based on the intensive surveys, see Bart et al., chapter 2, this volume). Corrected densities are applied across the subregions to generate the overall population estimate. For the analysis for Region 12, we used a single detection ratio (1.27) calculated from all intensive plots in Canada (see Bart and Smith, chapter 14, this volume). Data were analyzed for each subregion separately. Totals for Region 12 were calculated by summing subregional estimates and pooling their variances; this approach yields slightly different estimates than would be obtained by the combined approach used in the range-wide PRISM analyses (see Bart et al., chapter 2, this volume). Since the MGP subregion is nested within the MD West subregion, they are sometimes discussed together as a whole ("Mackenzie Delta").

Other Surveys

Aerial surveys for shorebirds in general, or Whimbrel (for scientific names, see Appendix C) specifically, were conducted in 2005, 2006, and 2008. Detailed discussion of the methods and results for aerial surveys in this region are presented in Elliott and Smith (chapter 9, this volume) and Pirie and Johnston (chapter 10, this volume). Auxiliary data, such as weather, insect emergence, bird habitat use, nest success, bird and plant phenology, and predator cycles, were collected at the Region 12 intensive camps (which also comprise an Arctic PRISM Tier 2 site). These data are described in Pirie et al. (chapter 11, this volume).

RESULTS

Weather and Timing of Breeding

The weather conditions in Region 12 for June and July were fairly consistent between years, though conditions were wettest in 2006 (Table 5.3). In 2006 there was a storm surge flood on 3 July which covered much of the lowland area of Niglintgak Island with 20–100 cm of water and destroyed 26 of 65 Red-necked Phalarope nests that were part of a parallel research project, just days before hatch (Beveridge 2007).

The first shorebird nests with complete clutches were found on 11 June in 2005 and 2006, 9 June in 2007, and 6 June in 2008. The peak of egg hatch occurred from 4 to 6 July in 2005, 5 to 7 July in 2006, and 2 to 4 July in 2007 and 2008. Based on an average incubation period of 21 days, most clutches were completed 11–16 June.

Surveys

Over the three years, we surveyed 171 "rapid" plots covering 19.56 km^2 (0.1% of Region 12). Rapid surveys in this region took 0.75–3.0 hours to complete

TABLE 5.4

Number of indicated pairs of shorebirds present during rapid surveys in Region 12.

	Subregion				
Species	Yukon NS	MD West	MGP	MD East	Total
Black-bellied Plover	0	0	0	1	1
American Golden-Plover	5	3	4	2	14
Whimbrel	8	5	20	2	35
Hudsonian Godwit	0	0	6	0	6
Semipalmated Sandpiper	8	14	7	23	52
Least Sandpiper	3	4	3	2	12
Baird's Sandpiper	2	0	0	0	2
Pectoral Sandpiper	30	8	12	22	72
Stilt Sandpiper	8	4	5	5	22
Long-billed Dowitcher	11	2	2	0	15
Wilson's Snipe	7	11	16	4	38
Red-necked Phalarope	38	28	54	27	147
Red Phalarope	1	0	0	0	1
Total	121	79	129	88	417

depending on ease of walking through the terrain and how many bird territories were present. Surveyors recorded 13 species of shorebirds (Table 5.4). The eight most abundant (Red-necked Phalarope, Pectoral Sandpiper, Semipalmated Sandpiper, Wilson's Snipe, Whimbrel, Stilt Sandpiper, American Golden-Plover, and Least Sandpiper) were present in all four subregions. We also recorded 40 species of non-shorebirds during rapid surveys. The most abundant songbirds were Savannah Sparrow (658 indicated pairs), Lapland Longspur (644), and American Tree Sparrow (159). The most abundant waterfowl were Greater White-fronted Goose (85) and Northern Pintail (64).

An average of 32 person-hours per plot (range 20–43 hours) was devoted to intensive surveys. Rope drags of intensive plots averaged 12 person-hours per survey.

Intensive survey crews found 113 shorebird breeding territories or nests on the intensive plots. Red-necked Phalarope, Pectoral Sandpiper, Semipalmated Sandpiper, Stilt Sandpiper, Whimbrel, Hudsonian Godwit, and American Golden-Plover (in order of abundance) all nested on intensive plots. Apparent hatching success, including nests found surrounding and en route to the intensive plots, ranged from 26% (2007, $n = 103$) to 57% (2006, $n = 111$). The main cause of nest failure was predation by birds (primarily Parasitic Jaegers) and ground predators (e.g., arctic fox).

We completed 41 rapid surveys on the 20 intensive plots. Of the 113 breeding territories or nests that were actually present on the intensive plots, the rapid surveys recorded 110. Although this would suggest a detection ratio for Region 12 closer to 1.0 than the Canada-wide detection ratio (1.27) used, at this interim stage the sample sizes do not justify using region- or species-specific detection ratios.

Population and Density Estimates

The total estimated population of shorebirds breeding in Region 12 is approximately 512,500 birds (Table 5.5). Not surprisingly, the highest shorebird densities were recorded in the wetland habitat. Species for which densities (birds/km^2) were greater than 10.0 in at least one subregion were Red-necked Phalarope, Pectoral Sandpiper, Semipalmated Sandpiper, and Wilson's Snipe (Table 5.6).

TABLE 5.5

Population estimates with standard error (SE) and coefficient of variation (CV) for shorebirds in Region 12.

Species are given with AOU 4-letter codes. Refer to Appendix C, this volume for full names.

	Yukon NS		MD West		MGP		MD East		Total[a]	
Species	Estimate (SE)	CV	Estimate (SE)	CV	Estimate (SE)	CV	Estimate (SE)	CV	Estimate (SE)	CV
BBPL	—	—	—	—	—	—	3,477 (3,583)	1.03	3,477 (3,583)	1.03
AMGP	16,274 (8,708)	0.54	7,691 (4,200)	0.55	175 (119)	0.68	5,614 (5,748)	1.02	29,753 (11,248)	0.38
WHIM	14,036 (12,586)	0.90	5,135 (3,761)	0.73	1,143 (645)	0.56	1,918 (1,628)	0.85	22,232 (13,252)	0.60
HUGO	—	—	—	—	127 (130)	1.02	—	—	127 (130)	1.02
SESA	4,318 (2,575)	0.60	39,435 (16,578)	0.42	369 (199)	0.54	58,867 (18,728)	0.32	102,988 (25,145)	0.24
LESA	1,032 (763)	0.74	5,190 (4,063)	0.78	87 (62)	0.71	6,771 (6,511)	0.96	13,079 (7,713)	0.59
BASA	3,553 (3,257)	0.92	—	—	—	—	—	—	3,553 (3,257)	0.92
PESA	21,690 (9,696)	0.45	17,904 (10,358)	0.58	772 (807)	1.05	56,204 (18,524)	0.33	96,570 (23,347)	0.24
STSA	4,233 (2,207)	0.52	7,121 (4,014)	0.56	222 (121)	0.55	12,023 (4,619)	0.38	23,599 (6,507)	0.28
LBDO	6,315 (3,414)	0.54	3,647 (3,760)	1.03	42 (43)	1.02	—	—	10,004 (5,079)	0.51
WISN	5,486 (3,652)	0.67	38,071 (16,125)	0.42	1,694 (1,475)	0.87	10,894 (5,926)	0.54	56,145 (17,625)	0.31
RNPH	29,422 (10,504)	0.36	59,916 (21,688)	0.36	5,485 (4,264)	0.78	54,907 (21,038)	0.38	149,730 (32,272)	0.22
REPH	1,377 (1,426)	1.04	—	—	—	—	—	—	1,377 (1,426)	1.04
Total	107,737 (22,098)	0.21	184,109 (34,511)	0.19	10,115 (4,639)	0.46	210,674 (35,831)	0.17	512,634 (54,633)	0.11

[a] Totals for Region 12 were calculated by summing subregional estimates and combining their variances; this approach yields slightly different estimates than would be obtained by the combined approach used in the range-wide PRISM analyses (see Chapter 2).

TABLE 5.6

Shorebird density (birds/km²), by subregion and habitat with standard error (SE) for species present during rapid surveys in Region 12.

		Wetland		Vegetated upland		Dry		All	
Species	Subregion	Dens.	SE	Dens.	SE	Dens.	SE	Dens.	SE
Black-bellied Plover	MD East	1.26	1.30	—	—	—	—	0.79	0.82
American Golden-Plover	MD East	2.03	2.14	—	—	—	—	1.28	1.31
	MGP	0.46	0.32	1.22	0.89	—	—	0.58	0.40
	MD West	0.56	0.60	3.32	1.91	—	—	1.11	0.61
	Yukon NS	1.06	1.06	0.69	0.69	3.50	2.39	1.78	0.95
Whimbrel	MD East	0.69	0.62	—	—	—	—	0.44	0.37
	MGP	5.37	2.85	6.76	4.03	—	—	3.80	2.14
	MD West	1.78	1.37	—	—	—	—	0.74	0.54
	Yukon NS	1.42	0.75	2.68	2.68	—	—	1.53	1.37
Hudsonian Godwit	MGP	2.05	1.88	—	—	—	—	0.42	0.43
Semipalmated Sandpiper	MD East	11.94	4.41	—	—	15.96	7.57	13.43	4.27
	MGP	3.58	2.17	1.23	1.26	—	—	1.22	0.66
	MD West	3.15	2.14	12.98	7.06	3.00	3.12	5.70	2.40
	Yukon NS	2.61	1.83	0.30	0.31	—	—	0.47	0.28
Least Sandpiper	MD East	2.45	2.29	—	—	—	—	1.54	1.49
	MGP	0.86	0.80	0.28	0.31	—	—	0.29	0.21
	MD West	1.80	1.46	—	—	—	—	0.75	0.59
	Yukon NS	0.91	0.70	—	—	—	—	0.11	0.08
Baird's Sandpiper	Yukon NS	0.29	0.30	0.69	0.69	—	—	0.39	0.36
Pectoral Sandpiper	MD East	16.41	5.50	—	—	6.71	4.33	12.82	4.22
	MGP	2.73	2.50	1.84	1.90	3.20	3.60	2.56	2.68
	MD West	5.16	3.34	1.66	1.60	—	—	2.59	1.50
	Yukon NS	10.36	4.24	2.15	1.68	—	—	2.37	1.06
Stilt Sandpiper	MD East	3.24	1.65	—	—	1.89	2.05	2.74	1.05
	MGP	2.40	1.74	0.61	0.63	—	—	0.74	0.40
	MD West	1.41	0.92	1.66	1.60	—	—	1.03	0.58
	Yukon NS	3.75	1.84	—	—	—	—	0.46	0.24
Long-billed Dowitcher	MGP	0.68	0.63	—	—	—	—	0.14	0.14
	MD West	1.26	1.33	—	—	—	—	0.53	0.54
	Yukon NS	5.59	2.84	—	—	—	—	0.69	0.37

TABLE 5.6 *(continued)*

Species	Subregion	Wetland		Vegetated upland		Dry		All	
		Dens.	SE	Dens.	SE	Dens.	SE	Dens.	SE
Wilson's Snipe	MD East	1.69	1.68	—	—	3.84	1.92	2.48	1.35
	MGP	3.45	1.30	2.80	1.94	9.60	10.79	5.63	4.90
	MD West	10.95	4.76	1.66	1.60	1.57	1.62	5.51	2.33
	Yukon NS	1.89	1.22	0.72	0.71	—	—	0.60	0.40
Red-necked Phalarope	MD East	11.35	5.44	—	—	14.52	8.66	12.52	4.80
	MGP	16.60	7.54	2.27	1.94	35.11	33.62	18.22	14.17
	MD West	17.33	6.83	1.66	1.60	3.14	3.24	8.67	3.14
	Yukon NS	14.04	5.07	2.92	1.68	—	—	3.21	1.15
Red Phalarope	Yukon NS	—	—	0.30	0.31	—	—	0.15	0.16

Yukon NS Subregion

The shorebird species with the highest overall densities were Red-necked Phalarope and Pectoral Sandpiper (Table 5.6). The highest habitat-specific densities were in wetland habitat for all species, with the exception of American Golden-Plover (highest in dry habitat), Whimbrel (highest in vegetated upland habitat), and Red Phalarope (present only in vegetated uplands). Yukon NS is the only subregion where we recorded Red Phalarope and Baird's Sandpiper.

MD West Subregion

The shorebird species with the highest densities were Red-necked Phalarope and Semipalmated Sandpiper (Table 5.6). With the exception of American Golden-Plover and Semipalmated Sandpiper, whose densities were highest in vegetated upland habitat, shorebird densities were highest in wetland habitats. Stilt Sandpiper had equivalent densities in wetlands and vegetated uplands.

MGP Subregion

The shorebird species with the highest densities were Red-necked Phalarope and Wilson's Snipe (Table 5.6). The densities for Wilson's Snipe were highest in the dry habitat (which, notwithstanding the usually dry nature of this habitat, also includes habitats of willow–standing water in this PRISM region). MGP is the only subregion where we recorded Hudsonian Godwit during our rapid surveys.

MD East Subregion

The most abundant shorebird species were Semipalmated Sandpiper and Pectoral Sandpiper (Table 5.6). Nearly equal numbers of shorebirds were recorded in dry habitat and in wetland habitat with the exception of Pectoral Sandpiper, which had higher densities in wetland habitat. However, no vegetated upland habitat plots were surveyed in this subregion. MD East is the only subregion where we recorded Black-bellied Plover.

DISCUSSION

Shorebird Densities and Distribution

Yukon NS Subregion

Our estimated overall density of shorebirds in the Yukon NS (11.8 birds/km^2) was much higher than that found by Richardson et al. (1975; 0.23 birds/km^2), but their densities were in large part based on fixed-wing aerial surveys from an unknown height with no correction factor, so the data are not comparable.

Results of ground surveys conducted by Salter and Davis (1974) were much larger, ranging from a 2.5-fold difference between our density estimate and theirs for Red-necked Phalarope to a

TABLE 5.7

Shorebird densities (birds/km²) from studies in the Yukon NS subregion in the 1970s (Salter and Davis 1974, Schweinsburg 1974) and in the MD West/MGP subregions in the late 1980s-early 1990s (Dickson et al. 1989, Gratto-Trevor 1994).

	Yukon NS		MD West/MGP		
Species	Salter and Davis	Schweinsberg	Gratto-Trevor[a]	Dickson et al.[b]	Arctic PRISM[c]
American Golden-Plover	54	4.9	1.3	0.5	0.8
Whimbrel	—	—	1.7	3.0	2.3
Hudsonian Godwit	—	—	1.4	3.5	0.2
Semipalmated Sandpiper	187	6.5	4.8	7.5	3.5
Pectoral Sandpiper	136	7.0	0.9	7.5	2.6
Stilt Sandpiper	17	2.7	4.8	5.0	0.9
Long-billed Dowitcher	—	—	0.4	1.0	0.3
Wilson's Snipe	—	0.5	7.7	7.0	5.6
Red-necked Phalarope	8	13.5	25.7	30.5	13.4

[a] Adjusted (using our Canada-wide detection ratio of 1.27) and averaged between habitat types.

[b] Average of the two survey years (1985 and 1986).

[c] Unweighted average of total density for MD West and MGP subregions only.

400-fold difference for Semipalmated Sandpiper (Table 5.7). Their estimate of 187 Semipalmated Sandpipers/km² seems improbable when the highest density recorded in our surveys of Alaska (wetlands in Region 17—Yukon Delta; McCaffery et al. chapter 3, this volume) was less than 70 birds/km². Their survey methods were too different from ours to permit application of the PRISM Canada-wide detection ratio.

The results from ground surveys by Schweinsberg (1974) were similar to ours for all species except Semipalmated Sandpiper, for which he recorded much higher densities (Table 5.7). These results are likely because his surveys were not stratified by habitat and favored a disproportionately large amount of heath tundra (which could be classified as vegetated upland or dry in the Region 12 habitat classification depending on the type of heath; Table 5.2), which Semipalmated Sandpipers prefer (Gratto-Trevor 1992).

Mackenzie Delta (MD West and MGP Subregions)

Two in-depth shorebird studies were conducted in the outer Mackenzie Delta in 1985–1986 (Dickson et al. 1989) and 1991–1992 (Gratto-Trevor 1994). Results of these studies are compared with ours in Table 5.7. Gratto-Trevor's study area was approximately 50% of the total area covered by the MD West and MGP subregions. Gratto-Trevor (1994) used a "rapid" survey technique only, so we applied our Canada-wide detection ratio (1.27) to her results for comparison. The work by Dickson et al. (1989) covered a relatively small area, but the surveys were intensive, so the results are presented as originally published.

For three species (American Golden-Plover, Whimbrel, and Pectoral Sandpiper), our results were within the range of Dickson et al.'s and Gratto-Trevor's. The other results were close to the results of Gratto-Trevor with the exception of Red-necked Phalarope, for which our results are approximately half of hers. Her large estimate for Red-necked Phalarope may be an artifact of small plot size; congregations of phalaropes on ponds within the plot were counted as being on the plot (Gratto-Trevor 1994) and thus inflated density estimates. We did not count large congregations of non breeders, whether they were on the plot or not, because they do not reflect individual nesting pairs. They were instead recorded as incidentals and not used in the density and population estimate calculations. Dickson et al.'s (1989) estimates were greater than both Gratto-Trevor's and ours for all species except American Golden-Plover and Wilson's Snipe. They concentrated on areas of very high-quality wetland habitat at the expense of coverage for species which nest in vegetated upland or dry habitats. Using the adjusted densities and the area of each habitat type she surveyed, the estimated population size of shorebirds in Gratto-Trevor's study area was 95,334 shorebirds. If we double this to compensate for the difference in size of her study area and ours, our population estimates are nearly identical—191,000 birds versus 194,000 birds (MD West and MGP subregions). The comparison of the two sets of data supports a conclusion that shorebird densities and populations have not changed much in the Mackenzie Delta over the past 16 years.

MD East Subregion

There are no past studies specifically on shorebirds in this subregion. However, in his thesis on geese, Barry (1967) listed the shorebird species in the area and their qualitative nesting abundances. His ranking exactly matches our density results for the most abundant (Semipalmated Sandpiper), very common (Pectoral Sandpiper and Red-necked Phalarope), and common (Stilt Sandpiper and Wilson's Snipe) species. Thus, although we are unable to compare density or population sizes, we can say that *relative* abundance for these five species of shorebird has not changed in this subregion over the past 40 years. Of the four subregions in Region 12, this subregion has experienced the least amount of human influence.

An ongoing challenge in the attainment of Arctic PRISM population estimates with acceptable standard errors is the quality of available habitat classifications. The habitat classifications for the Yukon NS, MD West, and MGP subregions were created using a combination of aerial photographs, Landsat data, and thousands of ground control points, and we are confident that they are as accurate as can be expected for a deltaic environment. We are, however, less confident about the classification created for the MD East subregion, as it was an unsupervised classification with no ground control points. Our main concern is that the vegetated upland and dry classes do not always accurately represent what is seen on the ground. For example, in June 2009 we visited several plots in MD East that were classified as vegetated upland habitat, but they were in fact primarily shrub covered and should have been classed as dry habitat. Plots observed on the ground to be dry sparsely vegetated upland were accurately portrayed as dry in the classification. However, wetland habitat was actually a mix of wetland and vegetated upland habitat, and those need to be teased apart. If a more accurate habitat classification is ever produced, the data from the MD East subregion will be reanalyzed to refine the estimates.

Importance of Region 12 to Canada

Over half a million shorebirds are estimated to be in Region 12 (Table 5.5). The Region 12 population estimate for Whimbrel is 86% of the current North American estimate (26,000 birds) for the western subspecies of Whimbrel (*N. p. rufiventris*; Morrison et al. 2006). This population of Whimbrel is also found across Alaska, so the actual number of *N. p. rufiventris* is likely much higher than was originally thought. Using the population estimates of Morrison et al. (2006), Region 12 is also estimated to hold 19%, 6%, and 5%, respectively, of North American breeding populations of Pectoral Sandpipers, Red-necked Phalaropes, and Semipalmated Sandpipers.

Among regions surveyed to date, Region 12 ranks third in Canada for the most shorebirds per total land area (behind Regions 3 and 4). However, it is the most diverse region yet surveyed because it hosts both mid-arctic (e.g., Semipalmated Sandpiper, Pectoral Sandpiper, Red Phalarope) and low-arctic (e.g., Least Sandpiper) breeders and species with localized breeding ranges (e.g., Hudsonian Godwit, Long-billed Dowitcher, Whimbrel). Although samples were small, the density of Hudsonian Godwit in wetland habitat in Region 12 was the highest of anywhere in either Alaska or Canada. For Whimbrel, with the exception of Region 16 (Selawik, western Alaska), densities in all habitat types were greater than or equal to those in Alaska, showing that for these localized breeders, Region 12 is an important part of their breeding range. The other Canadian region that is expected to have high densities of Hudsonian Godwit and Whimbrel (based on the current range maps) is Region 6 (southwestern Hudson Barrens, southern Nunavut), which, like Region 12, follows the Taiga–Tundra ecotone. The Canadian contribution to the overall

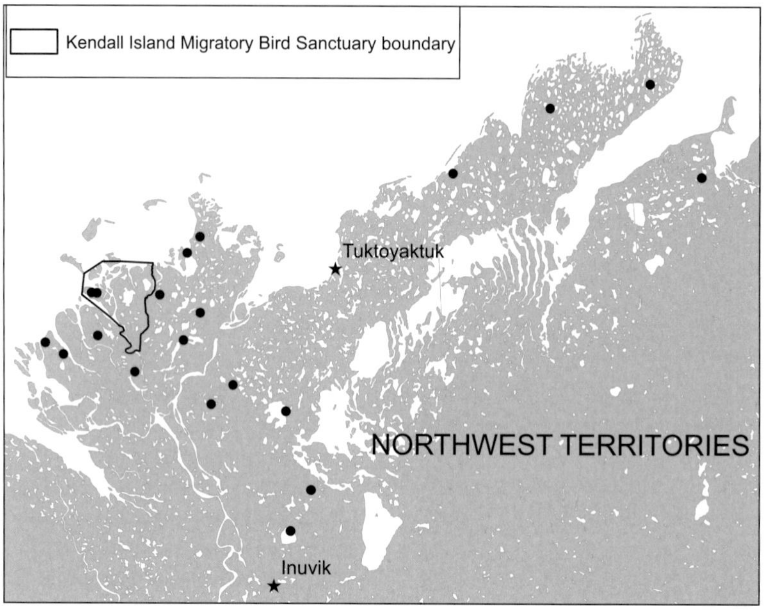

Figure 5.2. Sightings of Short-eared Owls collected during PRISM surveys in Region 12. Short-eared Owl is a species listed as Special Concern under the Canadian Species at Risk Act.

population size of these species is expected to remain small even after estimates are completed for all Canadian PRISM regions since the densities found in Alaska are greater over larger areas (see McCaffery et al., chapter 3, this volume, Bart et al., chapter 4, this volume, and Bart and Smith, chapter 14, this volume). While the density of Whimbrel within the MGP subregion is equal to that in western Alaska (see Table 14.6 in Bart and Smith, chapter 14, this volume), the MGP subregion is a fraction of the size of Selawik (50 times smaller). However, it is important to note that the birds breeding in central Canada (Region 6) are different populations/subspecies from those in Alaska. Conservation actions for these much smaller populations will need to reflect their distinct status.

Value-added Uses of Region 12 PRISM Data

Goal Five of PRISM is "to assist local managers in meeting their shorebird conservation goals" (Skagen et al. 2003). Data from Region 12 PRISM surveys have made essential contributions to species at risk reports and maps for the Mackenzie Delta. The Mackenzie Delta is an area that is under pressure from development; PRISM data have been critical instruments in the environmental assessment of the Mackenzie Gas Project.

Species at Risk

Since 2008, the Committee on the Status of Endangered Wildlife in Canada (COSEWIC) has been preparing a draft update of the status report for Eskimo Curlew, which is currently listed as Endangered under the Canadian Species at Risk Act (last assessed November 2009; Government of Canada 2010). Eskimo Curlew have historically bred in Region 12, but no nests have been found since 1866 (Government of Canada 2010). There were unconfirmed sightings in the 1980s and early 1990s, but despite extensive surveys of the breeding grounds by expert birders since then, including our PRISM surveys, the species has not been recorded (Government of Canada 2010, J. Obst and A. Spalding, unpubl. report). The status of Eskimo Curlew will be revisited in 2014 and at that point it will have been more than 50 years since the last confirmed sighting. It is expected that its status will be updated to Extinct.

Short-eared Owl is another COSEWIC listed species which is found in Region 12. It is listed as a Species of Special Concern under both COSEWIC and Species at Risk Act designations (Government of Canada 2009). Since 2004, we collected sightings of Short-eared Owl while doing Region 12 reconnaissance and surveys and have seen Short-eared Owl every year (n = 1, 8, 10, 1, 4, for 2004–2008, respectively; Fig. 5.2). In 2008 we also found one Short-eared Owl nest. They are reported to be scattered throughout the Northwest Territories and Nunavut, but no population status or trend information is available (COSEWIC 2008). We intend to continue collecting information about Short-eared Owl through our Tier 2 site activities in the Mackenzie Delta (Pirie et al., chapter 11, this volume).

Mackenzie Gas Project Environmental Assessment

There has been, and continues to be, much interest and activity in the Mackenzie Delta related to oil and gas exploration and development (Region 12 subregions MD West and MGP). A major undertaking of the Canadian Wildlife Service (Yellowknife) since 2005 has been the development of baseline population data for birds in the Mackenzie Delta. From the PRISM data, combined with other shorebird data, we concluded that:

- More than 1% of the estimated North American population of eight shorebird species utilize the outer Mackenzie Delta;
- Two species that have limited breeding ranges in Canada, Hudsonian Godwit and Long-billed Dowitcher, nest in and around the Kendall Island Bird Sanctuary (KIBS) in the outer Mackenzie Delta;
- KIBS alone (which is only 623 km^2) hosts more than 1% of the breeding populations of Hudsonian Godwit and Whimbrel; and
- The density of shorebirds in the Mackenzie Delta (especially in the potentially impacted gas extraction areas) is greater than at many other sites across the arctic (Table 5.8; Environment Canada 2006).

These conclusions, which highlight the importance of the Mackenzie Delta to continental shorebird populations, were submitted to the panel assessing the environmental impacts of the proposed Mackenzie Gas Project.

Implications of Region 12 Data to Research on Shorebird Declines

There are few locations in the arctic where there are current and historic data sets available for comparisons of bird densities and population sizes, something which Arctic PRISM hopes to remedy over time (refer to chapters 13–16, this volume). However, interest in baseline data collection and inventorying in Region 12, stemming from resource exploration and potential development, has afforded us the opportunity to compare our results with those of older studies. Despite most shorebird species experiencing significant or suspected population declines across North America, populations of shorebirds that nest in the Mackenzie Delta appear to have remained unchanged over more than 16 years. This information puts local managers in the unique situation of having a stable baseline against which to measure the effects of human development on shorebird populations in the arctic.

More broadly, it poses the question as to why populations in this region have not shown declines—or, alternatively, why populations in other regions of the arctic have, and continue to, show declines (e.g., the W–E gradient in Semipalmated Sandpiper declines, C. Gratto-Trevor pers. comm.). Examination of the differences between western and eastern arctic shorebird populations may help us tease apart the cause(s) of shorebird population decline.

CONCLUSIONS AND RECOMMENDATIONS

Based on the four years of data collected within Region 12, and past shorebird studies in the area, we have the following conclusions and recommendations:

1. Region 12 has a diverse breeding shorebird assemblage and contains approximately half a million breeding shorebirds;
2. The PRISM data set has been critical to local managers of species at risk and environmental assessments in Region 12. It is important for shorebird conservation that we make the Arctic PRISM data readily available and that local managers know that these data exist;
3. Shorebird densities, population estimates, and relative abundances have not changed over the past several decades, which is a promising sign that shorebird populations in this region are healthy; and
4. Research into why shorebird populations in this region are stable as compared to those in the east may help us answer questions about the causes of continental shorebird population declines.

TABLE 5.8

Shorebird densities (birds/km²) in the outer Mackenzie Delta (potentially gas extraction–affected areas) compared to other arctic locations.

Species	Outer Mackenzie Delta	Kent Peninsula/ Melbourne Island	West Baffin/ Prince Charles Island	Rasmussen lowlands	Somerset Island	Selawik, AK	Yukon Delta, AK	Churchill, MB	Phillips Bay/Stokes Point, YK	Kagloryuak River Valley	Colville Delta, AK
Hudsonian Godwit	1.4	—	—	—	—	0.1	—	—	0.7	—	—
Long-billed Dowitcher	0.7	—	—	—	—	0.3	1.1	—	—	—	2.2
Pectoral Sandpiper	5.0	4.3	0.0	3.3	0.8	—	1.8	—	28.7	4.3	7.4
Red-necked Phalarope	44.7	0.8	—	—	—	0.8	26.2	—	16.7	—	4.2
Semipalmated Sandpiper	8.7	1.7	1.2	1.7	—	4.2	24.9	—	19.0	13.2	9.5
Stilt Sandpiper	4.9	0.5	—	0.2	—	—	—	—	4.1	2.9	1.3
Whimbrel	1.2	—	—	—	—	2.1	—	5.5	0.6	—	0.1
Wilson's Snipe	1.3	—	—	—	—	0.2	1.2	—	0.6	—	—

SOURCE: Taken from Appendix I, Table 4 in Environment Canada (2006).

ACKNOWLEDGMENTS

J. Bart, A. Manning, and P. A. Smith carried out the analyses to estimate population size. S. Kennedy and C. Wood helped create maps. Fieldwork was funded by the Canadian Wildlife Service (Environment Canada). Thanks to our many field workers; AXYS Environmental/Westworth Consultants, J. Bart, J. Beaubier, R. Binder, R. Braden, K. Brady, V. Charlwood, L. Christenson, C. Cockney, K. Coddington, S. Davidson, C. Dockrill, S. Earnst, D. Ebner, K. Elliott, C. Francis, M. Gill, R. Greig, K. Hansen-Craik, G. Holroyd, A. Levesque, I. MacDonald, T. Marsh, S. McKenzie, W. Nixon, M. Noksana, E. Nol, L. Pirie, J. Ranger, S. Ripley, N. Senner, P. Sinclair, K. Sittler, R. Vittrekwa, A. and B. Walpole, R. Wilson, C. and C. Wood, and C. Yi; and the pilots at Aklak Air and Canadian Helicopters based in Inuvik, NT. This report is PCSP Contribution No. 00909.

ONLINE CONTENT

Abstracts are available in French and Spanish from www.ucpress.edu/go/sab.
Une traduction du résumé est disponible en français.
Una traducción del resumen está disponible en español.

CHAPTER SIX

Southampton and Coats Islands

Paul A. Smith, Victoria Johnston, and Jennie Rausch

Abstract. As part of the PRISM effort to generate arctic-wide estimates of shorebird populations, we conducted surveys on Southampton and Coats Islands, Nunavut, Canada, in June and July of 2004 and 2006. We surveyed 53 12-ha plots rapidly on foot, and recorded 310 breeding pairs of shorebirds. We conducted intensive nest searching on an additional 12 12-ha plots, and identified the nests or territories of 46 shorebirds. Our estimates suggest that approximately 880,000 shorebirds inhabit Coats Island and the southern 63% of Southampton Island. The most abundant species were Semipalmated Sandpiper, Red Phalarope, and Dunlin. We also report briefly on a study we conducted on the habitat preferences of loons.

Key Words: arctic, Coats Island, Dunlin, monitoring, Nunavut, population size, PRISM, Red Phalarope, Semipalmated Sandpiper, shorebirds, Southampton Island.

Shorebird populations across North America appear to be in a state of widespread decline. In particular, 19 of 26 species that breed in the North American arctic show signs of decline (Morrison et al. 2006, Johnston and Bart, chapter 1, this volume). This situation clearly warrants attention, yet our ability to respond is hindered by a poor understanding of basic parameters such as population size, distribution, and habitat relationships. To fill these knowledge gaps, Canada has embarked on a collaborative effort with the United States to census shorebirds across the continent. The objectives of the Program for Regional and International Shorebird Monitoring (PRISM) are described in Johnston and Bart (chapter 1, this volume).

Here we present the results of Arctic PRISM surveys undertaken between 2004 and 2006 at Southampton and Coats Islands, Nunavut, Canada. Habitat relationships, distributional information, and population estimates are provided for the region as a whole. Breeding ecology data are presented from intensively surveyed plots on both islands. We also present the results of a targeted survey for Pacific and Red-throated Loons (for scientific names, see Appendix C). Finally, we discuss our results in light of current knowledge of shorebird population sizes, and give recommendations for improving PRISM surveys in the future.

Smith, P. A., V. Johnston, and J. Rausch. 2012. Southampton and Coats Islands. Pp. 113–126 *in* J. Bart and V. Johnston (editors). Arctic shorebirds in North America: a decade of monitoring. Studies in Avian Biology (no. 44), University of California Press, Berkeley, CA.

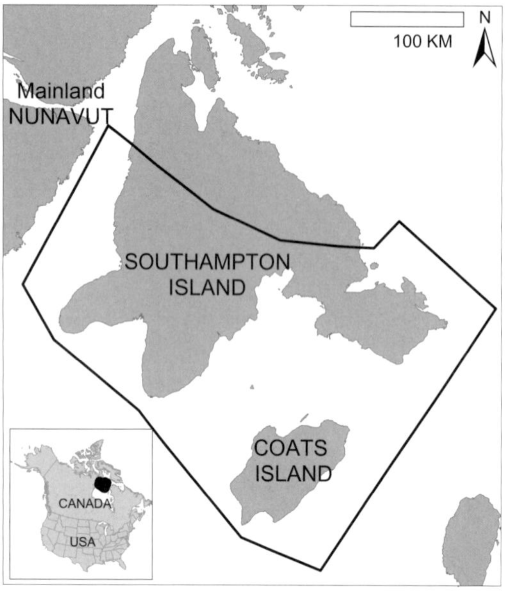

Figure 6.1. PRISM Region 4, comprising all of Coats Island and the southern 63% of Southampton Island, Nunavut.

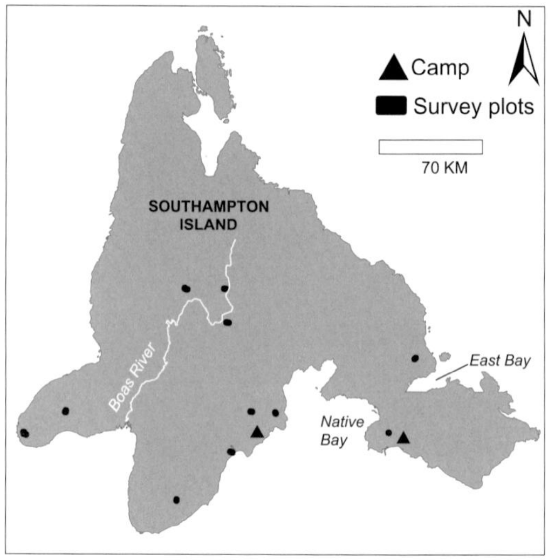

Figure 6.2. The location of pairs of rapid plots on Southampton Island. Four intensive plots were monitored at each of the camps.

METHODS

Study Area

The study area, defined as Region 4 in the PRISM sampling scheme (Bart et al., chapter 2, this volume), encompasses all of Coats Island and the southern 63% of Southampton Island, Nunavut (Fig. 6.1). The northern portion of Southampton Island, along with the Melville Peninsula and the Wager Bay area, comprise PRISM Region 5, and will be surveyed separately.

Coats Island lies at the north end of Hudson Bay, with an area of approximately 5,500 km^2. Exposed outcroppings of Precambrian metamorphic rock dominate the northeast corner, while the remainder of the island is composed primarily of lowland tundra and exposed Palaeozoic sedimentary rocks (Heywood and Sanford 1976). Smaller areas of upland heath tundra and raised beach deposits are also common across the island.

Southampton Island, at the mouth of Hudson Bay, has a total area of approximately 41,000 km^2, though only the southern 26,000 km^2 of the island are included in PRISM Region 4. This portion of the island consists of extensive tracts of coastal lowland tundra, with large expanses of unvegetated sedimentary rock farther inland. Raised beach deposits that resulted from (ongoing) isostatic uplift following the retreat of the Pleistocene ice sheets are common (Innes et al. 1968).

A PRISM camp was established in 2003 at East Bay, Southampton Island (63°59′N, 81°40′W), but work in this season was complicated by poor weather and logistical difficulties, and few useful data were collected. All Southampton Island data presented here were collected between 13 June and 10 July 2004. Two camps were established at 63°44′36″N, 82°00′30″W and 63°44′10″N, 84°01′07″W (Fig. 6.2). Work at Coats Island was conducted from 2 June to 21 July 2006 at a camp located at 62°51′06″N, 82°29′04″W (Fig. 6.3).

Weather data were recorded at the Coats Island camp with a Davis Vantage Pro weather station. At the Southampton Island camp, we used weather data from the Environment Canada weather station at Coral Harbor.

Previous Work in the Region

Reports on the avifauna of the region include early accounts by Halkette (1904), Sutton (1932), Bray (1943), and Parker and Ross (1973). More recent accounts include those of Abraham and Ankney (1986), Gaston et al. (1986, 2007), and Gaston and Ouellet (1997). However, the work reported on here represents the first extensive, systematic, ground-based survey targeting shorebirds.

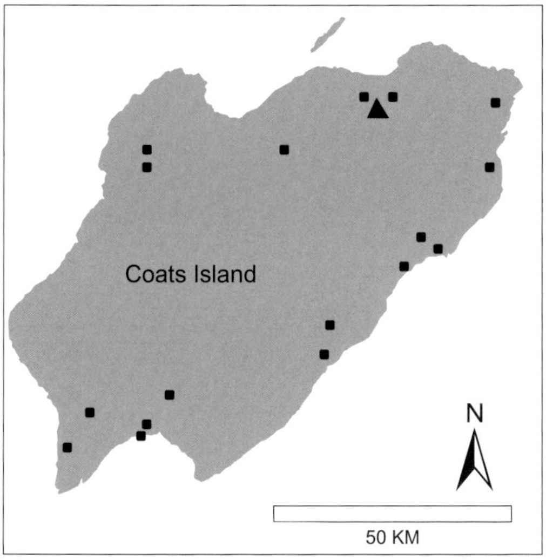

Figure 6.3. Plot locations on Coats Island. Two plots were surveyed at each square. The triangle denotes the location of the camp and the four intensive plots.

Previous work either targeted larger species or was more opportunistic in terms of design.

A number of intensive studies of shorebird ecology are ongoing at both Southampton Island and Coats Island (Smith 2003, 2009; Perkins et al. 2007; Smith et al. 2007a, 2007b).

PRISM Survey Strata

In Region 4, we defined three strata on Southampton Island and three on Coats Island. In both instances, these strata comprised wetlands, vegetated uplands, and sparsely vegetated uplands/barren areas (Table 6.1). For both sites, larger bodies of water constituted an additional stratum, but these are not included in analyses. On Coats Island, we also excluded large areas of bare rock. The methods used to delineate these strata differed between the islands due to the quality of GIS data available.

On Southampton Island we had access to a supervised classification, with ground-truthing completed by Alain Fontaine of the Canadian Wildlife Service. Complete methods are available from A. Fontaine and M. Mallory (unpubl. data), but in brief, three scenes of Landsat Enhanced Thematic Mapper (ETM+) imagery were classified into 25 habitat classes. The three images were combined to form a base image covering the entire study area. The Landsat images were captured between 20 July and 4 September 2000, and ground-truthing for the classification was done between 20 and 30 July in 2001 and 2002. The 25 classes developed by Fontaine and Mallory were lumped into the five habitat classes used for plot selection following Table 6.2.

The coverage of these habitat classes within a grid of 12-ha plots was used to delineate three strata: wetlands (good-quality shorebird habitat), vegetated uplands (habitats used extensively by some shorebird species), and sparsely vegetated or barren uplands (used by a small number of species; Table 6.1). Grid cells containing more than 70% water were excluded, and the remaining grid cells comprised the sample of potential survey plots. We used classification rules to assign plots to a single stratum: plots containing more than 35% of wetland habitat classes were designated as wetland plots, plots with more than 25% of barren habitat classes were designated as barren plots, and the remainder of plots were designated as vegetated uplands. The area of the plots assigned to each stratum was summed to provide the region-wide estimates of stratum size (Table 6.1).

TABLE 6.1

Area (km^2) and description of strata on Coats Island, southern Southampton Island (SHI), and PRISM Region 4 as a whole.

		Stratum area (km^2)		
Stratum	Description	Coats	SHI	Region 4
Wetlands	Wetlands, moist grasslands	1,375	5,104	6,479
Vegetated uplands	Heavily vegetated uplands, heaths, drier grasslands	960	13,041	14,001
Sparsely vegetated/barren	Sparsely vegetated uplands, barren areas, bare gravel	835	6,732	7,567

TABLE 6.2

Relationship between habitat classes of Fontaine and Mallory (2011) and those used in PRISM plot selection.

Fontaine and Mallory class	PRISM class
Deep to mid-depth clear waterbodies	Water
Shallow and/or turbid waterbodies	Water
Ice and snow ridges	Unsuitable
Drained waterbodies (thaw lakes)	Vegetated uplands
Coastal foreshore flats	Dry or barren habitats
Coastal backshore flats	Dry or barren habitats
Active deposits	Unsuitable
Bedrock outcrops	Dry or barren habitats
Boulder ridges	Dry or barren habitats
Lichen–heath tundra	Dry or barren habitats
Mix of lichen–heath and boulder ridges	Dry or barren habitats
Hand size or larger fragments deposits	Dry or barren habitats
Gravel-size fragments deposits	Dry or barren habitats
Algae-covered lag	Dry or barren habitats
Heath mats	Vegetated uplands
Heath nets	Vegetated uplands
Heavy heath–shrub nets	Vegetated uplands
Mixed tundra (heath/graminoid/shrub)	Vegetated uplands
Exposed peat and sediments	Vegetated uplands
Hydric moss and peat carpet	Dry or barren habitats
Hydric moss carpet	Dry or barren habitats
Graminoid meadows	Wetlands
Low-center polygons	Wetlands
Unclassified	No data
Clouds	No data

On Coats Island, the method for delineating strata was somewhat different, as no supervised habitat classification was available. Steven Mamet, of the University of Alberta, performed an unsupervised classification of Landsat data. Two Landsat 7 images were downloaded from the Global Land Cover Facility (GLCF) at the University of Maryland (http://glcf.umiacs.umd.edu/index.shtml). ETM+ imagery for path 25, row 16 and path 27, row 16 were downloaded for 20 July and 19 August 2000, respectively. The individual bands were stacked using Erdas Imagine 8.7 (ERDAS Inc., Madison, AL) and the normalized difference vegetation index (NDVI) was computed for each image using ENVI 4.1 (Exelis Visual Information Solutions, Boulder, CO). The NDVI was used to separate habitat classes based on vegetation vigor. Classes were initially separated based on water content using the normalized difference water index (NDWI), but this method did not produce results as accurate as the NDVI.

The NDVI imagery was then separated into 25 classes using an unsupervised classification in Erdas Imagine, with 15 iterations (*M*) and a convergence threshold (T) of 0.99. The value of T represents the maximum percentage of pixels whose class values are allowed to be unchanged between iterations (for a more detailed discussion, see Schrader and Pouncey 1997). Water and extensive unvegetated areas of exposed rock were isolated and excluded. The remaining habitat classes were grouped into one of three shorebird habitat classes based on vegetation vigor—wetlands, vegetated uplands, and sparsely vegetated or barren areas—and were very similar to the same classes used on Southampton Island. The classes were then evaluated by examining the results in areas with which the authors were familiar. The classification was then rerun to produce the expected results for these known habitat features. The resulting three classes comprise the strata for Coats Island (Table 6.1). Plot assignments to habitat classes and calculation of stratum sizes were carried out as for Southampton Island.

Survey Plots

In PRISM, a large number of plots are selected for rapid, helicopter-supported surveys, and a smaller number are selected for intensive surveys ("rapid" plots and "intensive" plots, respectively). At Coats Island, both rapid and intensive plots were 12 ha, with dimensions of 400 m N–S and 300 m E–W. At Southampton Island, plots had dimensions of approximately 350 m × 350 m, for an area of approximately 12.25 ha.

In all regions and years, we select more plots than we expect to sample and do not survey plots that are determined in the field to be unsuitable for sampling. The Landsat imagery available for arctic regions is frequently captured in late summer, and many of the shallow wetland habitats used by shorebirds are poorly represented by late summer imagery. Often, areas that appear to be wet tundra on late summer imagery are inundated in early June and unavailable to shorebirds. These plots are assigned densities of 0 in the analyses, and additional randomly selected plots are then surveyed.

As for all PRISM regions, the sampling plan involved stratification by subregion and habitat (wetlands, vegetated uplands, barrens), followed by selection of "zones" of potential survey plots and then selection of plots (see Bart et al., chapter 2, this volume). On Southampton Island, 18 zones of potential rapid plots were systematically selected to represent the study area. Zones were subdivided into 30 plots. Two plots were randomly selected from each of these zones. To keep the habitat class within plots as uniform as possible, and avoid inclusion of large water bodies, some plot boundaries were adjusted in the GIS. The sample that was surveyed comprised three wetland plots, 13 vegetated upland plots, and eight barren plots.

Zones of rapid plots at Coats Island were selected to conform to a ratio of six wetland plots:two vegetated uplands:one barren. The resultant sample of surveyed plots comprised 24 wetland, three vegetated upland, and two barren plots. Six additional plots were surveyed from the bare rock class to confirm that no shorebirds were present.

Selection of intensive plots was less objective. Although a simple random sample would be statistically preferable, the logistics of erecting a camp near the plots must be considered. Also, plots must contain a reasonable number of nesting shorebirds in order to generate detection ratios. To locate the intensive camps on Southampton Island, a random sample of zones of wetland habitat were identified. Areas near these zones were scrutinized for landing strips, and two zones with suitable camp locations were selected (Fig. 6.2). Surveyors non-randomly selected four plots at each camp from the wetland habitat stratum in order to maximize the number of nesting birds found in the plots.

The PRISM work at Coats Island was based out of an existing camp (Fig. 6.3), and intensive plots were by necessity selected to be within walking distance of that camp. Four plots were selected non-randomly: three in wetland habitat and one in mixed upland/wetland habitat (vegetated upland class).

Rapid Surveys

Rapid surveys were conducted by two observers walking line transects, covering 7–10 ha per hour. Observers were separated by 25 m and recorded birds within 12.5 m on either side, for a transect width of 50 m. Observers used GPS to ensure that the plot was surveyed completely. Sightings were recorded on plot maps using predetermined codes for nests, probable nests, pairs, males, females, unknown sex birds, and groups. These plot maps were used to minimize the number of birds that were double-counted. All species detected were recorded. Immediately after each survey, a summary was prepared and an estimate was made of the number of pairs plus the individual breeding birds whose mate was not seen ("indicated pairs") present in the plot. Basic habitat information, including percent cover of various habitat classes and landforms, was recorded on prepared data sheets at each plot. Original data sheets are archived at the Canadian Wildlife Service office in Yellowknife.

Intensive Surveys

Intensive plots were surveyed primarily by individual observers, but rope drags were also employed. Each plot was surveyed by two observers using area search methods (systematic searching on foot); although these observers worked independently, they shared information freely. No information was shared between crews conducting intensive and rapid surveys. The intensive plots are also surveyed rapidly to generate detection ratios, and complete independence between these two survey types is an important assumption of our sampling design.

Observers visited intensive plots every one to two days and spent approximately four hours per visit throughout the late courtship and early incubation stages. Locations of territories were marked on plot maps, and nest locations were recorded with GPS. After each visit to the plots, territory and nest locations were compared to previous plot maps to determine whether new territories had been identified. Plots were searched throughout the breeding season, and when no new territories were discovered for three consecutive visits, search effort was reduced to once- or twice-weekly visits to monitor nests.

We also surveyed each intensive plot by rope drags once on Southampton Island and twice on Coats Island. Rope-drag teams consisted of two to three people, dragging a 25-m-long, 0.6-cm-diameter rope with dropper lines of thin cord tied along its length. Plots were covered exhaustively, in straight transects, causing incubating birds to flush as the rope approached. Territories and nests were mapped as for single observers.

Eggs from the Coats Island intensive plots were floated to estimate developmental stage, and followed throughout the breeding season to monitor hatching success (Liebezeit et al. 2007). Basic habitat information was recorded for each nest using a standardized data sheet.

Regional Population Estimates

The population estimation procedure is based on estimators derived from standard survey sampling methods (Cochran 1977). Details of the procedure are found in Bart et al. (chapter 2, this volume). In brief, population is estimated by using observed densities in surveyed plots to calculate a mean density. Detection ratios are calculated by comparing the rapid surveys of intensive plots to the true number of nesting birds known to be there from the intensive surveys. These detection ratios are then used to correct the observed densities on rapid plots. Corrected densities are applied across the area of interest to generate the population estimate.

The detectability of birds on surveys can differ between sites, species, or years. Given the small number of nests found in the intensive plots, it is not practical to generate detection ratios for an individual PRISM region. The precision of the detection ratios would be so low that their use would be unjustified statistically. In this analysis, a single detection ratio (±SE) of 1.27 ± 0.21 is used for all species. The detection value is based on data from across the Canadian arctic, including this region.

Lake Surveys for Loons

Surveys of nearby lakes were conducted at both of the Southampton Island intensive camps to gain more information on arctic-breeding loon nesting habitat preferences and food sources. In brief, crews surveyed all lakes within 3 km of their camp and recorded size, depth, and habitat characteristics. Lakeshores were searched for nests, and habitat descriptions of loon nests were recorded. Minnow traps were deployed in a subsample of ponds.

Other Surveys

Rapid survey crews undertook extensive aerial surveys between rapid survey plots. Methods and results of these surveys are presented in Elliott and Smith (chapter 9, this volume).

TABLE 6.3

Weather data from community weather station at Coral Harbor, Southampton Island, and Intensive Camp on Coats Island, Nunavut.

Year	Date of 50% snow cover	Mean temperature (°C) June (min, max)	July	Total precipitation (mm) (June–July)	Rain events (field season duration)
Environment Canada weather station at Coral Harbor, Nunavut (surrogate for Southampton Islands camps)					
2004		2.0 (−1.3, 5.2)			
29-year average (1974–2003)		2.8 (−0.8, 6.3)			
Coats Island Intensive Camp					
2004	23 June	1.3	6.7	27.9	18 events (9 June–24 July)
2005	pre-June 7	4.5	7.3	49.3	24 events (7 June–22 July)
2006	5 June	3.6	8.4	28.9	21 events (2 June–21 July)

RESULTS

Weather and Timing of Breeding

The spring and summer of 2004 were late in the Southampton Island study area, where June temperatures (at the Coral Harbor weather station) were cooler than the 29-year average (Table 6.3). Snow cover on Southampton Island in 2004 was heavy at the time of the intensive crews' arrival. At one camp, snow cover was estimated at 95% on 15 June, while at the other, the intensive plots were 80–100% covered by snow and standing water on 16 June. By 20 June, snow cover at both camps had receded considerably, but standing water persisted, with 60–80% of the intensive plots flooded. Data from the East Bay field station (approximately 60 km northeast of one of our intensive camps) indicate that the summer of 2004 was later than usual, but similar snow conditions and colder temperatures in late June had been encountered in 1999.

Timing of breeding in 2004 was similar at the two Southampton Island camps. At one, the first shorebird pairs were seen on 17 June, and the first nest, a Red Phalarope nest with one egg, was found on 21 June. At the other Southampton Island camp, shorebird copulations were noted on 16 June and a Dunlin nest with four eggs was found on 21 June. Dates of nest initiation for shorebirds ranged from 17 June to 5 July, with a peak in late June.

Weather at the Coats Island camp is poorly represented by the Environment Canada weather station at Coral Harbor, 130 km away. Weather and timing of breeding data are available from this camp for 2004 and 2005, and these are used as a benchmark to assess the phenology of the 2006 breeding season.

Snow cover at the Coats Island camp was 80% upon our 2 June 2006 arrival. June temperatures were warm, however (Table 6.3), and snow cover was reduced to 50% by 5 June, and none remained on flat areas by 11 June. Weather throughout June and early July was generally warm and dry, and the flow of small rivers in the area was much lower in 2006 than in 2004 or 2005. Landfast sea ice retreated in mid-June 2006, roughly two weeks earlier than in the previous two years at this site.

As is typical in the arctic, there were several extreme weather events in 2006 that could have affected breeding birds. There was a snowfall of 5–10 cm on 18 June; this snow melted within 12 hours. A storm on 23 June brought 0°C temperatures, 72 km/hr winds, and snow. On 7 July, there were extreme winds (high of 98 km/hr) and light rain. This coincided roughly with peak hatch, and may have affected chick survival.

Timing of breeding in 2006 was approximately ten days earlier than in the "late" year of 2004, but similar to 2005. Dunlin and American Golden-Plover were present upon our arrival on 2 June, and Red Phalarope, Ruddy Turnstone, Semipalmated Plover, Semipalmated Sandpiper, and White-rumped Sandpiper had arrived by 4 June. The first shorebird nests found were one Dunlin and one Semipalmated Sandpiper nest with three eggs each on 11 June. The peak date of clutch completion was approximately 15 June.

Timing of breeding and weather patterns are explored more completely in Smith et al. (2010). We present these indications of timing here to demonstrate that our rapid surveys coincided with late courtship/early incubation, the desired timing for Arctic PRISM surveys (Bart et al., chapter 2, this volume).

Rapid Surveys

At Southampton Island, we conducted rapid surveys on 24 plots from 24 to 29 June 2004 (Fig. 6.2). At Coats Island, we conducted rapid surveys at 34 plots from 18 to 20 June 2006 (Fig. 6.3), with 29 in "suitable" habitat types used in analyses (the remainder were bare rock, surveyed to confirm the absence of shorebirds). Survey times for the 12-ha plots were typically 60–90 minutes, depending on terrain and bird densities.

On Southampton Island, seven species of shorebirds and 11 other species were recorded on rapid survey plots. A total of 203 indicated pairs of all species was recorded on the surveys, with Lapland Longspurs being the most abundant overall (Table 6.4). Among shorebirds, Red Phalarope, Dunlin, and Semipalmated Sandpiper were the most abundant. The former two species had highest densities in wetland habitats, while the latter was associated with medium-quality, drier habitats (Table 6.5). The density for all shorebird species combined was highest in wetland habitats.

On Coats Island, we recorded eight species of shorebirds and nine other species. Though only five more plots were surveyed on Coats Island than on Southampton Island, we recorded more than twice the number of shorebirds at Coats Island (Table 6.4). The main reason is that Semipalmated Sandpipers were much more abundant on Coats Island, where they comprised over 50% of the 217 shorebirds recorded. Red Phalaropes and Dunlin were also abundant. The habitat-specific densities show a strong association with wetlands for all species except the White-rumped Sandpiper (Table 6.5).

Intensive Surveys

Four plots were intensively surveyed at each of the two camps on Southampton Island. Intensive surveys were conducted from 16 June to 10 July 2004, with an average observer effort of 9.1 h/plot at one camp, and 34.9 h/plot at the other. Rope drags were completed once on each plot by a team of three. Although secretive individuals may always be missed, a lower level of effort was expended at the former camp (Camp no. 1) because shorebirds were scarce, not vice versa.

We identified 26 shorebird territories within the plots (Table 6.6), but found only 18 nests in total: one American Golden-Plover, seven Dunlin, three Red Phalarope, and seven Semipalmated Sandpiper. Many nests were still active when we left the area, and rates of nest predation are thus approximate. Overall, we estimate that 44% and 32% of nests were lost to predators at camps 1 and 2, respectively.

At Coats Island, we worked out of an existing shorebird research camp, and intensively monitored four plots. Although we were present at this site from 2 June to 21 July, we monitored the intensive plots following PRISM protocols from 12 June to 3 July. Two single observers shared information, spending an average of 24.8 h in each plot. Rope drags were completed once on each plot by a team of two (the same observers as the single observer surveys), for an additional six person-hours per plot.

We found 19 shorebird nests, and believed that one additional bird nested within the plot, but failed before we found its nest, for a total of 20 territories within the plots (Table 6.7). These comprised five Dunlin, one Pectoral Sandpiper, five Red Phalaropes, six Semipalmated Sandpipers, and three White-rumped Sandpipers. Analyses of nest survival were completed for the study area as a whole, as part of a Ph.D. thesis (Smith 2009). The estimated daily survival rate for all species was 0.952 ± 0.007, or approximately 34% nest success over an incubation period of 22 days. Both the nest success and nesting densities of polygamous species were significantly higher at this site in 2006 than in 2004 or 2005.

TABLE 6.4

The total number of pairs recorded on rapid surveys at Coats Island and Southampton Island, Nunavut[a].

Species[b]		Total number recorded: Southampton Island (24 plots)	Coats Island (29 plots)
Shorebirds			
AMGP	American Golden-Plover	5	6
BBPL	Black-bellied Plover	1	4
DUNL	Dunlin	24	27
PESA	Pectoral Sandpiper	1	0
PUSA	Purple Sandpiper	0	1
REPH	Red Phalarope	33	51
RUTU	Ruddy Turnstone	0	4
SESA	Semipalmated Sandpiper	18	115
WRSA	White-rumped Sandpiper	11	9
	All Shorebirds	93	217
Non-shorebirds			
ARTE	Arctic Tern	4	0
CAGO	Canada Goose/Cackling Goose	24	31
HOLA	Horned Lark	5	0
KIEI	King Eider	2	16
LALO	Lapland Longspur	58	83
LTDU	Long-tailed Duck	3	4
NOPI	Northern Pintail	0	1
PAJA	Parasitic Jaeger	0	3
PALO	Pacific Loon	4	3
RTLO	Red-throated Loon	0	1
SAGU	Sabine's Gull	2	0
SNBU	Snow Bunting	2	2
SNGO	Snow Goose	5	0
WIPT	Willow Ptarmigan	1	0
Total		203	361

[a] Surveys of Coats Island carried out 18–20 June 2006 and Southampton Island, 24–29 June 2004.

[b] Species names are displayed with AOU four letter codes.

Rapid Surveys of Intensive Plots

The eight intensive plots at Southampton Island were surveyed with rapid methods once, between 24 and 28 June. Agreement between the rapid survey estimates of the number of breeding territories and the intensive survey results was poor (Table 6.6). Estimates from the rapid surveys suggested that 70 shorebird pairs were breeding in the plots, while the intensive surveys concluded that only 26 territories were present, for a

TABLE 6.5

Shorebird densities (birds/km^2 ± SE) in the three habitat strata on Southampton and Coats Islands, Nunavut.

	Southampton Island			Coats Island		
Species	Wetland	Veg. upland	Barren	Wetland	Veg. upland	Barren
AMGP	3.0 ± 2.5	4.2 ± 2.7	—	2.6 ± 1.3	—	6.6
BBPL	—	1.6	—	2.2 ± 1.0	—	—
DUNL	12.5 ± 5.8	7.4 ± 3.4	—	12.9 ± 3.8	—	—
PESA	—	1.9	—	—	—	—
PUSA	—	—	—	0.6	—	—
REPH	21.8 ± 6.1	5.1 ± 3.3	—	27.5 ± 9.3	—	—
RUTU	—	—	—	2.0 ± 1.4	—	—
SESA	5.8 ± 4.6	13.4 ± 4.3	—	60.8 ± 15.3	7.9 ± 2.5	6.6
WRSA	10.5 ± 5.7	—	—	4.4 ± 2.8	5.3 ± 2.3	—
All	53.6 ± 11.4	33.6 ± 7.4	—	113.0 ± 18.7	13.2 ± 3.4	13.2 ± 9.3

NOTE: Densities are missing for species and habitats with no records (—). Some SEs are missing because of too few records.

TABLE 6.6

The number of shorebird pairs estimated to be present on eight rapid plots, and the number determined to be nesting (actual) through intensive surveys on Southampton Island, Nunavut.

	Estimated					Actual				
Plot	DUNL	REPH	SESA	WRSA	Other	DUNL	REPH	SESA	WRSA	Other
1-A	0	2	0	0	0	0	1	0	0	0
1-B	2	3	0	1	0	1	1	0	0	0
1-C	0	4	0	0	REKN-1	0	2	0	1	0
1-D	0	0	0	0	0	0	0	0	0	0
2-A	6	10	3	0	AMGP-1	2	1	0	0	0
2-B	3	13	1	0	RUTU-1	1	2	0	0	0
2-C	1	7	0	0	0	1	2	1	0	0
2-D	3	0	7	1	0	2	0	6	1	AMGP-1
All	15	39	11	2	3	7	9	7	2	1

NOTE: Four rapid plots (A–D) were selected at each camp (1–2).

detection ratio of 2.69. The discrepancy was especially large for Red Phalarope, for which rapid surveys estimated more than four times the number of birds found during the intensive surveys.

Each of the four intensive plots at Coats Island was surveyed by rapid methods twice, between 14 and 15 June, and 22 and 25 June. There was better agreement between intensive surveys and rapid surveys at this site. Intensive surveys identified 20 territories within the plots; the first round of rapid surveys estimated 33 territories, while the second estimated 26 (detection ratios of 1.65 and 1.30, respectively; Table 6.7). The poorest agreement was for Semipalmated Sandpiper in the first

TABLE 6.7

The number of shorebirds estimated (Est) to be present through rapid surveys, and the actual (Act) number determined to be nesting through intensive surveys of the same plots at Coats Island, Nunavut.

	AMGP		DUNL		PESA		REPH		SESA		WRSA	
Plot (survey)	Est.	Act.	Est.	Act.	Est.	Act.	Est.	Act.	Est.	Act.	Est.	Act.
A	0/1	0	2/2	3	2/0	1	5/3	2	4/2	0	3/2	1
B	0	0	1/1	1	0	0	1/2	2	5/2	2	1/0	2
C	0	0	0	0	0	0	1/1	1	7/6	3	0	0
D	0/1	0	0/2	1	0	0	0	0	1/1	1	0	0
Mean total[a]	1	0	4	5	1	1	6.5	5	14	6	3	3
Detection ratio[b]	—		0.80		1.00		1.30		2.33		1.00	

[a] Plots A–D were each surveyed rapidly twice. Both results are presented here, but the mean of the two surveys was used to calculate detection ratios.

[b] The detection ratio is calculated by dividing the estimated number of birds by the actual number present.

TABLE 6.8

The estimated population size, measured in individual birds ± SE, for PRISM Region 4, including all of Coats Island and 63% of Southampton Island, Nunavut.

	Southampton Island		Coats Island	
Species	Est. pop. ± SE	CV	Est. pop. ± SE	CV
AMGP	65,039 ± 27,177	0.42	9,798 ± 2,853	0.29
BBPL	16,200 ± 16,216	1.00	2,247 ± 1,129	0.50
DUNL	167,907 ± 62,468	0.37	13,290 ± 4,547	0.34
PESA	19,809 ± 20,119	1.02	—	—
PUSA	—	—	585 ± 603	1.03
REPH	213,055 ± 61,198	0.29	28,237 ± 11,417	0.40
RUTU	—	—	2,052 ± 1,491	0.73
SESA	180,662 ± 66,862	0.37	78,064 ± 22,294	0.29
WRSA	77,639 ± 43,565	0.56	10,153 ± 4,814	0.47
All[a]	740,310 ± 124,187	0.17	144,426 ± 26,138	0.18

[a] Calculated by summing species-specific estimates and combining their variances.

round of surveys, for which rapid surveys estimated 2.83 times the number of territories that intensive surveys eventually identified.

Population Estimates

We estimated a total population of approximately 880,000 breeding shorebirds for PRISM Region 4 (Table 6.8). At Southampton Island, Red Phalaropes were most abundant, followed by Semipalmated Sandpipers, Dunlin, and White-rumped Sandpipers (Table 6.8). In contrast, Dunlin and White-rumped Sandpipers were less common on Coats Island, and Semipalmated Sandpipers were nearly three times more abundant than any other shorebird (Table 6.8).

Lake Surveys for Loons

On Southampton Island, 22 water bodies were surveyed between 27 June and 8 July. Pacific Loons were observed on five lakes, and nested on four lakes, while Red-throated Loons were observed on ten lakes, and nested on four lakes. Lakes hosting Pacific Loons tended to be deeper and to have emergent vegetation and islands. In contrast, lakes hosting Red-throated Loons tended to be shallow, and some had islands or emergent vegetation while others did not.

The nests of both Pacific Loons and Red-throated Loons were predominantly on islands; this frequent use of islands is consistent with findings from other studies of both species in Canada and Alaska (Barr et al. 2000, Russell 2002). Nests were also typically located on the south or east side of the islands, and thereby were sheltered from the prevailing winds and waves. At Southampton Island, minnow traps were set in ponds with nesting loons but no minnows were captured.

Other species were observed on the 22 surveyed lakes including Long-tailed Ducks (10 lakes), Canada/Cackling Geese (8), King Eiders (3), Herring Gulls (3), and Long-tailed Jaegers (2). Arctic Tern, Atlantic Brant, and Parasitic Jaeger were observed on one lake each.

DISCUSSION

Shorebird Densities and Distribution

The PRISM program is designed primarily to provide arctic-wide estimates of population across a sampling interval of several years; the regional estimates are of necessity a snapshot in time. Densities of shorebirds are known to fluctuate between years (for example, in response to broad-scale patterns in weather; Meltofte 1985), yet long-term data sets of shorebird densities are exceedingly rare for arctic Canada. Studies of shorebirds have been undertaken at East Bay, Southampton Island, since 1999 (Smith 2003; Perkins et al. 2007; Smith et al. 2007a, 2007b, 2010), providing the rare opportunity to compare the density estimates from PRISM surveys to long-term averages. In the present study, we estimated densities of 53.6 ± 11.4 shorebirds/km^2 in wetland habitats and 33.6 ± 7.4 birds/km^2 in drier, vegetated upland habitats on Southampton Island

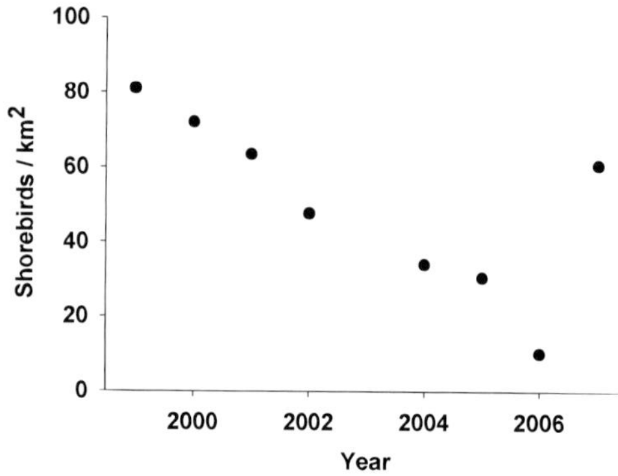

Figure 6.4. The density of shorebirds (all species combined) in a 1.6 km × 1.63 km plot at East Bay, Southampton Island, surveyed in late June, 1999–2010. No surveys were completed in 2003 (P. A. Smith, unpubl. data).

(zero in barren habitats). The long-term average from a fixed plot at East Bay was 52.1 ± 8.3 birds/km^2, ranging from 81 birds/km^2 in 1999 to 10 birds/km^2 in 2006 (Fig. 6.4). This 2.56-km^2 fixed plot is composed of a mix of wetland and beach ridge habitats, and is situated on the northeast corner of Southampton Island.

Habitat-specific densities and comparisons between the sites can yield further insight into species' distributions and habitat preferences. Semipalmated Sandpipers, for example, were observed at much higher densities on Coats Island than on Southampton Island; the latter site is near the northern limit of the species' range. American Golden-Plovers on Southampton Island were observed at the highest densities in drier, medium-quality habitats. The association of this species with heath tundra has been documented previously (Johnson and Connors 1996a).

Population Estimates

In general, species for which we estimate large populations in the region also have large estimates for range-wide population (e.g., Semipalmated Sandpiper and Red Phalarope), while species for which we estimate smaller populations are also thought to have smaller populations range-wide (e.g., Black-bellied Plover and Purple Sandpiper; Morrison et al. 2006). Thus, the relative patterns of population estimates between species are consistent with the current estimates of population size. Further, relative abundances of Semipalmated Sandpiper are consistent with our knowledge of the species' range. Semipalmated Sandpiper occurs in far higher densities in the southern part of Region 4 (Coats Island) than in the northern part (Southampton Island), near the northern limit of its breeding range.

That said, in many cases, population estimates for this comparatively small part of the arctic represent a large proportion of the existing estimates for populations range-wide. For example, our estimate for American Golden-Plover in Region 4 represents 37% of the previous range-wide estimate (Morrison et al. 2006). This proportion was large for other species as well, including 81% for the Dunlin *hudsonia* subspecies, 19% for Red Phalarope, and 13% for Semipalmated Sandpiper. As acknowledged by Morrison et al. (2006), many of these previous estimates were based on limited data. Our sampling design requires numerous assumptions (including in this case a single detection rate of 1.27), but is based on survey data. That our estimates of density (the basis of our population estimates) are in agreement with a long-term study at East Bay lends credence to our belief that shorebird populations are larger than previous estimates suggest.

We were surprised by the scarcity of observations of Ruddy Turnstone and Black-bellied Plover, and in particular the near absence of records from Southampton Island. In a few km^2 study area at East Bay, on the northeastern coast of Southampton Island, Ruddy Turnstones nest at the highest densities observed to date in North America (Perkins et al. 2007). They also nest in coastal regions of Coats Island, though at much lower densities (P. A. Smith, pers. obs). As no Ruddy Turnstones were observed on plots on Southampton Island, we were unable to generate an estimate of population for this portion of the region; however, we had a confirmed nest near an intensive camp. Similarly, our population estimates for Black-bellied Plover are low, despite the fact that they are not uncommon around the East Bay area and were observed on several occasions during aerial surveys. These examples highlight the effect of sampling error on our population estimates and provide an argument for increasing the sample size of rapidly surveyed plots when regional estimates of population are desired.

Factors Affecting Accuracy of the Surveys

With few exceptions, we identified the greatest densities of shorebirds in areas that we classified as wetland habitat. That shorebirds prefer wetland habitats is perhaps not surprising. However, this result also demonstrates that our habitat classification was effective at identifying the habitats with the highest densities of shorebirds, a desirable condition for reducing variance in the population estimates. Unfortunately, the classification failed to separate degraded wetlands (that had been heavily grazed by geese) from intact wetlands of high value to shorebirds. Degraded wetlands accounted for only a small proportion of the entire wetland habitat on Southampton Island, but this classification error could have influenced the variance of shorebird densities in wetland plots. The classification also had difficulty distinguishing between areas of frost-shattered rock commingled with standing water and wetlands.

The PRISM program depends heavily on remote sensing, and supervised habitat classifications for the Canadian arctic are rare. Moreover, available imagery was often captured in late July, August, or even early September, at a time when conditions on the tundra are substantially drier than at the time of PRISM surveys. This may have contributed to one type of misclassification that we witnessed. Flooded areas in the Boas River and Native Bay areas of Southampton Island were classed as "wetland" habitat, whereas in reality much of it was so wet as to be "sparsely vegetated/barren." Because plots that were truly unsuitable were assigned values of zero for shorebird abundance, these misclassifications did not bias results. However, more accurate habitat classifications would improve precision.

Our detection rates were higher than desired. At both study sites, we counted more birds on rapid surveys than were found breeding in the plots during intensive surveys. The agreement was especially poor for Southampton Island in 2004, where we overcounted by as much as a factor of four in some cases. This is unlikely due to observer error on the rapid surveys as these were conducted by experienced observers (J. Bart, S. Earnst, V. Johnston). Nor is it likely to be due to poor nest finding in the intensive plots; these were also completed by experienced observers and the detection ratio differed little between the two camps (2.17 vs. 2.85). Instead, we suggest that extensive and late snow cover during the first few weeks of the 2004 breeding season caused numerous individuals to display and exhibit courtship behaviors in areas where they did not subsequently nest.

If we assume that these high detection values are reflective of Southampton Island as a whole, using a detection rate of 1.27 to estimate population would overestimate the true breeding population. We do not feel that our population estimates are grossly inflated, but we do recognize that some individuals that did not breed in the late year of 2004 may have been included in our estimates of breeding population.

Although our sample of intensive plots is small, comparison of the detection rates between species and camps reveals several patterns. At Coats Island, two rapid surveys of each plot were completed, and the mean was used to calculate detection rate. Detection rates at this camp were much closer to 1 than at Southampton Island, where a single survey was conducted. Further, our results suggest a tendency to overcount the more abundant species. Detection rates play a crucial role in population estimation, and understanding how detection rates vary with density or among species, sites, and years will allow us to refine the PRISM methodology. An investigation of the factors influencing detection rates is warranted.

CONCLUSIONS AND RECOMMENDATIONS

We are able to make the following recommendations, based on the results of this study:

1. The unsupervised classification of Coats Island, using the normalized difference vegetation index, was effective. This method could be employed in the future when supervised habitat classifications are not available.
2. Supervised classifications that are based on late summer imagery can lead to inaccurate quantification of usable shorebird habitat (usually by ranking flooded wetlands as "wetland" habitat, or by excluding wetland areas that have dried later in the season). We recommend the development of a classification "mask" that would identify and properly classify areas that are subject to regular flooding or snow cover during the June breeding season.
3. Each intensive plot should be surveyed twice with the rapid method.
4. Two shorebird species (Red Knot and Semipalmated Plover) as well as a few non-shorebird species (Atlantic Brant, Tundra Swan) known to breed in the region were not captured on rapid surveys, leading to population estimates of zero. Where region-specific estimates of population are desired, our sample of rapid plots should be increased to reduce the influence of sampling error.

ACKNOWLEDGMENTS

Many hands went into the development of this report. J. Bart created the PRISM sampling plan, and J. Leger and A. Manning contributed to GIS plot selection. We thank A. Fontaine and S. Mamet for completing the landcover classification of the satellite images. Fieldwork was funded by the Canadian Wildlife Service, and logistical support was provided by the Polar Continental Shelf Program. We are grateful for assistance in the field from J. Bart, G. Beyersbergen, V. Charlwood, S. Davies, L. Dickson, S. Earnst, K. Elliott, M. Kelly, M. Kotierk, J. McKay, K. Monaghan, C. Parker, M. Setterington, and A. Skok. This report is PCSP Contribution No. 00709.

ONLINE CONTENT

Abstracts are available in French and Spanish from www.ucpress.edu/go/sab.
Une traduction du résumé est disponible en français.
Una traducción del resumen está disponible en español.

CHAPTER SEVEN

Prince Charles, Air Force, and Baffin Islands

Victoria Johnston and Paul A. Smith

Abstract. As part of an effort to generate arctic-wide estimates of shorebird populations, we conducted surveys on Prince Charles and Air Force Islands in 1996–1997 and the western portion of Baffin Island in 2003–2004. Using the Program for Regional and International Shorebird Monitoring (PRISM) methodology, we surveyed 151 plots of 12–16 ha rapidly on foot, recording 1,546 breeding pairs of shorebirds. We conducted intensive nest searching on an additional eight plots, and identified the nests or territories of 40 shorebirds. Our estimates suggest that approximately 1.6 million shorebirds inhabit this 16,815-km^2 region, which encompasses the Foxe Basin islands south of Rowley Island and the lowlands of Baffin Island west of Nettilling and Amadjuak Lakes. The most abundant species were Red Phalaropes and White-rumped Sandpipers, totaling 1.5 million individuals (91% of all shorebirds in the region). We also report briefly on a study of habitat preferences of loons.

Key Words: Air Force Island, arctic, Baffin Island, monitoring, Nunavut, population size, Prince Charles Island, PRISM, Red Phalarope, shorebirds, White-rumped Sandpiper.

The portion of Baffin Island that is included in PRISM Region 3 is an important area for breeding migratory birds (Soper 1940). It was established as the Dewey Soper Migratory Bird Sanctuary in 1957, and designated a Ramsar Wetland of International Importance in 1982. Prince Charles and Air Force Islands are recognized as a key habitat site for migratory birds in general, and previous surveys have suggested that shorebirds are particularly abundant (Morrison 1997, Latour et al. 2008, Johnston and Pepper 2009). Here, we present the results of surveys of Prince Charles and Air Force Islands in 1996–1997, and the western portion of Baffin Island in 2003–2004. We discuss bird distribution and provide population estimates for the region as a whole. Breeding ecology data are presented from studies undertaken by J.-L. Martin at Prince Charles Island in 1996 and 1997 and from plots intensively surveyed by us on Baffin Island in 2003 and 2004. Last, we present briefly the results of a pilot study that described nest sites used by Pacific Loons (for scientific names, see Appendix C).

Johnston, V., and P. A. Smith. 2012. Prince Charles, Air Force, and Baffin Islands. Pp. 127–140 *in* J. Bart and V. Johnston (editors). Arctic shorebirds in North America: a decade of monitoring. Studies in Avian Biology (no. 44), University of California Press, Berkeley, CA.

METHODS

Study Area

The study area is part of Region 3 in the PRISM sampling scheme (Bart et al., chapter 2, this volume; Fig. 7.1), which encompasses a number of the Foxe Basin Islands, including Prince Charles, Air Force, Foley, Bray, South Tweedsmuir, and North and South Spicer Islands. It also includes a portion of the west coast of central Baffin Island, from the coast adjacent to Bray Island, south to Bowman Bay, and inland to Nettilling and Amadjuak Lakes. Our surveys were conducted on Prince Charles, Air Force, and Baffin Islands (Figs. 7.2 and 7.3).

The largest of the Foxe Basin islands, Prince Charles and Air Force Islands have areas of 9,900 km^2 and 1,700 km^2, respectively. The portion of Baffin Island included in Region 3 has an area of 7,765 km^2; much of this is a flat wetland known as the Great Plains of the Koukdjuak. Coastal areas throughout the region are characterized by wide intertidal flats and low-lying salt marshes. Further inland are large expanses of marshy tundra. The entire region is very flat and comprises relatively young landforms undergoing isostatic uplift since the retreat of the Pleistocene ice sheets (Gaston et al. 1986). As a result of this uplift, the center of Prince Charles Island is primarily unvegetated, broken shale. A portion of the region farthest inland on Baffin Island is also barren. Raised beach features are common, particularly on the west coast of Prince Charles Island and inland from the Baffin coast. The flora of Prince Charles and Air Force Islands is similar to that of the Great Plains of the Koukdjuak. A collection of vascular plants from Prince Charles and Air Force Islands identified 54 species, compared to 73 species on the Great Plains of the Koukdjuak, Baffin Island (Baldwin 1953, Kerbes 1969).

The region's climate is influenced greatly by the cold waters of the Foxe Basin. Landfast ice persists until July, and drifting pack ice precludes navigation until late August or September (Markham 1986). The coastal areas in the region are thus much cooler than inland areas, and colder than expected based on latitude. Snowmelt typically occurs in the last half of June. Temperatures are above freezing on most days in the latter part of June, July, and the first half of August.

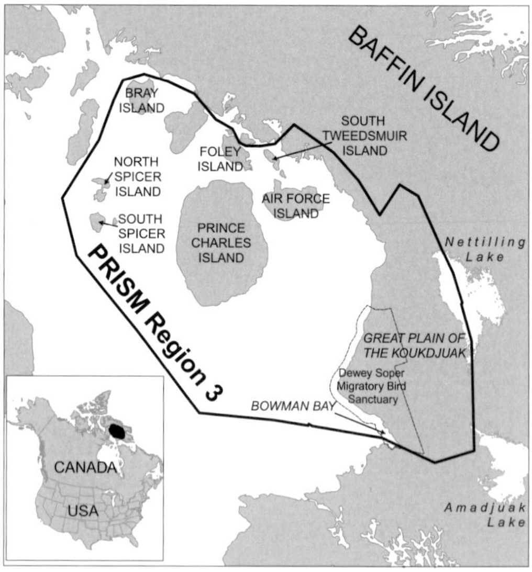

Figure 7.1. Arctic PRISM Region 3.

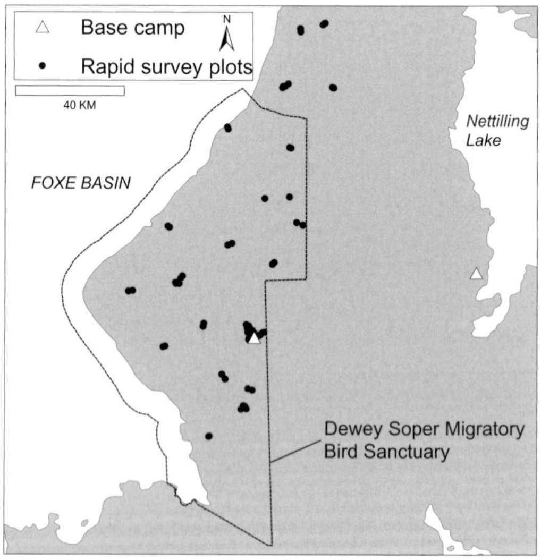

Figure 7.2. Camp and plot locations on western Baffin Island, 2003 and 2004.

Our work in the region spanned two periods. Prince Charles and Air Force Islands were surveyed during 19 June–16 July 1996 and 24 June–17 July 1997 (Johnston and Pepper 2009). Baffin Island was surveyed during 15 June–10 July 2003 and 16 June–10 July 2004. On Prince Charles Island, camps were established in the northwestern (1996) and east-central (1997) parts of the island (Fig. 7.2). On Baffin Island, camps were located in the Great Plains of the Koukdjuak, south of the Koukdjuak River (2003)

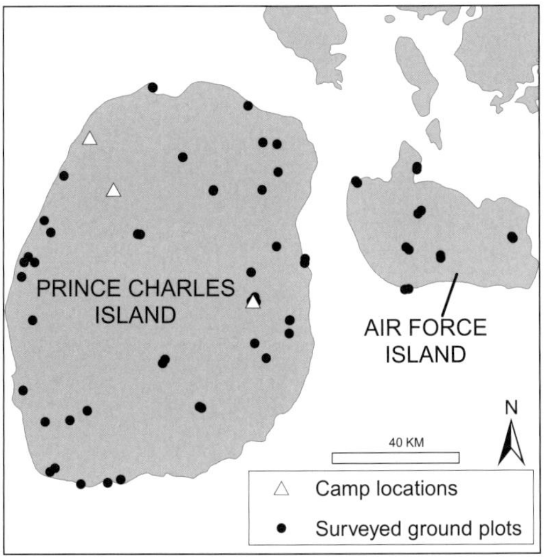

Figure 7.3. Camp and plot locations on Prince Charles and Air Force Islands, 1996 and 1997.

and further inland, east of the Great Plains near Nettilling Lake (2004; Fig. 7.3). Weather data were recorded daily at both locations using hand-held instruments at Prince Charles Island, and a small Davis Instruments weather station on Baffin Island.

Previous Work in the Region

Reports on the avifauna of the area include numerous and ongoing studies of Lesser Snow Geese (Kerbes 1969, D. Caswell and J. Leafloor, unpubl. data) and quantitative and qualitative surveys of shorebirds (Soper 1940).

The 1996–1997 surveys on Prince Charles and Air Force Islands were later incorporated as part of the arctic-wide PRISM surveys and are discussed further in Johnston and Pepper (2009). The studies of breeding behavior and rates of predation carried out by Jean-Louis Martin and colleagues remain largely unpublished (but see Smith 2009, Smith et al. 2010).

PRISM Survey Strata

The PRISM sampling protocol is described in Bart et al. (chapter 2, this volume). On Prince Charles and Air Force Islands, as well as on Baffin Island, we defined three habitat strata: vegetated lowlands/wetlands, vegetated uplands, and sparsely vegetated or barren uplands. These can be interpreted in general as good, medium, and poor shorebird habitat (Table 7.1). For both areas, larger bodies of water were an additional stratum, but are not included in analyses and were removed from the "surveyable area" in Table 7.1. The methods used to delineate these strata differed between areas, due to the quality of GIS data available.

For Prince Charles and Air Force Islands, a supervised habitat classification was available that was designed specifically for shorebird surveys (Morrison 1997). Because this classification did not cover the eastern 30% of Air Force Island, we estimated coverage of each habitat type in the eastern portion by extrapolating proportionally from the remainder of the island. Overflights of the eastern section of the island confirmed that it was similar in habitat composition to the western portion, with the exception of a unique rock outcrop habitat on the Fee Peninsula.

Morrison's classification was based upon a July 1985 Landsat 7 image and ground control points that were surveyed in July 1989. The classification aggregated a number of habitat types into four broad classes similar to those used in PRISM: water, wet graminoid habitats, vegetated tundra, and barren habitats (see Morrison 1997 for details).

A supervised habitat classification was available for a small portion of western Baffin Island at the time of our study, and its 22 classes were aggregated into the four PRISM classes (good, medium, poor, water) and overlaid on an orthorectified LandSat7 false-color composite. The relationships between the existing four classes and the false-color pixel values were visually assessed and applied to the rest of the false-color image beyond the area of overlap.

For Baffin, Prince Charles, and Air Force Islands, stratum sizes were estimated in ArcGIS (ESRI 2005) by overlaying the region with a plot-sized grid. Plots with 70% or greater water coverage were excluded, and the sample of potential survey plots comprised the remaining plots. We used classification rules to assign plots to a single stratum: plots containing greater than 50% of wetland habitat types were designated as good, plots with greater than 50% of poor habitat types were designated as poor, and the remainder were designated as medium. The area of the plots assigned to each stratum was summed to provide the region-wide estimates of stratum size (Table 7.1).

TABLE 7.1

Area, description of strata, and number of plots surveyed on Prince Charles and Air Force Islands (PCI/AFI), western Baffin Island (WBI), and PRISM Region 3 as a whole.

		Stratum size (km^2)		
Stratum	Description	Baffin (no. of plots)	PCI/AFI (no. of plots)	Region 3 (no. of plots)
Good	Wetlands, moist grasslands	4,559 (32)	5,039 (31)	9,598 (63)
Medium	Vegetated uplands, heaths, drier grasslands	2,904 (17)	50 (1)	2,954 (18)
Poor	Sparsely vegetated uplands, barren areas, bare gravel	301 (1)	3,961 (17)	4,262 (18)

Survey Plots

To carry out the double sampling method employed by PRISM, a large number of plots are selected for rapid, helicopter-supported surveys, and a smaller number are selected for intensive surveys ("rapid plots" and "intensive plots", respectively). At Baffin Island, rapid and intensive plots were 16 ha (400m × 400m) in 2003 and 12.25 ha (350m × 350m) in 2004. At Prince Charles and Air Force Islands, all rapid plots were 16 ha (400m × 400m), and no intensive plots were surveyed.

To select the plots, we divided the region into plot-sized cells, aggregated these into similar sized units called zones, and selected two to four plots randomly from within the zone. Plots within a zone may be comprised of different habitat types, and thus may be in different strata. Zones were selected using stratified random sampling, with more sampling effort placed in wetland areas where shorebird densities were expected to be highest. However, because plots within a zone are randomly selected, the ratio of good, medium, and poor PRISM classes within plots cannot be controlled entirely. To keep the habitat class within plots as uniform as possible, and to avoid inclusion of large water bodies, some plot boundaries were manually adjusted.

On Baffin Island, zones were 2.1 km × 2.1 km and contained a sample of 36 possible plots (note that plot size was different in 2003 vs. 2004). Within selected zones, two (2003) or two to four (2004) plots were drawn randomly for sampling. In 2003, we sampled 17 groups of two plots, with 23 plots in the good stratum, 10 plots in the medium stratum, and one in the poor stratum. We sampled an additional 14 plots in wetlands near the intensive camp. Because these plots were not randomly drawn from the region, they are not included in the region-wide estimates of density or population size. In 2004, we sampled 16 plots from five groups: nine plots from the good stratum and seven from the medium stratum (Fig. 7.2).

The surveys on Prince Charles and Air Force Island were completed before PRISM was developed. The plot locations were simply a random sample stratified by habitat type. This sample of plots was later assigned to our three habitat strata using the methods described above. The "zones" were therefore plot-sized. Plots were surveyed in 1996, however many of these were repeated in 1997 and given the poor weather in 1996, we excluded all 1996 data from the analysis. The sample consisted of 49 plots (Fig. 7.3). Of these, 31 were in the good habitat stratum, 1 medium, and 17 poor.

Intensive plots were selected non-randomly because of logistical constraints. At Baffin Island in 2003, four intensive plots were selected non-randomly near the camp; three were in wetlands (good stratum) and one was in vegetated uplands (medium stratum). On Baffin Island in 2004, four wetland plots were selected near camp (good stratum). No intensive surveys were carried out on Prince Charles Island.

Rapid Surveys

Rapid surveys were conducted by two observers, separated by 25 m, walking line transects, covering 7–10 ha per hour. We recorded all individual birds seen or heard on the plot; observers communicated with each other to avoid double-counting. Additional details of the methods used are presented in Smith et al. (chapter 6, this volume, Rapid Surveys).

Intensive Surveys

Throughout the late courtship and early incubation stage, single observers visited plots every one to two days and spent approximately four hours per visit. During the visits, we mapped territories and attempted to find and mark all shorebird nests. In addition, we covered each intensive plot by rope dragging three times in 2003 and twice in 2004. Rope drag teams consisted of three people, dragging a 25-m-long, 0.6-cm-diameter rope with dropper lines of thin cord tied along its length. In 2003, eggs were floated to estimate developmental stage and followed throughout the breeding season to monitor hatching success (Liebezeit et al. 2007). See Smith et al. (chapter 6, this volume, Intensive Surveys) for more details.

Regional Population Estimates

Complete details of the population estimation procedure are found in Bart et al. (chapter 2, this volume). In this analysis, a single detection ratio ($\pm$SE) of 1.27 $\pm$ 0.21 is used for all species. The detection value is based on data from across the Canadian arctic.

Lake Surveys for Loons

At Baffin Island in 2004, surveys of nearby lakes were conducted to gain more information on nesting habitat preferences and food sources of Pacific, Red-throated, and Common Loons. Crews surveyed all lakes within 3 km of their camp and recorded size, depth, and habitat characteristics. Lakeshores were searched for nests, and habitat descriptions of nests were recorded. Minnow traps were deployed in a subsample of ponds.

Other Data

The study of Prince Charles and Air Force Islands, 1996–1997, produced a wide variety of general natural history and phenological data (see Johnston and Pepper 2009). For Prince Charles Island, data on breeding behavior of shorebirds, rates of predation, and the effects of local weather events is also available (J.-L. Martin, unpubl. data).

On Baffin Island, intensive survey crews completed 84 Canadian Wildlife Service NWT/Nunavut bird checklist surveys. Another 141 checklists were completed on Prince Charles and Air Force Islands. A plant collection was made at the Baffin Island intensive camps. Data from the checklist surveys contribute to PRISM Tier 3 (Armer et al., chapter 12, this volume).

RESULTS

Weather and Timing of Breeding

Weather at the camps on Prince Charles Island differed markedly between 1996 and 1997. In 1996 the survey period was cooler, sunnier, and less windy than in 1997 (Table 7.2). Snow cover in 1996 was greater than 90% upon arrival to the field site (16 June), and persisted until the end of June, when rapid melting caused flooding over much of the island. In contrast, there was very little (less than 5%) snow cover upon arrival (24 June) in 1997, and meltwater had already drained off of the land. In 1996, the pre-lay period was much colder than the 26-year mean, while in 1997 this period was much warmer than the 26-year mean. Temperatures during the incubation period were not significantly different from the 26-year mean in either year (Johnson and Pepper 2009). Data from J.-L. Martin suggest that timing of laying on Prince Charles Island was as much as two weeks later in the cold, late snowmelt year of 1996, as compared to the warm, early snowmelt year of 1997 (Table 7.3).

The spring and summer of 2004 were considered late across most arctic and boreal regions. Our weather data from Baffin Island are incomplete for 2004 (Table 7.2), but from these data, and from discussions with goose biologists familiar with the area, we suspect that breeding was not as delayed in 2004 at the west Baffin Island site as it was at many other arctic locations. A comparison between our 2004 weather data and data from Environment Canada's nearby Dewar Lakes weather station supported this suspicion.

At Baffin Island in 2003, the mean date of nest initiation was 21 June, and 12 of the 19 nests monitored were initiated between 15 and 30 June (Table 7.3). In 2004, eggs were not floated

TABLE 7.2

Weather conditions on Prince Charles Island (PCI)[a] and on West Baffin Island (WB)[b] during the survey periods.

Year	Daily mean minimum temperature (°C)	Daily mean maximum temperature (°C)	Mean daily wind speed (km/hr)	Mean daily cloud cover (%)
1996 (PCI)	1.9	7.8	5.9	20.6
1997 (PCI)	5.0	9.3	19.1	54.0
2003 (WB)	6.3	12.0	19.3	n/a
2004 (WB)	4.7	15.9	8.6	n/a

[a] 19 June–16 July 1996 and 24 June–17 July 1997. Data from J.-L. Martin, unpublished.

[b] 21 June–8 July 2003 and 17 June–3 July 2004.

TABLE 7.3

Nest initiation dates at Prince Charles and Baffin Islands, observed directly, back-calculated from hatch, or estimated by floating eggs.

	Prince Charles Island		Baffin Island
Species	1996 ($n = 9$)	1997 ($n = 68$)	2003 ($n = 19$)
Black-bellied Plover	—	14 June	16 June
American Golden-Plover	—	16 June	—
Ruddy Turnstone	29 June	15 June	—
Dunlin	—	—	22 June
White-rumped Sandpiper	—	15 June	10 June
Semipalmated Sandpiper	—	—	14 June
Red Phalarope	30 June	15 June	13 June

to estimate nest age, and our data for timing of breeding are less informative. Of 14 shorebird nests monitored, ten were initiated no later than 20 June. White-rumped Sandpipers were seen copulating on 17 June, and the first shorebird nests were found with one and two eggs on 18 June.

Rapid Surveys

The mean duration of a rapid plot survey at Prince Charles and Air Force Island was 150 minutes, which is substantially longer than a typical PRISM rapid plot. At Baffin Island, survey times were more typical, and were similar for the 16-ha plots in 2003 and the 12-ha plots in 2004 (mean duration: 96 minutes and 106 minutes, respectively).

On Prince Charles Island, we recorded 11 species of shorebirds and 16 other species on the rapid surveys, with 2,700 bird sightings in total. Snow cover was heavy at the beginning of our surveys in 1996, and we were forced to select plots in snow-free areas. The densities of birds in these areas were much higher than normal breeding densities, and consequently, we used only 1997 data to derive the estimates of habitat-specific densities and regional population sizes. In the 49 plots surveyed in 1997, we recorded 497 indicated pairs of shorebirds (Table 7.4).

TABLE 7.4

Total number of indicated pairs of birds recorded on rapid surveys at Prince Charles Island (PCI), Air Force Island (AFI), and western Baffin Island (WBI), Nunavut.

For PCI/AFI, only the 1997 data are presented.

	Total shorebirds			Total non-shorebirds	
Species	PCI/AFI (49 plots)	WBI (50 plots)	Species	PCI/AFI (49 plots)	WBI (50 plots)
BBPL	7	2	RTLO	2	—
AMGP	7	2	PALO	2	—
RUTU	29	—	SNGO	1,162	488
REKN	2	1	BRAN	22	—
WRSA	193	83	CAGO	8	46
PESA	4	—	NOPI	1	—
PUSA	2	—	KIEI	8	6
DUNL	11	6	LTDU	9	—
REPH	242	194	LTJA	11	4
			PAJA	2	—
			POJA	13	—
			HERG	—	9
			SAGU	59	22
			ARTE	5	—
			WIPT	2	—
			HOLA	—	1
			LALO	99	42
			SNBU	6	—
Total	497	288	Total	1,411	617

Red Phalaropes were the most commonly recorded shorebird, followed by White-rumped Sandpipers, representing 49% and 39% of all shorebird detections, respectively. Among the other species, Lesser Snow Geese were extremely abundant on Air Force Island, and Sabine's Gulls and Lapland Longspurs were also common on Prince Charles and Air Force Islands (Table 7.4).

Our estimates suggest that the density of shorebirds on Prince Charles and Air Force Islands in 1997 was highest in the medium-quality habitats (Table 7.5), but only a single plot from this stratum was sampled. Our samples of plots from the wetland (good) and sparsely vegetated (poor) strata were adequate, and we also observed extremely high densities of shorebirds in these habitats. We observed as many as 33 White-rumped Sandpipers and 28 Red Phalaropes in single 16-ha plots.

Shorebird densities on western Baffin Island were lower than on Prince Charles and Air Force Islands. Overall, 905 indicated pairs of birds were recorded; 288 of these were shorebirds, comprising six species (Table 7.4). As on Prince Charles and Air Force Islands, Red Phalarope and White-rumped Sandpiper were the most abundant species, representing 67% and 29% of shorebird detections, respectively. Though relatively common on Prince Charles and Air Force Islands,

TABLE 7.5

Number of shorebirds recorded on the surveys, and the estimated density of shorebirds (individual birds/km^2) in the three habitat strata on Prince Charles and Air Force Islands and on western Baffin Island.

	Prince Charles and Air Force Islands[a]						Western Baffin Island					
	Good		Medium		Poor		Good		Medium		Poor	
Species	*n* obs.	Birds/km^2	*n* obs.	Birds/km^2	*n* obs.	Birds/km^2	*n* obs.	Birds/km^2	*n* obs.	Birds/km^2	*n* obs.	Birds/km^2
BBPL	3	1.21 ± 0.74	1	11.23 ± 1.86	3	3.08 ± 2.35	1	0.46 ± 0.450	1	0.55 ± 0.58	0	0
AMGP	1	0.55 ± 0.55	0	0	6	2.12 ± 1.25	2	1.22 ± 1.02	0	0	0	0
RUTU	19	6.37 ± 3.33	2	22.46 ± 3.72	8	4.90 ± 3.14	0	0	0	0	0	0
REKN	2	1.20 ± 1.21	0	0	0	0	1	0.91 ± 0.960	0	0	0	0
WRSA	101	47.30 ± 20.41	10	112.31 ± 18.60	82	58.48 ± 36.83	50	18.83 ± 6.16	33	22.11 ± 8.48	0	0
PESA	3	0.71 ± 0.50	0	0	1	0.06 ± 0.070	0	0	0	0	0	0
PUSA	2	0.37 ± 0.38	0	0	0	0	0	0	0	0	0	0
DUNL	8	3.95 ± 2.77	0	0	3	1.18 ± 0.79	0	0	6	3.40 ± 2.86	0	0
REPH	187	71.68 ± 21.65	2	22.46 ± 3.72	53	40.32 ± 19.25	126	42.15 ± 11.98	68	39.59 ± 12.99	0	0
All	326	133.34 ± 30.11	15	168.46 ± 22.36	156	110.14 ± 41.77	180	63.57 ± 16.54	108	65.65 ± 15.78	0	0

[a] Data collected on Prince Charles and Air Force Islands in 1996 were not used in these analyses.

Ruddy Turnstones were not observed in the plots on Baffin Island. Lesser Snow Geese, Cackling Geese, and Lapland Longspurs were the most common non-shorebird species in our sample. Densities of shorebirds were similar in the habitats that we classified as good and medium (Table 7.5).

Intensive Surveys

In 2003 at western Baffin Island, we surveyed three 16-ha intensive plots in wetland habitats and one in vegetated uplands. From 23 June to 9 July, single observers made eight visits to the intensive plots, for an average of 29.2 h of observation per plot. Three rope drags were conducted on each plot, requiring an average of 7.1 person-hours per survey (21.3 person-hours per plot). These were independent from the single observers with no shared information.

In total, 20 nests and seven additional territories were found, including 19 of the White-rumped Sandpiper, five of the Red Phalarope, one of the Black-bellied Plover, one of the Dunlin, and one of the Semipalmated Sandpiper. Our ability to find nests was hindered somewhat by a high rate of nest predation and an early start to nesting. Eight nests were known to be active for two or more visits by single observers, and eight nests were active for only a single rope drag. An additional two nests were depredated before the first rope survey. One Red Phalarope nest found by a single observer consisted of one egg laid on moss with no scrape, attended by an apparently unpaired female for several days. This nest was later abandoned and was not included in the above total or in analyses.

Nest success in 2003 was less than 40%. While we were present, 12 of 20 nests failed, seven were still incubating when we left the area, and a single nest had hatchlings. We estimated a hatch success of 7% for all species combined using the Mayfield method, but with only 92 exposure days, the precision of this estimate is poor (daily survival rate $\pm$SE $= 0.88 \pm 0.03$, Mayfield 1961).

On Baffin Island in 2004, we monitored four wetland plots between 16 June and 10 July, with an average of 17.4 h of search effort per plot by single observers. In addition, the plots were covered by three-person rope drag teams twice. A total of seven White-rumped Sandpiper, five Dunlin, and one American Golden-Plover nests or territories were found. Most nests were still active when we stopped monitoring them on 7 July, and this, coupled with a small sample size, made derivation of an accurate estimate of nest success difficult.

No intensive plots were monitored on Prince Charles or Air Force Island. However, in the study of shorebird breeding ecology conducted by Jean-Louis Martin, the Mayfield estimate of nest success for a 21-day incubation period was 70% in 1996 and 83% in 1997 (all species combined, 484 and 566 exposure days, respectively).

Rapid Surveys of Intensive Plots

In 2003, the intensive plots were surveyed rapidly twice: 22–23 June and 6–8 July. These surveys yielded substantially different results for the estimates of the number of birds breeding in the intensive plots. The first rapid survey recorded 18 pairs of Red Phalaropes, while the second survey recorded only two. However, intensive surveyors concluded that there were only five phalarope territories within the plot (Table 7.6). Similarly, the first rapid survey recorded five American Golden-Plover, and none were recorded in either the second rapid survey or through the intensive surveys. Intensive surveyors found a single Black-bellied Plover nest on the border of plot number three; this bird was not recorded on either of the rapid surveys of this plot. The detection ratio (estimated/actual) was 1.96 for the first round of rapid surveys and 0.52 for the second.

In 2004, the intensive plots were surveyed rapidly only once, on 20 June. For White-rumped Sandpiper and Dunlin, more pairs were recorded on rapid surveys than nests or territories were found (Table 7.7). For all shorebirds, the detection rate was 1.77.

Population Estimates

Our estimates suggest that 1.6 million shorebirds breed in PRISM Region 3 (Table 7.8). The most abundant species in the region were the Red Phalarope and White-rumped Sandpiper, together contributing nearly 1.5 million individuals (91% of the total estimated population for all shorebirds). Ruddy Turnstones were the next most abundant species on Prince Charles and Air Force Islands, but none were counted in plots on Baffin Island, leading to a population estimate of zero for the latter area. The CVs of

TABLE 7.6

Number of pairs of birds estimated to be breeding in the intensive plots based on two (A and B) rapid surveys (estimated), and on intensive surveys (actual), at western Baffin Island in 2003.

	Estimated (A/B)						Actual					
Plot	BBPL	AMGP	SESA	WRSA	DUNL	REPH	BBPL	AMGP	SESA	WRSA	DUNL	REPH
1	0/0	1/0	0/0	8/3	2/0	5/2	0	0	0	4	0	2
2	0/0	1/0	0/0	5/5	3/0	8/0	0	0	0	8	0	1
3	0/0	0/0	1/0	3/1	0/0	1/0	1	0	1	2	0	0
4	0/0	3/0	0/0	7/3	1/0	4/0	0	0	0	5	1	2
Mean total	0	2	1	17.5	3	10	1	0	1	19	1	5

TABLE 7.7

Number of pairs of birds estimated to be breeding in the intensive plots based on rapid surveys (estimated), and the number of pairs found through intensive surveys to have nests or territories within plot boundaries (actual), at western Baffin Island in 2004.

	Estimated			Actual		
Plot	AMGP	WRSA	DUNL	AMGP	WRSA	DUNL
A	1	8	1	1	2	1
B	0	7	4	0	2	1
C	0	0	1	0	1	1
D	0	1	0	0	2	2
Total	1	16	6	1	7	5

our population estimates are large for all but the most common species; the PRISM program is designed to ensure acceptable precision of range-wide, but not necessarily regional, estimates of population.

Lake Surveys for Loons

On Baffin Island, 81 water bodies were surveyed. Pacific Loons were present on 25% (20 lakes) and nested on 10% (eight lakes), and Red-throated Loon and Common Loon were observed on one water body.

In brief, lakes having Pacific Loons tended to be medium/deep rather than shallow, and have emergent vegetation and islands. No analysis was possible for Red-throated Loon or Common Loon as only one of each species was observed.

We found ten Pacific Loon nests on Baffin Island in 2004; six were on islands, two on peninsulas, and two on shorelines. The frequent use of islands is consistent with findings from other studies. Nests were also typically oriented to the south or east, and sheltered from the prevailing winds and waves.

In 2004, minnow traps were set on six lakes used by Pacific Loons and four lakes on which no loons had been seen. Of the six Pacific Loon lakes, sticklebacks or unknown minnows were recorded in all six, with an average of 28.8 minnows per lake per 24-hour trapping period. Of the four non-loon lakes, only half had minnows, with an average of two minnows per lake per period. Of the 181 minnows captured, the vast majority (173, or 96%) were sticklebacks, and the remainder are yet to be identified.

TABLE 7.8

Estimated population size, measured in individual birds ± SE, for Prince Charles Island, Air Force Island (PCI/AFI), and western Baffin Island, Nunavut.

The coefficient of variation (CV) is also presented for each estimate.

	PCI/AFI		Western Baffin Island	
Species	Est. pop. ± SE	CV	Est. pop. ± SE	CV
BBPL	19,118 ± 9,567	0.50	3,698 ± 2,694	0.73
AMGP	10,927 ± 5,694	0.52	5,327 ± 4,356	0.82
RUTU	53,546 ± 23,919	0.45	—	—
REKN	6,158 ± 6,303	1.02	3,971 ± 4,145	1.04
WRSA	477,220 ± 188,304	0.39	150,517 ± 44,714	0.30
PESA	3,891 ± 2,617	0.67	—	—
PUSA	1,917 ± 1,954	1.02	—	—
DUNL	24,766 ± 14,609	0.59	10,467 ± 9,259	0.88
REPH	523,825 ± 154,194	0.29	306,498 ± 75,560	0.25
All	1,121,368 ± 245,344	0.22	480,479 ± 88,531	0.18

Other waterbirds observed on the surveyed lakes include Long-tailed Ducks (22 lakes), King Eiders (12), Herring Gulls (5), Arctic Terns (4), and Long-tailed Jaegers (3).

DISCUSSION

Shorebird Densities and Distribution

The most striking result of our surveys in PRISM Region 3 was the high density of shorebirds on Prince Charles and Air Force Islands. Our estimate of density in good habitats exceeds all published estimates of shorebird density in the Canadian arctic archipelago and is far higher than what the authors have seen elsewhere in this part of the arctic (see Johnston and Pepper 2009 for comparison with other arctic sites). Only in the Yukon Territory and Alaska have higher densities of shorebirds been reported. The high densities estimated for medium-quality habitat may in part be due to sampling error, but the number of birds observed in all plots on these islands leaves no question that the area is important for shorebirds.

A large proportion of the shorebirds observed on Prince Charles and Air Force Islands were White-rumped Sandpipers and Red Phalaropes. Although data are limited, the densities of these and other polygamous species are thought to be more variable than those of monogamous species (Meltofte 1985, 2007b, and references therein). In a nine-year study at East Bay, Southampton Island, densities of Red Phalaropes and White-rumped Sandpipers ranged from 0 to 36 birds/km^2, and 7 to 21 birds/km^2, respectively (P. A. Smith, pers. obs.). The cause of these fluctuations is not yet understood. It is interesting to note that Jean-Louis Martin (pers. comm.) reported hatching success of 83% for shorebirds, primarily Red Phalaropes and White-rumped Sandpipers, on Prince Charles Island in 1997. This nest success is also much higher than reported elsewhere for these species in the Canadian arctic (Smith 2009), and the high densities and high nest success may reflect atypically good breeding conditions. A return visit to Prince Charles Island is warranted to determine whether the densities observed by us are typical or were the product of "irruptive years" for these polygamous shorebirds.

Our results also provide insight into the limits of the range of some species, including Semipalmated Plover and Dunlin. Semipalmated Plover is a common breeder at Southampton Island (Smith et al. 2007a), less than two degrees of latitude south of the southern reaches of Region 3. In the western part of their range, they breed as far north as Rasmussen Lowlands and Victoria and Banks Island (NWT-Nunavut Checklist Survey Database, Johnston et al. 2000). Fewer than five were seen during our surveys (outside of plots), suggesting that this area may be at the northern range limit for this species. The species has also been recorded at Clyde River, but with no evidence of breeding (V. Johnston, pers. obs.). In contrast, Dunlin were present throughout the region during our surveys; these sightings and those from Sarcpa Lake represent the most northerly breeding records for the eastern Canadian arctic (Montgomerie et al. 1982, 1983). No Dunlin were observed on Prince Charles Island during surveys in 1989, suggesting that they may be present only in some years, as is the case at East Bay, Southampton Island (Morrison 1997, Smith 2009). Alternatively, the species may be a newly established breeder in the region (Johnston and Pepper 2009). The first breeding record for Dunlin on Baffin Island was in 1986 (Martin et al. 1988).

We failed to observe any Ruddy Turnstones during our surveys on Baffin Island, but they were common on Prince Charles Island. From our transits of the area, it appeared that the beach ridge–wetland habitat that turnstones favored on Prince Charles Island was almost entirely missing from the coastal portion of the Great Plains of the Koukdjuak (Johnston and Pepper 2009). Alternatively, their apparent absence from the area could simply be a result of a small sample size of coastal plots. Ruddy Turnstones in some arctic locations are semi-colonial (Perkins et al. 2007) and associated with the coast (Johnston and Pepper 2009). If a species has a clumped distribution, our small sample of coastal plots is problematic.

We were also surprised by the observation of similar densities of shorebirds in high- and medium-quality habitats on Baffin Island. Many of the areas identified as good habitat on satellite imagery were in fact heavily grazed by Snow Geese, and our observations may instead reflect depressed densities in the "good" areas. We recorded the density of goose nests on the survey plots, and a more thorough analysis of the relationship between goose density and shorebird density is warranted.

Population Estimates

Our population estimates demonstrate that the shorebird community of the region is dominated by Red Phalaropes and White-rumped Sandpipers. The population estimate for White-rumped Sandpipers may be artificially inflated by a high density in the undersampled medium strata (see below), but notwithstanding, an extremely large number of both species were present in the region during our surveys. Our estimates for Red Phalaropes and White-rumped Sandpipers in this region comprise the majority of their range-wide populations (66% and 56% of range-wide estimates, respectively; Morrison et al. 2006). Most other shorebirds recorded were Ruddy Turnstones, Dunlin, and Black-bellied Plovers. Our population estimates for these species still represent a significant proportion of the range-wide estimates presented in Morrison et al. (2006), but not to the extent of the species noted above (Ruddy Turnstone: 23%, Dunlin: 2%, Black-bellied Plover: 11%).

Semipalmated Sandpiper and Pectoral Sandpiper are typically abundant birds where they are present, but this was not the case in our surveys. Our only observations of Semipalmated Sandpipers were off-plot; consequently, we estimated a population of zero for this species in this region. Semipalmated Sandpipers breed as far north as Creswell Bay, Somerset Island (Latour et al. 2005). They have been observed breeding on Igloolik Island, but do not appear on Bylot Island (Forbes et al. 1992, L. McKinnon, pers. comm.). The northeastern limit of their range has not been established. Similarly, we estimated a population of zero for Pectoral Sandpipers, although we noted their presence off-plot on Baffin, Prince Charles, and Air Force Islands. This species is common at sites of this latitude further west, including the Boothia Peninsula (Patterson and Alliston 1978). It is much less common, however, in the eastern arctic.

Factors Affecting Accuracy of the Surveys

The PRISM population estimates for shorebirds on Prince Charles and Air Force Islands are 1.9 to 2.6 times higher (with the exception of Dunlin, which was only half as large) than Johnston and Pepper's (2009) estimates. Both studies used three broad habitat categories: wet graminoid lowlands, vegetated tundra, and unvegetated barrens. Johnston and Pepper (2009) found densities of shorebirds to be much higher in the wet lowlands than in the latter two categories. Our classification assigned a single plot to the medium stratum and 14 plots to the poor stratum. Because the poor plots contained many shorebirds, and the area classified as poor was fairly large, a substantial fraction of the population estimates presented here come from habitats classified as poor. Thus, for these islands, the GIS-based classification did a poor job of distinguishing good- from poor-quality habitat, but for other arctic regions it did quite well (e.g., Region 8 [Queen Maud], Coats Island in Region 4 [Southampton], J. Rausch unpubl. data). Although a less than optimal stratification will produce a larger than desirable SE, it does not produce a biased estimate and thus does not account for the difference between studies. The coefficients of variation (CVs) for population estimates are also larger in this study; for the six most abundant species, the average CV was 0.46, or 2.25 times larger than CVs in Johnston and Pepper (2009). This difference may in part relate to differences in habitat classifications, but also reflects the large differences in analytic methods, such as consideration of variability in rates of detection in this study, but not Johnston and Pepper (2009). Regardless, the discrepancies between the two studies highlight the fact that PRISM is designed to produce unbiased and precise arctic-wide estimates and that more intensive region-specific surveys might be needed for regions of particular importance.

Another critical component of the PRISM program is the detection rate. The ratio of birds observed to those actually nesting is used to correct the survey results and to generate a population estimate. We used a single detection rate of 1.27, which is based on data from many species across arctic Canada. For Baffin Island in 2004, our detection rate of 1.77 was somewhat higher than this value, although imprecision of the estimates precludes formal testing. For the 2003 plots, the detection rate was high in the first round of surveys (1.96, 22–23 June) and low in the second (0.52, 6–8 July). This large difference may in part be due to the high rate of depredation. Anecdotally, over half of the nests that we monitored in 2003 were lost to predators, and some nests failed quite quickly. If during the first round of rapid surveys we counted individuals whose nests failed before they could be found by our intensive surveyors, our detection rates would tend to be more than 1 on the first round of rapid surveys. Likewise, if nests that were found by intensive surveyors then failed prior to the second round of rapid surveys, and those individuals left the study areas to prepare for migration along the coast (as is common in shorebirds), then the detection rate would tend to be less than 1 on the second round of rapid surveys.

With a sample of only four intensive plots, it is difficult to understand the true source of the discrepancy. However, it is clear that two rapid surveys of intensive plots are desirable; the overall detection rate for both surveys at western Baffin Island was 1.24, vs. 1.27 for the Canada-wide detection rate. The timing of these surveys also requires careful consideration; it would appear that 6–8 July is too late for the second round of rapid surveys, but an objective rule (e.g., based on snowmelt, see also Smith et al. 2010) is required. It is also critical to find all or nearly all nests present. On Baffin Island in 2003, we found several nests using the rope drag technique that would have otherwise been missed. The value of rope dragging in other situations, and the fraction of nests that observers can expect to discover, is explored in Smith et al. (2009).

CONCLUSIONS AND RECOMMENDATIONS

The densities of birds encountered on Prince Charles and Air Force Islands were extremely high, which is consistent with the findings of previous studies there (Morrison 1997, Johnston and Pepper 2009). A return trip is warranted to determine if these numbers have been maintained when so many species are showing declines in migration surveys. The potential for a long-term (PRISM Tier 2) field site on Prince Charles Island could also be evaluated. In a future survey, truly barren habitat such as expanses of frost-shattered rock, far from water, which comprise 15–20% of Prince Charles Island, could be removed from strata prior to plot selection.

During our surveys on Baffin Island, we observed no Ruddy Turnstones, though they undoubtedly nest in the area. Ruddy Turnstones were also missed in parts of PRISM Region 4 (see Smith et al., chapter 6, this volume). In Regions 3 and 4 this species appears to be associated with coastal beach ridge complexes, and may have an aggregated distribution. For rare species of interest or those that occupy specific and relatively patchily distributed habitat, additional stratification, or targeted surveys, may be warranted.

Similarly, we had small numbers of plots from some habitat strata such as one medium plot on Prince Charles and Air Force Islands, leading to questionable point and interval estimates of density. While an even distribution of plots among strata is not necessarily desirable, we should strive for adequate samples from all strata. This may be difficult, however, when the GIS classification is *post hoc*. However, it should be noted that PRISM is not designed to give estimates for individual strata. If regional results are desired, more intensive habitat studies need to augment the rapid surveys.

The variability in detection rates can be large, and we have an inadequate understanding of its source. We should conduct an analysis using data from all PRISM regions to identify relationships between detection and factors such as species, timing, survey effort, and weather conditions.

Timing of the rapid surveys appears to have a large effect on detection ratios, and we could benefit from objective rules for when these surveys should take place. Meltofte et al. (2007a) and Smith et al. (2010) linked timing of breeding to snow cover. Local snow cover could be used to predict median nest initiation dates, which in turn could be used to determine optimal timing for the rapid surveys. If logistic constraints prevent flexibility in the timing of the surveys, it would still be instructive to determine what the optimal timing would have been, based on observed or predicted breeding schedules.

ACKNOWLEDGMENTS

J. Bart developed the PRISM sampling plan and assisted with analyses to estimate population size. J. Leger assisted with GIS plot selection. We thank M. Fuller and S. Earnst for reviewing a draft of this chapter. Fieldwork was funded by the Canadian Wildlife Service, and logistical support was provided by the Polar Continental Shelf Program. The rapid surveys in 2003 were led by B. A. Andres. We also thank J. Bart, S. Earnst, K. H. Elliott, G. Fernandez, M. Kotierk, H. Krumbholz, B. McBride, R. Book, D. Kaleta, and Q. Quvianaqtuliak for assistance in the field. This report is PCSP Contribution No. 00609.

ONLINE CONTENT

Abstracts are available in French and Spanish from www.ucpress.edu/go/sab.
Une traduction du résumé est disponible en français.
Una traducción del resumen está disponible en español.

CHAPTER EIGHT

Small-scale and Reconnaissance Surveys

Jonathan Bart, Brad A. Andres, Kyle H. Elliott, Charles M. Francis, Victoria Johnston, R. I. G. Morrison, Elin P. Pierce, and Jennie Rausch

Abstract. This chapter describes small-scale surveys at seven locations in arctic Canada. At Kent Peninsula, the standard double sampling method (Bart et al., chapter 2, this volume) was used to estimate densities and population sizes. Shorebird densities were low except on Melbourne Island. At the northern tip of Ellesmere Island, densities were too low for intensive plots to be practical, and some of the species became extremely cryptic once incubation started. Methods are suggested for dealing with both problems. On Somerset Island, detailed surveys were made at Creswell Bay and estimated densities and population sizes were obtained. On the rest of the island, we lacked a good habitat map and densities were extremely low. We found a few scattered shorebirds but were not able to obtain estimates of density of population size. At Québec, shorebird density was strongly related to elevation. This relationship may provide an important basis for stratification in this region, especially since detailed landcover maps are lacking. On Melville and Prince Patrick Islands, shorebirds were found in the interior of the islands and in almost completely unvegetated areas, indicating that future surveys will need to cover the islands extensively rather than just the wetlands and adjacent bare areas. On Ellesmere and Axel Heiberg Islands, a combination of aerial and ground surveys showed that while species richness was low, areas of Axel Heiberg Island and the Fosheim Peninsula had surprisingly high numbers of breeding shorebirds. Rope drag surveys proved critical in identifying nesting Red Knots. In the Kivalliq region on the west side of Hudson Bay, a combination of aerial and ground surveys documented high diversity and numbers of breeding shorebirds and other species. In addition, large numbers of passage birds were observed on the coast, though the overall importance of this region to spring migrants remains unknown. These surveys demonstrate the value of conducting small-scale reconnaissance surveys in unfamiliar regions prior to beginning the full-scale PRISM surveys to estimate density and population size.

Key Words: aerial surveys, Alert, arctic, Axel Heiberg Island, Canada, Ellesmere Island, Fosheim Peninsula, Kent Peninsula, Kivalliq, Melville Island, monitoring, Northwest Territories, Nunavut, population size, Prince Patrick Island, PRISM, Québec, range maps, reconnaissance, Red Knot, shorebirds, Somerset Island.

Bart, J., B. A. Andres, K. H. Elliott, C. M. Francis, V. Johnston, R. I. G. Morrison, E. P. Pierce, and J. Rausch. 2012. Small-scale and reconnaissance surveys. Pp. 141–156 *in* J. Bart and V. Johnston (editors). Arctic shorebirds in North America: a decade of monitoring. Studies in Avian Biology (no. 44), University of California Press, Berkeley, CA.

This chapter contains brief reports on surveys of breeding shorebirds at six locations in arctic Canada. Surveys on Kent Peninsula used the standard double sampling method, lasted for two years, and provided estimated densities and population sizes. Methods in the other surveys varied depending on shorebird density and on how large the study area was. In the Québec study area, the standard Arctic PRISM method for rapid surveys was used. In the Alert study area, hybrid methods were developed to deal with an extremely low density of shorebirds. In the other sites, investigators did not confine their surveys to predefined, randomly selected plots. We did not estimate densities or population sizes from the reconnaissance surveys, but the results provided useful new information on distributions and some indication of abundance. These preliminary surveys will also help us design large-scale surveys to estimate density and population size.

KENT PENINSULA

This study area encompassed Kent Peninsula, Melbourne Island, and the adjacent mainland from Bathurst Inlet to the western boundary of the Queen Maud Gulf Migratory Bird Sanctuary (Fig. 8.1). The study area was stratified by the normal three PRISM habitats (wet, moist, upland) and four geographic areas. Rapid surveys using the standard methods (Bart et al., chapter 2, this volume) were made on 52 randomly selected plots during 17–29 June 2001 and 19–29 June 2002. Intensive plots were established in both years but had too few shorebirds to estimate detection rates. We therefore used the Canada-wide detection ratio of 1.27 to adjust results from rapid surveys.

Surveyors recorded 88 shorebirds of 11 species (Table 8.1); 78% of the records were of the five most common species: American Golden-Plover, Dunlin, Semipalmated Sandpiper, Pectoral Sandpiper, and Stilt Sandpiper (for scientific names, see Appendix C). Densities were consistently highest in wetlands and were about equal in moist areas and uplands (Table 8.1).

Although shorebird density was low compared to other mid-arctic sites, we found high densities at Melbourne Island. Only six plots were surveyed, but 33 shorebirds were recorded, and the density of observations was more than

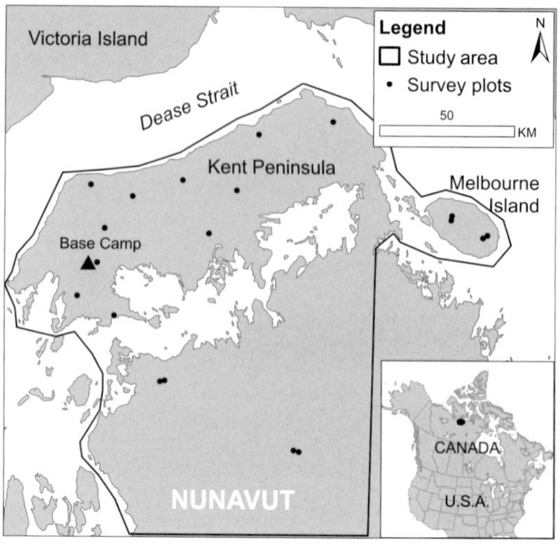

Figure 8.1. Kent Peninsula study area.

twice the density at the rest of the plots. A preliminary analysis of satellite imagery suggested that 15% of Melbourne Island is covered by wetlands, about three times more than in other parts of the study area. An important task for future surveys in the mid-arctic regions will be to identify small but habitat-rich locations like Melbourne Island that have extensive high-quality habitat.

ALERT

This study area (Fig. 8.2) covered 135 km^2 near Alert on the northeast coast of Ellesmere Island. The terrain was mainly frost-shattered rock, gravel, and bare clay with little vegetation. Most "barren" areas had less than 5% vegetation cover; "tundra" areas generally had 5–15% cover (mostly *Dryas, Salix,* and *Saxifraga*); and "wetlands," which occurred below persistent snow and ice banks, had up to 85% cover (mostly graminoids and mosses). Snow cover in spring varies between years but is usually extensive until the end of May or early June. Snow-free patches begin to occur in late May and early June, especially on south-facing slopes on higher ground, where wind action results in only a thin layer of snow.

We selected 75 plots (61 in 2001, 14 in 2007) in locations considered to be representative of the area and surveyed them during 10 June–15 July 2001 and 24–29 June 2007. Plot size varied from 1 to 50 ha; 43 covered 5–20 ha. Most

TABLE 8.1

Number recorded, estimated densities, and population sizes (with CVs) of shorebirds in the Kent Peninsula study area.

	Number by habitat type			Density (birds/km^2) (CV)			
Species	Wetlands	Moist areas	Uplands	Wetlands	Moist areas	Uplands	Population size
American Golden-Plover	1	1	9	0.18 (1.13)	0.99 (0.93)	4.13 (0.44)	29,794 (0.38)
Semipalmated Plover	0	0	1	0 (0)	0 (0)	0.43 (0.98)	2,590 (0.99)
Dunlin	8	3	1	9.21 (0.39)	2.97 (0.93)	0.37 (1.01)	23,588 (0.68)
Semipalmated Sandpiper	7	2	4	10.37 (0.63)	1.25 (1.02)	1.75 (0.76)	24,628 (0.67)
Least Sandpiper	0	2	3	0 (0)	1.34 (0.77)	2.29 (0.85)	20,239 (0.65)
White-rumped Sandpiper	1	0	1	1.11 (1.05)	0 (0)	0.27 (1.03)	2,491 (0.77)
Pectoral Sandpiper	15	3	4	16.24 (0.30)	2.22 (0.69)	2.75 (0.49)	39,822 (0.38)
Baird's Sandpiper	2	0	0	2.52 (0.52)	0 (0)	0 (0)	1,971 (0.76)
Stilt Sandpiper	3	0	2	3.47 (0.43)	0 (0)	1.86 (0.68)	13,964 (0.60)
Red-necked Phalarope	9	0	2	3.47 (1.00)	1.25 (1.02)	0 (0)	8,673 (0.84)
Red Phalarope	3	1	0	7.55 (0.65)	0 (0)	0.54 (1.03)	9,143 (0.56)
Total	49	12	27	54.12 (0.27)	10.04 (0.38)	14.40 (0.28)	176,903 (0.27)

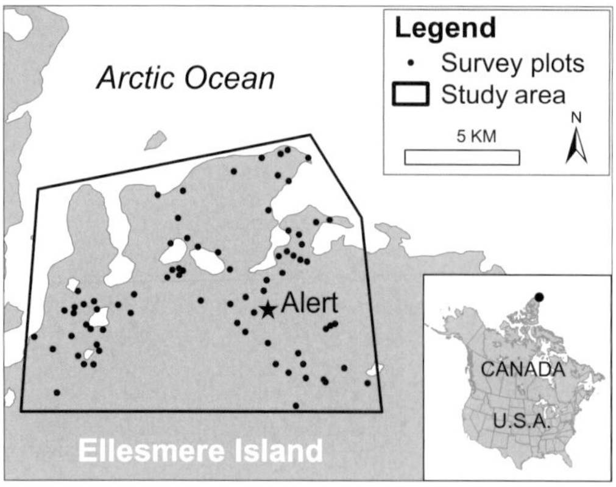

Figure 8.2. Alert study area.

surveys followed the rapid survey method but some involved rope dragging (Bart et al., chapter 2, this volume). We recorded only those birds thought to be nesting in the plot, as was done in Alaska. Due to the low density of shorebirds at this site, estimating detection rates using standard intensive plots was judged not to be practical. Instead, plots were surveyed up to six times, with more surveys on those plots thought to have more birds. The number of plots covered one, two, three, and four or more times were 38, 18, 10, and 9, respectively. Plots covered more than once were estimated to contain 30 territories, of which 25 were found after the first survey. This indicated a detection rate of 0.83 on the first visit, a value very close to the 0.81 obtained in Alaska using similar methods (see McCaffery et al., chapter 3, this volume, and Bart et al., chapter 4, this volume). This result suggests that most nesting birds were located during these surveys, so we have assumed a detection ratio of 1.0 while acknowledging that we lack rigorous proof for this absolute value.

In both years, Ruddy Turnstone, and Red Knot were common (Table 8.2). In 2001, a single Red Phalarope was recorded. In 2007, Red Phalaropes were more common and Baird's Sandpiper was also recorded. Among non-shorebirds, Snow Bunting was by far the most common species, though Long-tailed Jaegers were also regularly noted. Forty-seven of the 75 plots (63%) had bird territories (all species combined); 27 of the 75 plots (36%) contained shorebird territories. Off-plot birds were also recorded in 2007 and, as a result, this doubled the number of non-shorebird species recorded (8 vs. 4 in 2001). The density and population size of all shorebirds was estimated at 10.65/km^2 and 1,434, respectively (Table 8.3). These estimates should be treated with caution due to the lack of a formal plan for selecting plots.

Once incubation began, Red Knot and Sanderling ceased conspicuous activities on their territories, so their nests were difficult to locate. Ruddy Turnstones remained active on their territories and were easier to detect. Rope dragging did reveal nests, but generally did not reveal new nests on plots that had previously been searched. This was because densities were low and territories were initially searched during the territory-establishment period when birds were active. For example, during nine rope surveys of plots previously surveyed at least once, no new nests were discovered. However, later in incubation, and for Red Knot in particular, some nests would have been missed had we not expanded the rope dragging effort. Predation levels were recorded qualitatively as "high" during the 2001 surveys. In 2007, half the nests are known to have failed and some nests had not hatched by the time surveyors departed.

In other parts of the arctic, wet areas are often heavily used by birds. Most species occurring at Alert, however, favored "upland" types of tundra, and in the case of the Red Knot, nesting could occur in extremely barren rocky, almost unvegetated terrain. The only records of shorebirds nesting in "marsh" habitats at Alert were Baird's Sandpiper, which nested in the marsh at Kirk Lake over several years (though this species typically nests in drier habitats elsewhere), and a single Ruddy Turnstone, which nested unsuccessfully in the same marsh in 2002. Red Phalaropes and Red-necked Phalaropes have been recorded at Alert in wet areas in the spring, but no nesting records were recorded.

Although these high-arctic species prefer drier nesting habitats, presumably much of their foraging occurs in nearby wetlands. To determine how close nests were to wetlands, the distance from the nest to the nearest water body or edge of a wetland complex was recorded in 2007. The average distances to water were Sanderling (2 nests): 13.5 m, Ruddy Turnstone (8 nests): 34 m, and Red Knot (2 nests): 217 m. For all nests, the average was 61 m.

TABLE 8.2

Birds recorded on rapid surveys in the Alert study area.

Species	2001	2007	
	On plots	On plots	Off plots
Shorebirds			
Common Ringed Plover	1	0	0
Ruddy Turnstone	14	1	6
Red Knot	15	1	11
Sanderling	7	1	5
Baird's Sandpiper	0	3	1
Red Phalarope	1	1	3
Other species			
Red-throated Loon	0	0	2
Brant	0	0	3
King Eider	1	0	0
Long-tailed Duck	0	0	5
Rock Ptarmigan	2	0	0
Long-tailed Jaeger	6	3	15
Thayer's Gull	0	0	1
Glaucous Gull	0	0	3
Arctic Tern	0	0	7
Snow Bunting	43	9	7

TABLE 8.3

Estimated densities and population sizes of shorebirds in the Alert study area.

Species	Number recorded	Density (birds/km^2)	Population size	CV
Red Knot	23	4.80	647	0.30
Red Phalarope	1	0.21	28	1.03
Ruddy Turnstone	16	3.34	450	0.34
Sanderling	11	2.30	309	0.42
All species	51	10.65	1,434	—

SOMERSET ISLAND

This study area (Fig. 8.3) covered 4,359 km^2 and was divided into the Creswell Bay region 139 km^2 and the remaining area (4,220 km^2). Rapid surveys were conducted during 27 June to 5 July 2001 on 56 plots in Creswell Bay and 47 plots in the remaining area. Plots were searched a single time, in most cases by two observers walking slowly, 25 m apart, along parallel lines that completely covered the plot. We attempted to conduct intensive surveys but arrived too late for them to be effective.

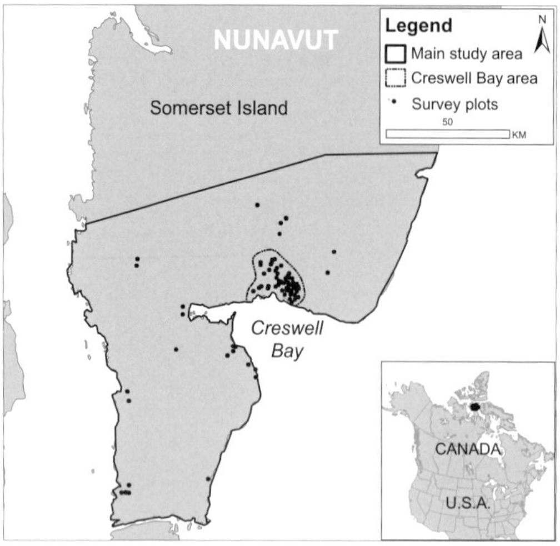

Figure 8.3. Somerset Island study area.

During the surveys at Creswell Bay, we recorded 201 shorebirds of eight species (Table 8.4). About half the birds recorded were Red Phalaropes and a third were White-rumped Sandpipers. All other species were much less common (less than 6%). The density of shorebirds was nearly 50 birds/km^2. The estimated population size was 6,751 shorebirds.

We lacked a useful landcover map for the rest of the study area, and many areas portrayed as wetlands were actually shadows. As a result, the randomly selected areas had virtually no birds. Eventually a decision was made to seek areas that might have suitable habitat and survey them to determine whether any appreciable number of birds occurred in the area. We found 28 indicated pairs of shorebirds, including Baird's Sandpipers (8), Sanderlings (7), Red Phalaropes (6), American Golden-Plovers (2), Black-bellied Plovers (2), White-rumped Sandpipers (2), and Red Knots (1). These data provide distributional information and show that shorebird populations in these areas are sparse but present. They do not provide a good basis for estimating density or population size.

We also recorded 14 species other than shorebirds. The species (and number of indicated pairs) at Creswell Bay were Lapland Longspur (36), King Eider (9), Horned Lark (7), Red-throated Loon (3), Canada Goose (1), Long-tailed Duck (1), and Parasitic Jaeger (1). On the rest of the study area the records were Lapland Longspur (15), Snow Bunting (13), Snow Goose (12), Horned Lark (6), Long-tailed Duck (3), Red-throated Loon (2), Long-tailed Jaeger (1), Pacific Loon (1), Redpoll (1), Snowy Owl (1), and Glaucous Gull (1).

QUÉBEC

This study area (Fig. 8.4) covered 19,003 km^2 on the eastern shore of Hudson Bay, including the village of Puvirnituq (60°00′N, 77°10′W). Rapid surveys were conducted during 9–17 June 2002 on 98 randomly selected 10-ha plots. We used the standard PRISM method for surveys and analyses (see Bart et al., chapter 2, this volume). Two intensive plots were established but no birds were found breeding on them, so the Canada-wide detection ratio (1.27) was used to estimate density. We did not calculate estimated densities of species other than shorebirds.

Semipalmated Sandpiper was by far the most common breeding shorebird, followed by Semipalmated Plover and Wilson's Snipe (Table 8.5). Several other species were abundant including Lapland Longspur, Canada Goose, American Pipit, American Tree Sparrow, and Savannah Sparrow.

Shorebird densities showed a strong inverse relationship with elevation. For example, the mean number of shorebirds/plot (and CV) was 0.77 (0.24) for plots less than 15 m above sea level, but only 0.20 (0.29) for plots greater than 15 m above sea level ($P < 0.01$). Plots at higher elevations were better drained and had fewer wetlands.

Our results suggest that breeding Dunlin were more abundant on the Ungava Peninsula than suggested by Warnock and Gill (1996). Breeding was confirmed for Dunlin (nest-building) and probable for American Golden-Plover (copulation). Although we found only one confirmed breeding record for the Black-bellied Plover, they may be more abundant along coastal segments of rivers on the Ungava Peninsula than reported by Paulson (1995).

MELVILLE AND PRINCE PATRICK ISLANDS

This study area (Fig. 8.5) included Melville (42,149 km^2), Prince Patrick (15,848 km^2), and Eglinton (1,541 km^2) Islands in the western Queen Elizabeth Islands. Prior to the survey, Landsat images and a vegetation map prepared by Edlund (1982) were used to select 77 clusters of three plots each (231 plots). The average plot size was 23 ha, and 80% of the plots were between 9 and 41 ha.

TABLE 8.4

Results of shorebird surveys on Somerset Island, 2001.

Species	Number recorded	Estimated density (birds/km^2)	CV	Estimated population size
Red Phalarope	97	22.32	0.22	3,092
White-rumped Sandpiper	64	15.89	0.22	2,201
American Golden-Plover	12	3.24	0.39	448
Baird's Sandpiper	9	2.51	0.37	348
Black-bellied Plover	6	1.90	0.39	262
Pectoral Sandpiper	6	1.23	0.41	170
Buff-breasted Sandpiper	5	1.03	0.53	142
Ruddy Turnstone	2	0.63	0.72	87
Total	201	48.7		6,751

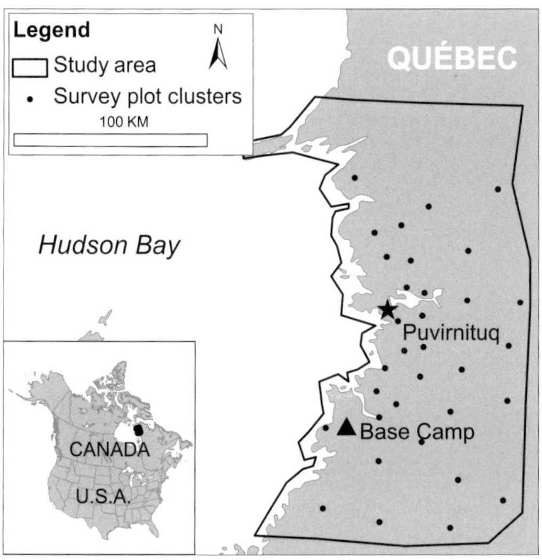

Figure 8.4. Québec study area.

Sites were concentrated in areas we thought likely to have the most birds, but a small sample of sites in the most barren areas was also selected.

Surveys were conducted during 18–29 June, 2007. We visited 208 plots in 64 clusters, including a few plots not in the original sample. When a plot was covered entirely by snow or water, we recorded zero birds for it and selected another nearby area to survey. If a plot was partially covered by snow or water, we adjusted the boundaries and later redrew the plot so that it covered the original plot plus the additional area surveyed. The survey of each plot was conducted by a single observer and lasted 30–45 minutes (longer for a few large plots located near base camps). During surveys we recorded indicated pairs and off-plot birds as usual for PRISM surveys (see Bart et al., chapter 2, this volume).

We recorded 12 shorebird species and 630 "indicated birds" (indicated pairs × 2 + birds seen off-plot), of which 36% were Red Phalarope. Species diversity was high, with ≥15 individuals of nine species recorded. An additional 97 shorebird individuals, most not identified to species, were recorded during helicopter flights.

Prior to our surveys, the avifauna of these two islands was poorly known. The results will be reported in more detail elsewhere, but here we note new information on distribution and abundance.

American Golden-Plover

The BNA range map excludes half of Melville Island and all of Prince Patrick Island, whereas our surveys show that both should be included (Johnson and Connors 1996a).

Ruddy Turnstone

The BNA range map includes all of Melville and the southeastern part of Prince Patrick Islands (Nettleship 2000). Our surveys suggest that the range may be limited to the southeastern quarter of Melville, though a few individuals could certainly occur elsewhere in the study area.

TABLE 8.5
Birds recorded on rapid surveys in the Québec study area.

Species	Number recorded	Estimated pairs	Estimated density (birds/km^2; CV)
Shorebirds			
Black-bellied Plover	1	0	0.00 (0.00)
American Golden-Plover	2	1	0.12 (1.02)
Semipalmated Plover	11	6	0.92 (0.41)
Dunlin	5	3	0.43 (0.59)
Semipalmated Sandpiper	40	23	3.59 (0.26)
Least Sandpiper	5	3	0.44 (0.59)
White-rumped Sandpiper	32	0	0.00 (0.00)
Wilson's Snipe	6	6	0.97 (0.48)
Other species			
Snow Goose	12	0	—
Canada Goose	387	134	—
Northern Pintail	8	7	—
Black Scoter	2	2	—
Long-tailed Duck	15	11	—
Rock Ptarmigan	6	4	—
Herring Gull	4	1	—
Horned Lark	48	46	—
American Pipit	134	107	—
American Tree Sparrow	84	76	—
Savannah Sparrow	70	64	—
White-crowned Sparrow	28	27	—
Lapland Longspur	304	294	—
Snow Bunting	16	7	—
Common Redpoll	5	3	—

Purple Sandpiper

The BNA range map includes eastern Melville Island and has a "?" for the rest of our study area (Payne and Pierce 2002). Our results show that this species is widely distributed throughout the study area and that it may be more common at Prince Patrick than on Melville Island.

Sanderling

The BNA range map shows this species nesting only close to the coast; our results suggest it probably also nests inland on these islands (MacWhirter et al. 2002).

White-Rumped Sandpiper

The BNA range map excludes Prince Patrick Island, whereas our results show that most, if not all, of the island should be included (Parmelee 1992b).

Baird's Sandpiper

The BNA range map includes the entire study area, whereas we found it only on Melville Island

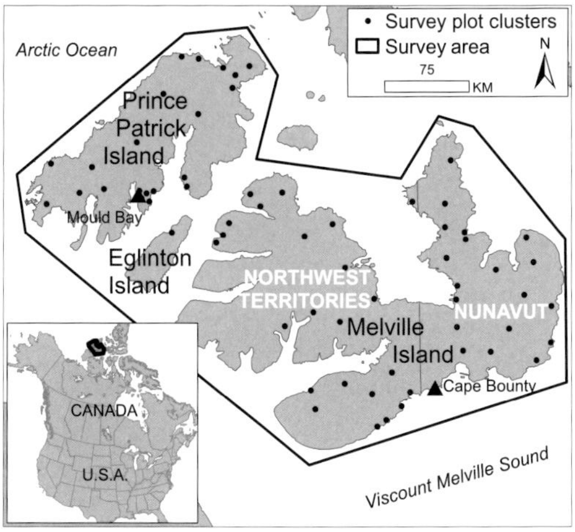

Figure 8.5. Melville and Prince Patrick Islands study area.

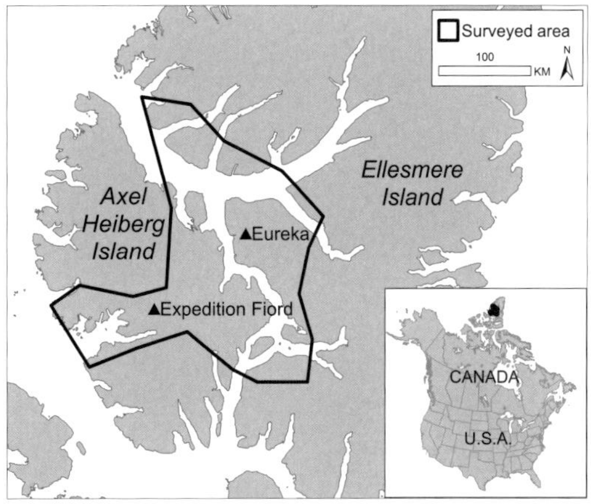

Figure 8.6. Central Ellesmere and Axel Heiberg Islands study area.

(Moskoff and Montgomerie 2002). The species may occur sparsely on Prince Patrick Island.

Pectoral Sandpiper

The BNA range map shows the species only on the eastern side of Melville Island, whereas we found it throughout Melville and at a few locations on Prince Patrick (Holmes and Pitelka 1998).

Buff-Breasted Sandpiper

The BNA range map includes only the southern quarter of Melville (Lanctot and Laredo 1994). We found it at one site in central Melville and, based on habitat and the distribution of other species, we suspect that it occurs more widely across the study area.

Red-Necked Phalarope

The BNA range map excludes all areas north of southern Victoria Island (Rubega et al. 2000). The single bird we recorded may have been due to an overflight or may suggest that a small population breed farther in our study area.

Red Phalarope

The BNA range map excludes Prince Patrick Island, whereas we found it breeding in the southern quarter of the island (Tracy et al. 2002).

The mean number of shorebirds per plot in different habitats showed that the highest numbers were in wetlands and vegetated uplands but that substantial numbers occurred in unvegetated soil and even rock (either bedrock or gravel). Some of these plots were far from highly vegetated areas. Shorebirds may thus be quite widely distributed across the study area.

CENTRAL ELLESMERE AND AXEL HEIBERG ISLANDS

This study area (Fig. 8.6) covered 44,000 km^2 on central Ellesmere and Axel Heiberg Islands. The survey crew was based at the Polar Shelf facility at Eureka and the research camp at Expedition Fiord from 17 to 26 June 2007. Three types of surveys were conducted: ground-based area searches, rope drags, and aerial transects. The locations of the areas searched and rope drag plots were non-randomly selected, while the general survey locations were selected using a coarse-scale habitat classification and topographic maps. We concentrated our surveys in wetlands. Aerial transects were conducted while flying between plots (Elliott and Smith, chapter 9, this volume).

Plots for area searches ($n = 5$) covered 4–16 ha and were surveyed by two people, who recorded all birds detected. Rope drag surveys of plots ($n = 6$) were carried out in wetlands. The surveyors circumnavigated the lake, pond, or wetland complex while spiraling outward from the edge of the wetland or water body. Distances from nests to the edge of the nearest wetland or water body were measured to test the hypothesis that shorebird nests would be found near wetlands. One hundred thirty-five shorebirds were seen during the ground surveys (area search and rope drag plots). We found 17 shorebird nests and recorded 89 indicated pairs of shorebirds (Table 8.6).

TABLE 8.6

Birds recorded on ground surveys (rapid surveys and rope drags) and aerial surveys in the central Ellesmere and Axel Heiberg Islands area.

Species	Ground surveys				Aerial surveys		
	Nests	Pairs	Singles	Total	Axel Heiberg Island	Ellesmere Island	Total
Shorebirds							
American Golden-Plover	0	0	1	1	1	0	1
Ruddy Turnstone	10	16	16	42	66	58	124
Purple Sandpiper	0	0	0	0	1	2	3
Red Knot	5	2	17	24	69	70	139
Sanderling	0	0	4	4	3	6	9
White-rumped Sandpiper	0	0	0	0	3	5	8
Baird's Sandpiper	1	0	0	1	2	0	2
Red Phalarope	1	11	5	17	20	73	93
Small shorebird	0	0	0	0	357	207	564
Medium shorebird	0	0	0	0	0	11	11
Other species							
Red-throated Loon	0	2	0	2	9	8	17
Snow Goose	0	0	10	10	44	29	73
Brant	0	1	0	1	42	10	52
Canada Goose	0	0	0	0	10	0	10
Common Eider	0	0	0	0	4	0	4
King Eider	0	5	1	6	156	81	237
Long-tailed Duck	0	8	3	11	39	45	84
Unidentified Duck	0	0	0	0	9	5	14

Glaucous Gull	0	0	0	0	4	3	7
Rock Ptarmigan	0	0	1	1	4	2	6
Parasitic Jaeger	0	0	1	1			
Long-tailed Jaeger	1	2	3	6	60	60	120
Arctic Tern	0	0	1	1	16	5	21
Common Raven	0	0	1	1	0	0	0
Lapland Longspur	1	13	28	42	3	6	9
Snow Bunting	0	1	7	1	61	64	125
Hoary Redpoll	0	0	3	42	0	0	0
Unidentified songbird	0	0	0	0	3	3	6

TABLE 8.7

Birds recorded in the Kivalliq region study area.

Species	Aerial surveys	Walkabouts	Total
Shorebirds			
Black-bellied Plover	0	1	1
American Golden-Plover	0	22	22
Semipalmated Plover	22	30	52
Lesser Yellowlegs	3	3	6
Whimbrel	8	10	18
Hudsonian Godwit	0	1	1
Ruddy Turnstone	39	109	148
Red Knot	0	9	9
Sanderling	6	51	57
Dunlin	9	101	110
Pectoral Sandpiper	0	7	7
White-rumped Sandpiper	12	87	99
Baird's Sandpiper	0	9	9
Semipalmated Sandpiper	0	234	234
Least Sandpiper	0	34	34
Stilt Sandpiper	2	55	57
Wilson's Snipe	0	4	4
Red Phalarope	0	24	24
Red-necked Phalarope	47	75	122
Small shorebirds	3,068	—	3,068
Medium shorebirds	218	—	218
Large shorebirds	4	—	4
Other species			
Red-throated Loon	8	5	13
Pacific Loon	16	8	24
Common Loon	0	1	1
Tundra Swan	124	7	131
Canada Goose	1,535	838	2,373
Greater White-fronted Goose	62	57	119
Ross's/Lesser Snow Goose	354	493	847
Mallard	1	0	1
Northern Pintail	442	99	541
Northern Shoveler	6	0	6
Green-winged Teal	54	16	70
Unidentified scaup	23	2	25
Common Eider	76	14	90

TABLE 8.7 (*continued*)

Species	Aerial surveys	Walkabouts	Total
King Eider	16	6	22
Long-tailed Duck	153	101	254
Surf Scoter	4	0	4
Black Scoter	8	0	8
Common Merganser	22	5	27
Red-breasted Merganser	2	6	8
Northern Harrier	2	2	4
Bald Eagle	2	2	4
Merlin	2	0	2
Willow Ptarmigan	189	73	262
Rock Ptarmigan	0	21	21
Sandhill Crane	67	39	106
Long-tailed Jaeger	0	21	21
Parasitic Jaeger	6	11	17
Herring Gull	346	107	453
Glaucous Gull	0	1	1
Arctic Tern	426	12	438
Common Raven	0	4	4
Horned Lark	0	70	70
Gray-cheeked Thrush	0	5	5
American Pipit	1	20	21
Yellow Warbler	0	1	1
Blackpoll Warbler	0	4	4
American Tree Sparrow	0	46	46
Savannah Sparrow	0	328	328
Harris's Sparrow	0	12	12
White-crowned Sparrow	0	26	26
Lapland Longspur	5	599	604
Snow Bunting	24	20	44
Redpoll	0	106	106

All nests were within 500 m of a wetland or water body. The rope drag was extremely effective, especially for Red Knots, which did not flush until the rope touched or passed over them.

Aerial transects (n = 36) were flown 30 m above ground at a speed of 80 km/hr. Observers recorded birds within a 400-m-wide transect, centered on the airplane. The total transect length was 3,500 km. Birds were identified to species where possible or to size category (Table 8.6). The small shorebird category included Baird's Sandpiper, Common Ringed Plover, Red Phalarope, Sanderling, and White-rumped Sandpiper. The medium shorebird category included American Golden-Plover, Black-bellied Plover, Purple Sandpiper, Red Knot, and Ruddy Turnstone. No large shorebird species were found in this region. We recorded 954 shorebirds on the survey, mainly unidentified small shorebirds, but substantial numbers of Red Knot, Ruddy Turnstone, and Red Phalarope. We also recorded more than 100 each of King Eider, Jaegers, and Snow Buntings. It is interesting to note that more shorebirds were seen on the surveys of Axel Heiberg Island than on Ellesmere Island. Axel Heiberg Island had more "good" wetland habitat than we expected. The density of birds (number/km^2) recorded on the aerial survey was impressive: shorebirds 0.68, geese 0.10, ducks 0.24, and all species 1.24. These results show the utility of aerial surveys for determining distribution of many species, including shorebirds and even a few small landbirds (e.g., Snow Buntings).

KIVALLIQ REGION

This study area (Fig. 8.7) covered 180,000 km^2 on the west coast of Hudson Bay. The survey crew was based in Arviat and Baker Lake from 10 to 28 June 2008. Surveys consisted of ground-based "walkabouts" and aerial transect surveys. The walkabouts were done by two observers walking in opposite directions from the helicopter for approximately 45 minutes. All species detected were recorded (Table 8.7). The aerial surveys were flown at 30 m above the ground at a speed of 80 km/hr with two observers on opposite sides of the helicopter (front left and back right), each recording birds within 200 m of the airplane. Shorebirds that could not be identified to species were classified by size category: Small (Baird's Sandpiper, Dunlin, Least Sandpiper, Red Phalarope, Red-necked Phalarope, Sanderling, Semipalmated Plover, Semipalmated Sandpiper, White-rumped Sandpiper), Medium (American Golden-Plover, Black-bellied Plover, Pectoral Sandpiper, Red Knot, Ruddy Turnstone, Stilt Sandpiper, Wilson's Snipe), and Large (Hudsonian Godwit, Lesser Yellowlegs, Long-billed Dowitcher, and Whimbrel).

The Kivalliq region has high species diversity for both shorebirds (n = 19) and other species (n = 44), as noted in the Northwest Territories–Nunavut Bird Checklist Survey (see Armer et al., chapter 12, this volume). The most common shorebirds recorded were Semipalmated Sandpiper, Ruddy Turnstone, Red-necked Phalarope, and Dunlin. The most common non-shorebirds were Canada Goose, Ross/Snow Goose, Lapland Longspur, Northern Pintail, Arctic Tern, and Herring Gull. We also noted hundreds of mid- and high-arctic–breeding

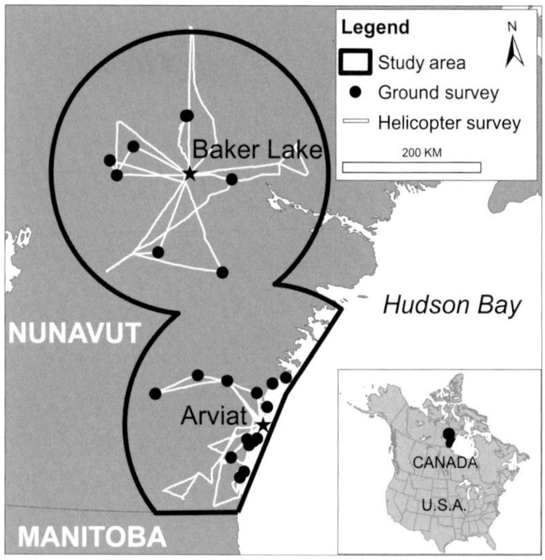

Figure 8.7. Kivalliq Region study area.

shorebirds (e.g., Ruddy Turnstone, Sanderling, White-rumped Sandpiper) feeding on the wrack lines along the western Hudson Bay coast. The importance of this region for northward migration of these species is still unknown.

ACKNOWLEDGMENTS

In the Kent Peninsula study area, L. Bourget, L. Dickson, G. Donaldson, A. Fontaine, V. Johnston, L. McIvor, N. Nashaooraitook, T. Ohokak, and J. Rausch conducted the surveys. Helicopter transport was provided by pilot J. Barry from Universal Helicopters and accommodations (Walker Bay Cabin) by the Government of Nunavut, Department of the Environment.

In the Alert study area, surveys were conducted by G. Morrison and E. P. Pierce in 2001 and by L. A. Armer and R. Braden in 2007. Funding and logistical support was provided by the Canadian Wildlife Service (Environment Canada) and the Polar Continental Shelf Program (Natural Resources Canada). We also thank the Commanding Officer and staff of Canadian Forces Station Alert (Department of National Defence) for their outstanding support.

In the Québec study area, V. Johnston, Y. Aubrey, and R. Cotter of the Canadian Wildlife Service (Environment Canada) provided funding and logistic support for this study. C. Perry, U.S. Fish and Wildlife Service, also provided travel funds. B. A. Andres, D. Buehler, G. Fernandez-Aceves, and J. Rausch conducted the surveys. Thanks to F. St.-Pierre, J. Lefebvre, S. Bachand, and V. Richard for their hospitality in the Canada Goose Tuksukatuk camp on the Polemond River. G. Tremblay helped with shorebird data collection. A. Tulugak aided us in Puvirnituq and arranged for transportation to the field camp. Special thanks go to our pilot D. Dubé, Canadian Coast Guard, who provided skillful and conscientious helicopter support.

In the Melville/Prince Patrick study area, surveys were conducted by J. Bart, C. Francis, and K. H. Elliott. Funding and logistical support were provided by the Canadian Wildlife Service (Environment Canada) and the Polar Continental Shelf Program (Natural Resources Canada). Hospitality at the Cape Bounty Camp was provided by Queen's University (S. Lamoureux).

In the Central Ellesmere study area, surveys were conducted by V. Johnston and J. Rausch. Funding and logistical support were provided by the Canadian Wildlife Service (Environment Canada) and the Polar Continental Shelf Program (Natural Resources Canada). Hospitality was provided by McGill University (W. Pollard) at the Expedition Fiord Camp and by the Canadian Forces Station and employees in Eureka. Thanks to our pilot, G. Greening of Universal Helicopters Newfoundland, for his flying skills and knowledge of the area.

In the Kivalliq study area, surveys were conducted by G. Gibbons, R. Illnik, V. Johnston, M. Kingaq, J.-L. Martin, J. Rausch, and L.-J. and J. van den Scott. Funding and logistical support were provided by the Canadian Wildlife Service (Environment Canada) and the Polar Continental Shelf Program (Natural Resources Canada). Accommodations in Arviat were provided by N. Lamoureux and L. Rollin (Mikilaaq Centre) and the Diocese of Churchill–Hudson Bay. Accommodations in Baker Lake were provided by B. and D. Cooper. Thanks to our pilot, J. Lafrenière of Helicopter Transport Services Canada, and our photographer, N. Lamoureux of the Mikilaaq Centre.

ONLINE CONTENT

Abstracts are available in French and Spanish from www.ucpress.edu/go/sab.

Une traduction du résumé est disponible en français.

Una traducción del resumen está disponible en español.

PART THREE

Methodology

CHAPTER NINE

Aerial Surveys

A WORTHWHILE ADD-ON TO PRISM SURVEYS, ESPECIALLY IN THE INTERIOR

Kyle H. Elliott and Paul A. Smith

Abstract. The ground-based PRISM surveys often require lengthy periods of transit between plots. Conducting aerial surveys during transit can decrease time available for ground surveys (by requiring slower flight speeds), but can provide additional data about birds' abundance and distribution. Previously, most surveys for shorebirds and waterbirds in the Canadian arctic focused on coastal regions. In contrast, our aerial surveys during transits often cover significant portions of island interiors. To examine the value that aerial surveys might add to the PRISM program, we conducted surveys between 2003 and 2008 of the Mackenzie Delta in the Northwest Territories, the Kivalliq region of Nunavut, and Prince Patrick, Melville, Coats, Southampton, Ellesmere, and Axel Heiberg Islands (all in Nunavut). Densities for all taxonomic groups were higher in the low-arctic (Mackenzie Delta, Kivalliq) and Foxe Basin compared to the high-arctic islands, but the bird communities were different. We found that gulls and terns were concentrated along the coast, but all other birds were equally abundant away from the coast. Our results demonstrate that aerial surveys conducted during transit between plots can increase knowledge of birds' relative density and distribution. For shorebirds, this knowledge could be used to refine stratification for future rounds of PRISM surveys. For waterbirds, we demonstrate that densities are high even in some inland areas; because these areas have been poorly monitored in the past, our surveys could add significantly to knowledge about population size and distribution. As observed densities were similar across flight speeds, we suggest that relatively high forward flight speeds be used to reduce the costs associated with slower transit speeds.

Key Words: aerial survey, arctic, Arctic Tern, coast, gulls, monitoring, Northwest Territories, Nunavut, PRISM, shorebirds, Southampton Island, waterbirds.

When conducting ground surveys over large regions, such as for Arctic PRISM, long periods are spent in transit. Carrying out aerial surveys during these transit periods requires reduced flight speeds, and therefore may increase costs or reduce the time available for ground surveys. Yet these aerial surveys may add value by providing

Elliott, K. H., and P. A. Smith. 2012. Aerial surveys: a worthwhile add-on PRISM surveys, especially in the interior. Pp. 159–176 *in* J. Bart and V. Johnston (editors). Arctic shorebirds in North America: a decade of monitoring. Studies in Avian Biology (no. 44), University of California Press, Berkeley, CA.

additional information on the distribution and abundance of shorebirds and other birds. The purpose of the current chapter is to examine the costs and benefits of conducting aerial surveys while traveling between Arctic PRISM plots in the Canadian arctic, using the six years of data collected to date.

Bird populations in the Canadian arctic are generally poorly known, with the exception of a few high-profile groups, such as geese (Béchet et al. 2004, Gauthier et al. 2007) and colonial waterbirds (Gilchrist and Mallory 2005, Gaston et al. 2006, Raven and Dickson 2009), which are easily counted at their breeding colonies. Detailed studies have provided robust population estimates for small regions, but it is unclear how representative these are of the entire arctic. Surveys in Alaska have more comprehensive coverage (McCaffery et al., chapter 3, this volume, Bart et al., chapter 4, this volume), but in Canada, estimates for larger regions are often based on coastal aerial surveys that assume most waterbirds occur close to the coast (Gaston et al. 1986). For example, earlier population estimates for geese and waterbirds in the Foxe Basin, Nunavut, were based on transects flown 200 m from the coast, with birds counted 200 m on either side of the aircraft (Gaston et al. 1986). Even surveys that take place further inland often miss large areas at the in-terior of islands in the Canadian arctic (Groves et al. 2009) and may not detect all species. Although large birds, such as waterfowl and raptors, can be accurately censused from the air (Smith 1995, Anthony et al. 1999, Barnhill et al. 2005), aerial censuses are seldom used for small birds and there is little information about the accuracy of aerial surveys for smaller birds such as shorebirds (but see Nebel et al. 2008). It is likely that the detection rates are low, but this has not yet been assessed.

During 2003–2008, we conducted aerial surveys at six sites across the Canadian arctic, including part of the Kivalliq region (near Arviat and Baker Lakes), several high-arctic islands (Axel Heiberg, Ellesmere, Melville, and Prince Patrick Islands), the Mackenzie Delta, and parts of the Foxe Basin (Coats and Southampton Islands; Fig. 9.1). The aerial surveys were conducted as an adjunct to Arctic PRISM rapid surveys (see Bart et al., chapter 2, this volume). We relate observed densities to habitat, region, and distance from

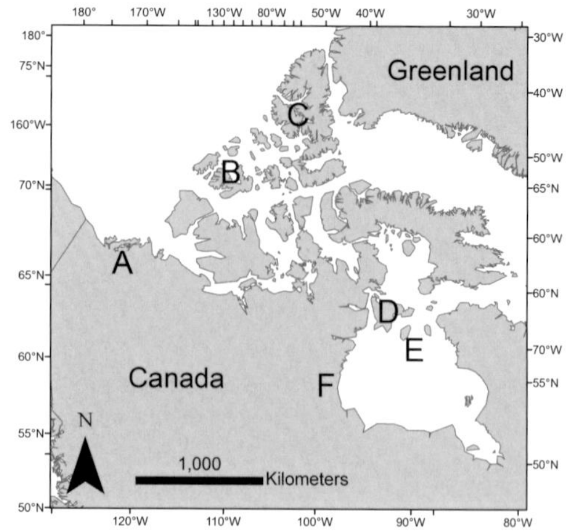

Figure 9.1. Six locations where aerial surveys were conducted: a) the Mackenzie Delta, b) Melville and Prince Patrick Islands, c) Ellesmere and Axel Heiberg Islands, d) Southampton Island, e) Coats Island, and f) the Kivalliq region near Arviat.

the coast. We provide preliminary indications of detection rates and provide suggestions for improving aerial surveys in the future. Implications for monitoring waterbirds in the Canadian arctic are discussed, as is the use of aerial surveys to improve Arctic PRISM's monitoring of shorebirds.

METHODS

We conducted aerial surveys when flying between plots where we conducted ground-based PRISM rapid surveys, except when poor visibility or extreme weather made this impractical. Plots for ground surveys were selected in a stratified random fashion, with more plots located in wetlands than in barren areas (typically chosen to meet a 6:3:1 wetlands:vegetated uplands:dry areas shorebird habitat ratio). Beyond this stratification by habitat, plot selection was random; distance to coast or geographic arrangement of plots was not considered. Although plots were randomly located, the order in which they were surveyed, and thus the transect length and orientation, was selected for maximum efficiency. Surveys should not therefore be considered a purely random sample of all habitats.

To measure the bias associated with our non random design, we placed a polygon around all the 25-km–separated points during the aerial surveys (birds seen along each 25-km section were binned into a single point halfway along that segment) that form the basis for the maps presented here. We then randomly selected the same number of points as the number of points sampled. We compared the distance to coast and proximity to wetland for each randomly selected point with that for the actual points.

Surveys were flown in a Bell 206L helicopter at 90 km/h, in June 2003–2008 (Fig. 9.1; Tables 9.1–9.2). All birds within 200 m of either side of the aircraft were recorded, except at the Mackenzie Delta in 2008 and Coats Island, where birds within 100 m were recorded. Whether birds were within this threshold was determined by placing reference markers on the windows of the helicopter. Surveys were flown at an altitude of 25–30 m, except on Coats Island and at the Mackenzie Delta in 2006, where surveys were flown at an altitude of 15 m. Additional aerial surveys (not part of the PRISM program, but conducted by PRISM crew) were also flown along the coast of Coats Island in 2005 using a Twin Otter aircraft. For all surveys, the location of bird sightings was established by logging a timed location of the aircraft using GPS, and recording the time of bird sightings on the voice recorders. Birds were identified to species where possible. Unidentifiable shorebirds were assigned to broader classes based on size (small, medium, large).

For all helicopter surveys, one observer was located in the front passenger (left) seat and one in the rear right seat. Habitat information was recorded along each transect by a third observer, at 2-km or 5-km intervals, and consisted of the estimated percent coverage (±5%) of habitat types and landforms. In the Kivalliq, the high-arctic islands, and Mackenzie Delta, photographs were taken and habitat type and landforms were recorded whenever ground cover appeared to change.

The data were subdivided into roughly 25-km transects by removing all transects less than 10 km in length and subdividing any transect longer than 35 km into 25-km pieces such that the final piece was between 15 and 35 km in length. The area covered (transect width × length) was calculated, and the number of birds divided by the area covered (i.e., density, in birds/km^2) is reported here. If transects were less than 25 km apart, we averaged values.

RESULTS

Randomly selected points tended to be farther from the coast ($\chi^2 = 217$, df = 162, $P = 0.005$) and farther from the closest wetland ($\chi^2 = 203$, df = 162, $P = 0.02$) compared to the actual aerial surveys. The aerial transects surveyed here were therefore biased toward coastal and wetland habitats. Despite this bias, it is noteworthy that our surveys include substantial areas of inland habitat, whereas previous surveys in the Canadian arctic typically have not.

The total number of birds seen in each location, with times and distances surveyed, is presented in Tables 9.1 and 9.2. Many mammals were seen, including one grizzly bear feeding on a muskox (for scientific names, see Appendix C) at 74°54.458′N, 109°35.483′W on Melville Island (well outside of the bears' normal range). There was no relationship between aircraft speed and the density of birds recorded by observers ($R^2 = 0.01$, $P = 0.56$). Apart from gulls and terns at Kivalliq and the Mackenzie Delta and shorebirds at Kivalliq (Figs. 9.2–9.8), no other bird groups at any location showed a relationship with distance from coast ($P > 0.50$).

Low-arctic and Foxe Basin Aerial Surveys

Mackenzie Delta and Kivalliq Region

At both the Mackenzie Delta and the Kivalliq region, gulls and terns showed an association with the coast (Fig. 9.8). Similarly, shorebirds were more abundant near the coast in Kivalliq (Fig. 9.8). Other bird groups were distributed throughout the study areas (Figs. 9.2–9.3). Ptarmigan were present in highest concentrations at the Tuktoyaktuk Peninsula (Fig. 9.2). The bird communities of the Mackenzie Delta and Kivalliq regions were similar (Tables 9.1–9.2), and included a number of species more typical of the treeline and subarctic. Waterfowl were the most numerous birds. Species included Tundra Swans, Cackling Goose, Northern Pintail, Long-tailed Duck, and, at the Mackenzie Delta, White-winged Scoter. These regions were also the only regions where significant numbers of raptors were seen, especially raptors with more southerly distributions, such as Bald Eagles. The Mackenzie Delta, known to be one of the regions with the highest waterbird densities in the Canadian arctic,

TABLE 9.1

Total number of birds and mammals seen on aerial surveys in the Canadian arctic.

	Prince Patrick Island	Melville Island	Ellesmere Island	Kivalliq	Coats Island	Southampton Island
Survey time (min)	301	843	248	1,292	520	585
Survey distance (km)	452	1,265	372	1,938	775	878
Tundra Swan				120	179	14
Greater White-fronted Goose				62		
Snow Goose		22	25	38	3,971	902
Brant	24	29	5		34	12
Cackling Goose		18	10	1518	781	160
Green-winged Teal				54		
Mallard				1		
Northern Shoveler				6		
Northern Pintail				442	13	1
Greater Scaup				15		
Unknown scaup				8		
Common Eider			4	76	6	
King Eider	33	38	139	16	310	84
Long-tailed Duck	4	3	37	153	113	79
Black Scoter				8		
Surf Scoter				4		
Common Merganser				22		
Red-breasted Merganser				2		
Unknown waterfowl	1		14	8		7
Red-throated Loon		3	5	8	10	26
Pacific Loon				16	10	17
Bald Eagle				2		
Rough-legged Hawk		3				
Northern Harrier				2		
Gyrfalcon		4				
Merlin				2		
Willow Ptarmigan				189		7
Rock Ptarmigan	22	41				4
Sandhill Crane				67		1
Black-bellied Plover					6	3
American Golden-Plover	1				1	1
Semipalmated Plover				22		1

TABLE 9.1 *(continued)*

	Prince Patrick Island	Melville Island	Ellesmere Island	Kivalliq	Coats Island	Southampton Island
Baird's Sandpiper			1			
Dunlin				9		
Lesser Yellowlegs				3		
Whimbrel				8		
Whimbrel/Hudsonian Godwit				58		
Ruddy Turnstone			59	39		2
Red Knot	2	4	36			4
Sanderling			9	6		
White-rumped Sandpiper		1	8	12		2
Purple Sandpiper	4		3			
Stilt Sandpiper				2		
Red Phalarope		17	82		14	196
Red-necked Phalarope				47		
"Large" shorebird				4	18	
"Medium" shorebird		18	10	218		17
"Small" shorebird		25	351	3068	133	518
Unknown shorebird	19	77				
Parasitic Jaeger			4	6	26	4
Pomarine Jaeger			1?			
Long-tailed Jaeger	1	24	32			6
Unknown jaeger	2	31	49			5
Arctic Tern			8	422	284	50
Glaucous Gull	2	17	1			2
Herring Gull				313	304	20
Sabine's Gull					318	15
Snowy Owl		2				
American Pipit				1		
Lapland Longspur		2	7	5		
Snow Bunting	6	10	28	24	5	11
Unknown songbird			6			85
Caribou	15	76		572		
Arctic fox		2		2		
Arctic hare		4		6		
Muskox	15	121				
Grizzly bear		1				

TABLE 9.2

Birds seen on aerial surveys of the Mackenzie Delta.

Bird	Delta (2005)	Delta (2006)	Delta (2008)
Survey time (min)	304	1,012	133
Survey distance (km)	414	1,380	181
Tundra Swan	86	260	146
Greater White-fronted Goose	154	863	122
Snow Goose	18	10	2
Cackling Goose	11	164	125
Unknown goose		37	
Green-winged Teal	8	18	2
American Wigeon	3	99	4
Mallard	16	51	29
Northern Shoveler	16	47	118
Northern Pintail	137	471	184
Canvasback	31	2	144
Common Goldeneye		19	
Greater Scaup		143	
Unknown scaup	133	150	151
Common Eider		23	
King Eider	13	49	
Unknown eider		2	
Long-tailed Duck	42	269	20
Surf Scoter		31	
White-winged Scoter	173	530	179
Unknown scoter		26	
Common Merganser			1
Red-breasted Merganser	10	29	
Unidentified waterfowl		193	2
Red-throated Loon	31	43	
Pacific Loon	29	55	66
Common Loon		9	
Yellow-billed Loon		2	
Unknown loons	3	11	
Bald Eagle		3	
Rough-legged Hawk	5	18	3
Northern Harrier	3	24	8
Golden Eagle	1	3	
Willow Ptarmigan	51	323	125

TABLE 9.2 *(continued)*

TABLE 9.2 (CONTINUED)

Bird	Delta (2005)	Delta (2006)	Delta (2008)
Rock Ptarmigan	16	20	
Unknown ptarmigan		36	
Sandhill Crane	33	54	116
American Golden-Plover	4	3	
Semipalmated Plover		1	
Baird's Sandpiper			
Whimbrel	10	34	12
Hudsonian Godwit		1	1
Dunlin			
Stilt Sandpiper		3	
Wilson's Snipe	3	2	1
Red Phalarope	1		
Red-necked Phalarope	41	106	88
"Small" shorebirds	70	706	138
"Medium" shorebirds	4	30	33
"Large" shorebirds		6	
Parasitic Jaeger	1	4	7
Long-tailed Jaeger	2	15	4
Arctic Tern	185	236	144
Common Raven	5	2	3
Glaucous Gull	33	96	42
Herring Gull		11	1
Mew Gull	1		
Unknown gull	1	2	
Short-eared Owl	7	10	3
Yellow Warbler			1
Unknown songbird	7		196

had relatively high densities of Sandhill Crane, Hudsonian Godwit, Whimbrel, Red-necked Phalarope, Arctic Tern, and Glaucous Gull. The only songbird identified to species, a Yellow Warbler, was also a species typical of more southerly locations (Table 9.2).

Southampton Island

We completed 878 km of surveys on Southampton Island. The most abundant bird was the Snow Goose, with 902 recorded, and we flew an additional 12 km of transects where geese were too numerous to count. Snow Geese were abundant wherever our surveys neared the coast, and were present inland in moderate abundance as well (Fig. 9.4c). Unidentified small shorebirds were the next most commonly recorded group of birds, followed by Red Phalaropes and Cackling Geese (518, 196, and 160 individuals counted, respectively). The aerial surveys did not reveal a pattern of higher density near the coast for shorebirds; densities were uniform across all

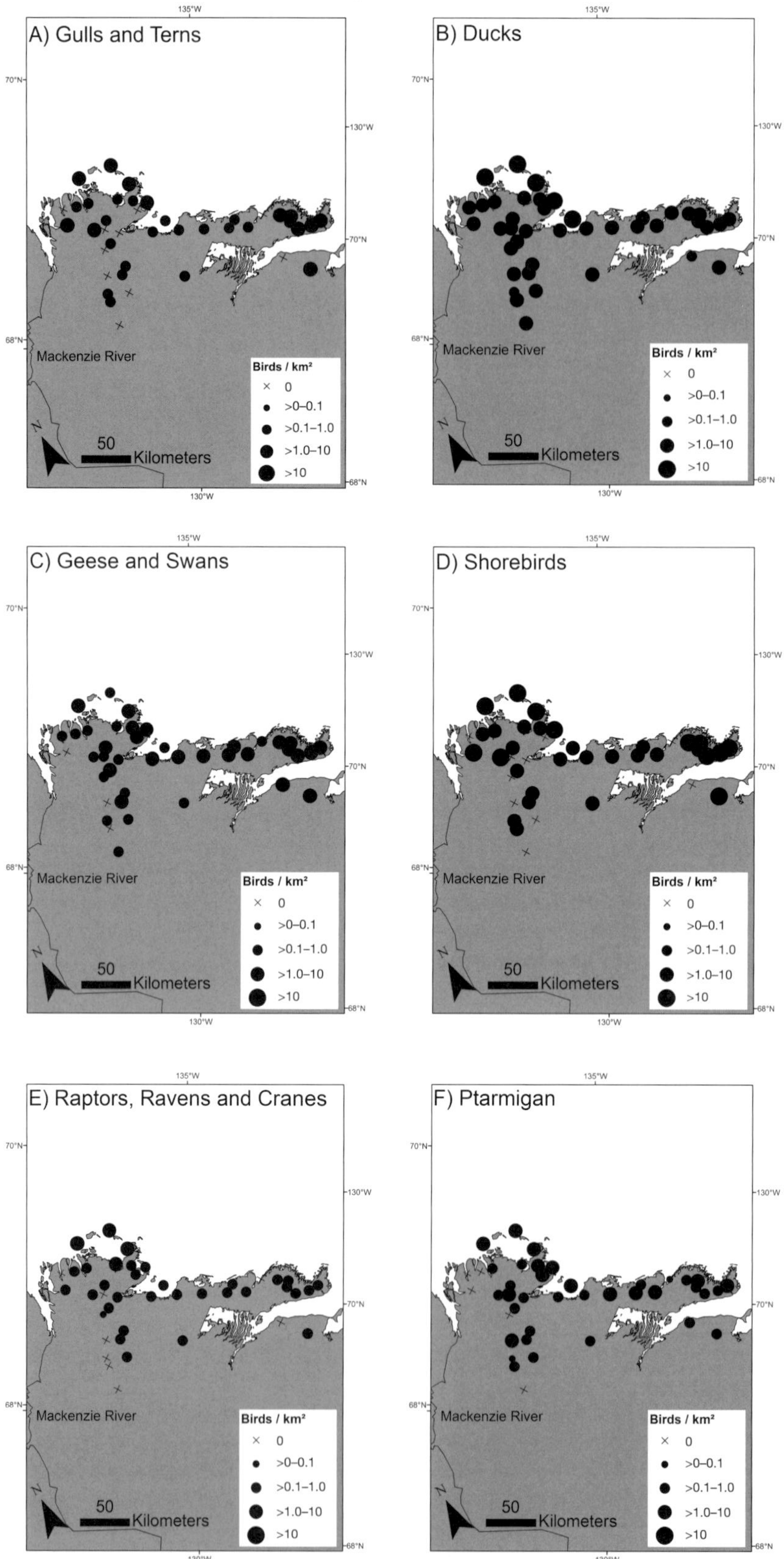

Figure 9.2. Densities of various birds groups counted during aerial surveys in the Mackenzie Delta.

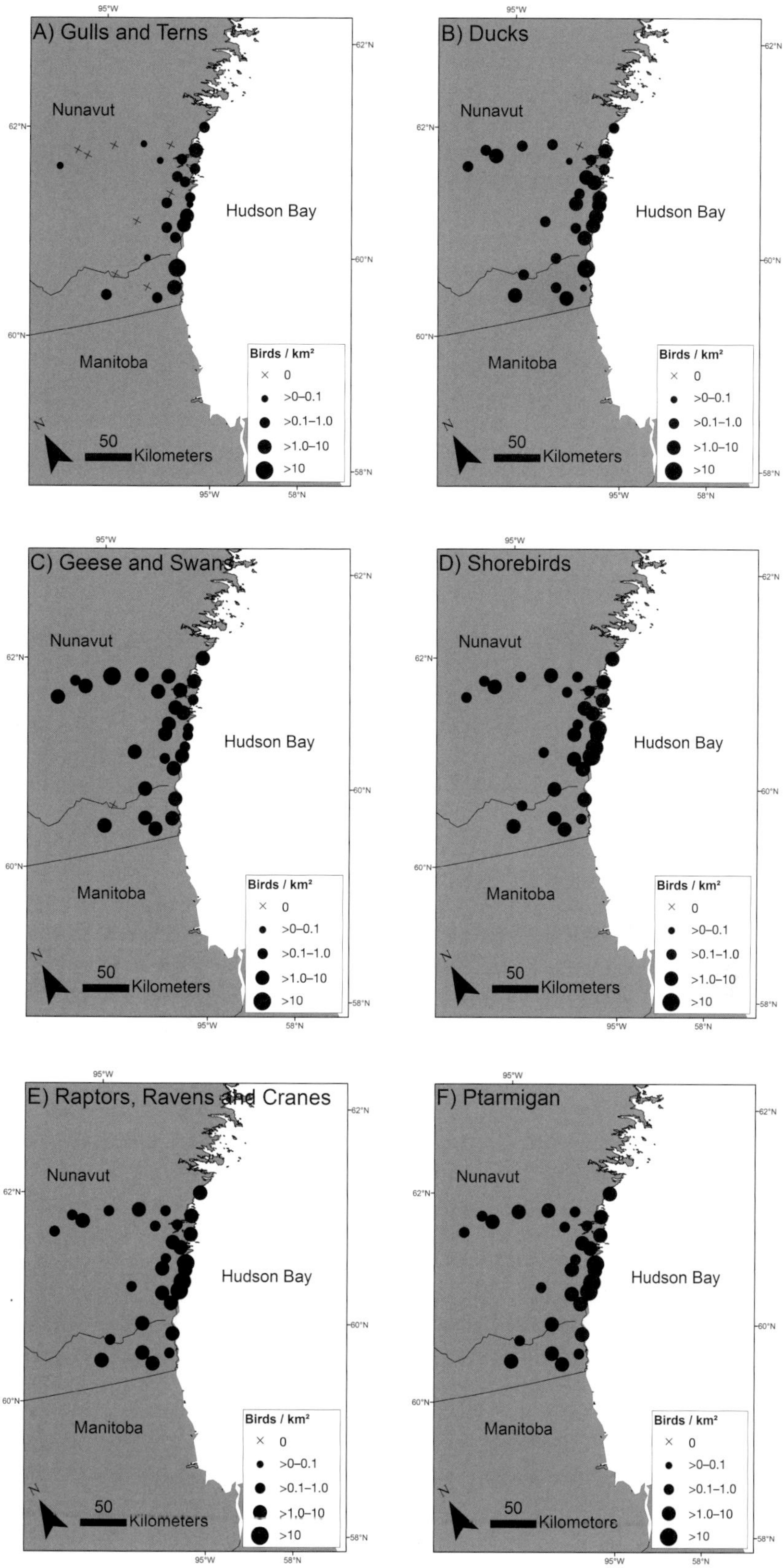

Figure 9.3. Densities of birds recorded during aerial surveys in the Kivalliq region.

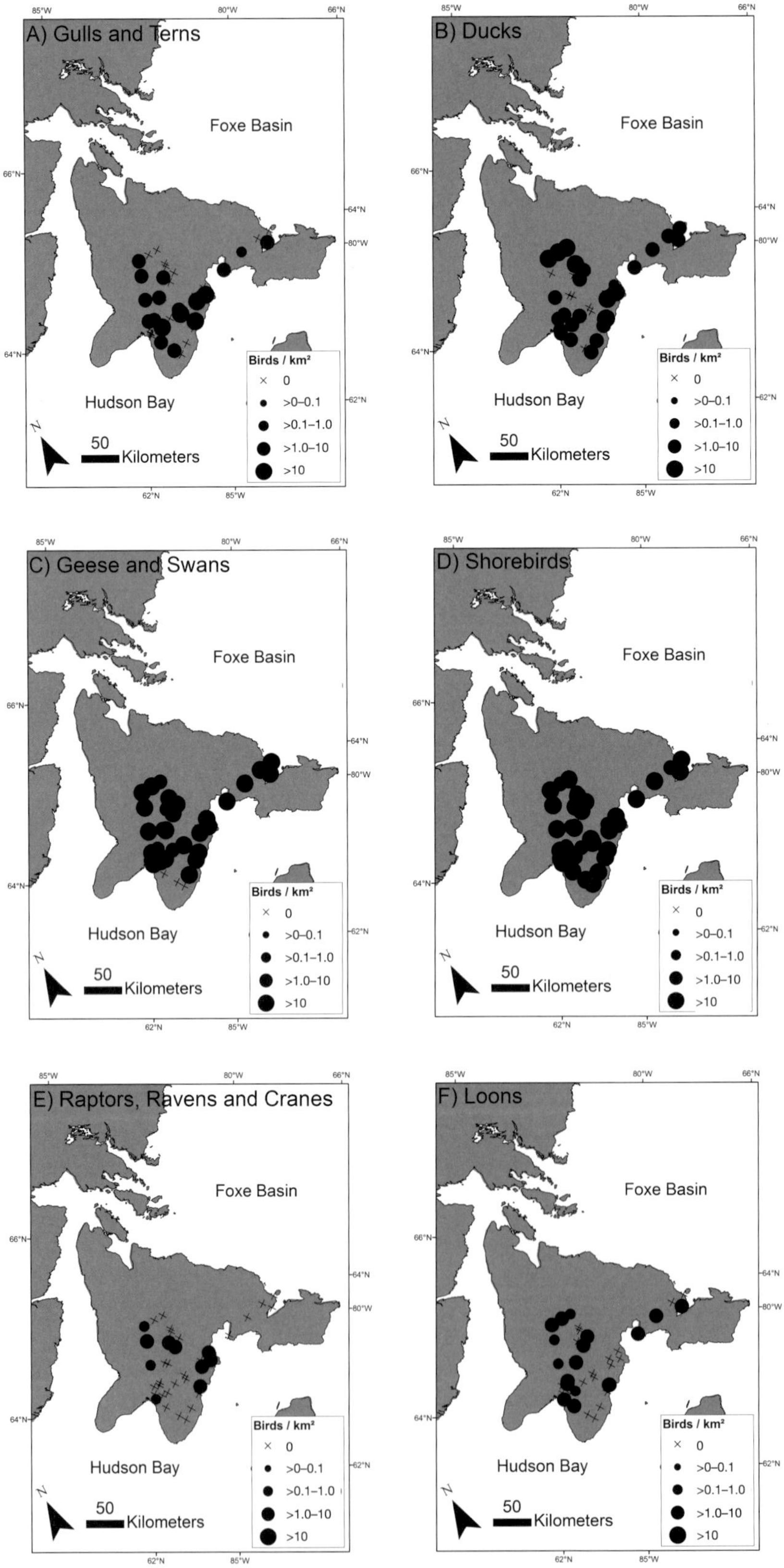

Figure 9.4. Densities of birds recorded during aerial surveys at Southampton Island.

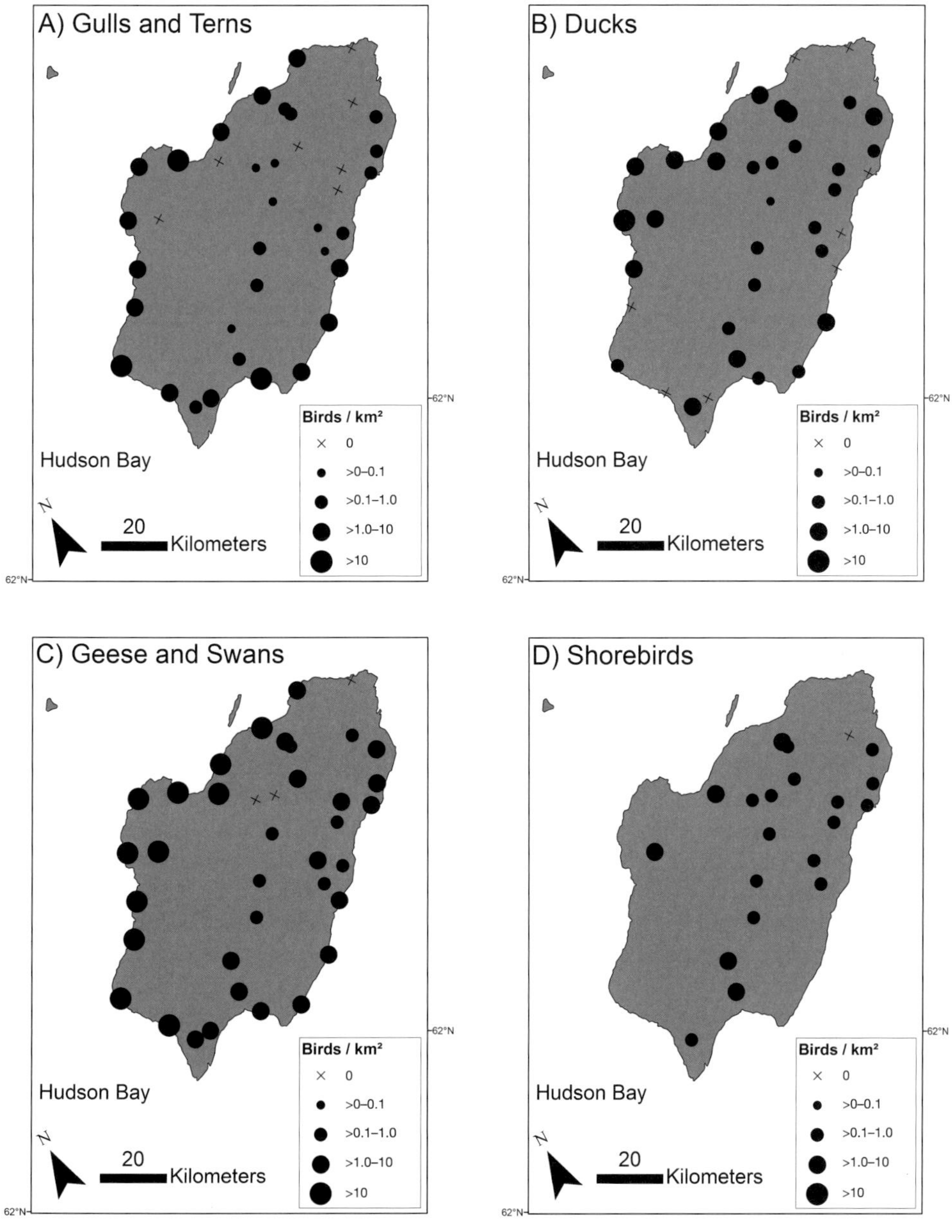

Figure 9.5. Densities of birds on Coats Island. Inland points were surveyed by helicopter in 2006 while the coastal points were surveyed by fixed-wing aircraft in 2005. Coastal surveys do not include shorebirds.

areas surveyed (Fig. 9.4d). Along with the Mackenzie Delta and Coats Island, shorebird aerial survey densities in the arctic were highest on Southampton Island (Fig. 9.4d).

Coats Island

At Coats Island, we completed 515 km of aerial surveys by helicopter in 2006 and 260 km of coastline surveys by fixed-winged aircraft in 2005 (Fig. 9.5). The most abundant bird recorded on the 2006 surveys was the Snow Goose, with 1,067 individuals counted. Nearly 1,000 of these were counted in a 30-km length of transect south of Calanus Bay at 62°49′N, 82°58′W. Although Snow Geese did not historically breed on the island (Gaston et al. 1986, Gaston and Ouellet 1997), surveys in 2005 identified large numbers of Snow Geese breeding on the northwest coast. The aerial surveys in 2006 rarely came close to

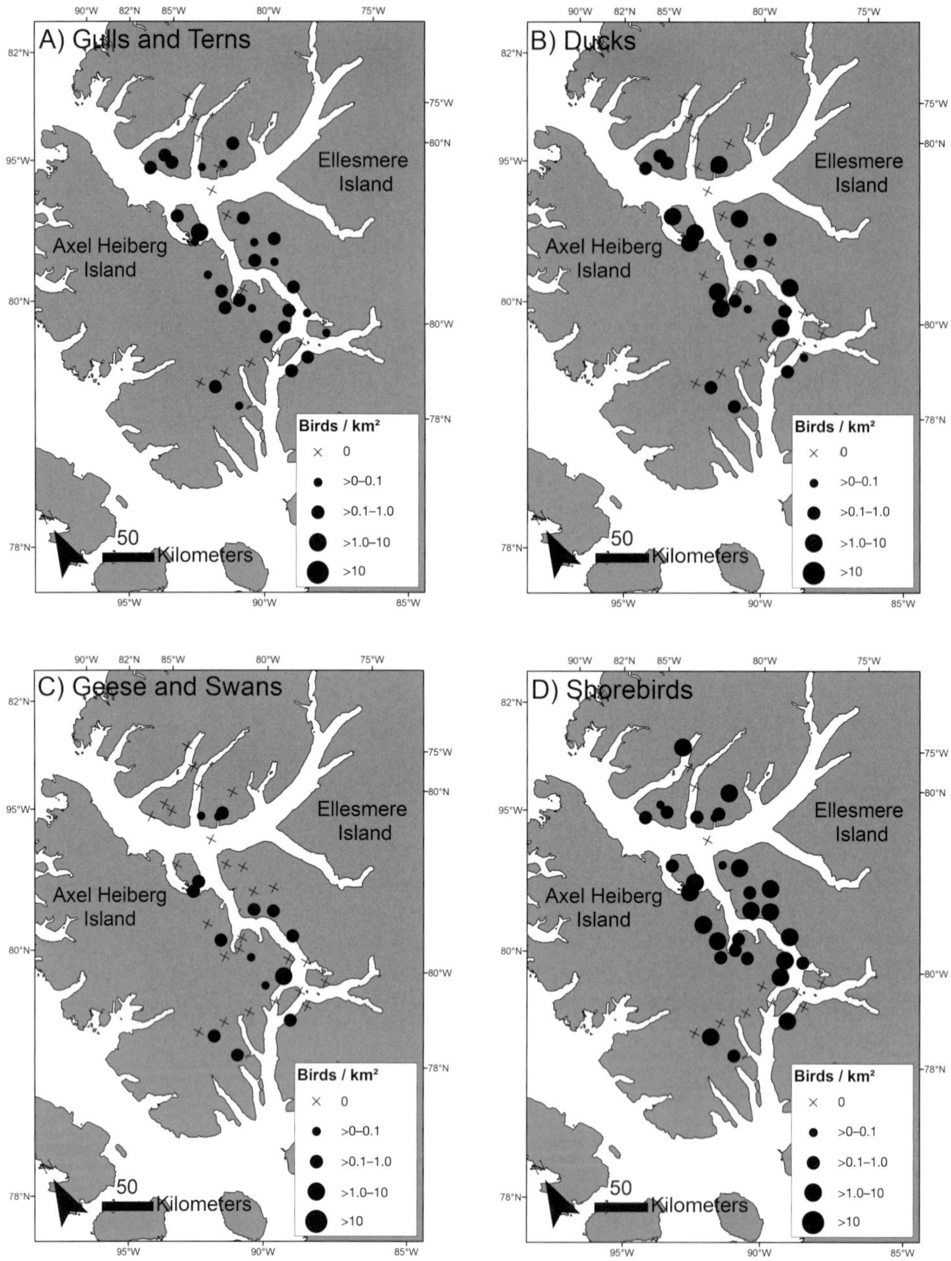

Figure 9.6. Densities of birds recorded during aerial surveys at Ellesmere and Axel Heiberg Islands.

this portion of the coastline, and the high counts of Snow Geese in 2006 were recorded during the short period when we did approach this portion of the coast. Cackling Geese and King Eiders were also abundant across the island, with 359 and 164 recorded, respectively. Both species were found breeding far inland.

We recorded shorebirds to species where possible, and felt reasonably confident with our identification of Red Phalaropes, Ruddy Turnstones, Black-bellied Plovers, and American Golden-Plover. Still, the majority (133/164) of our shorebird observations were "small shorebirds"; based on the ground surveys, we assume that these are primarily Semipalmated Sandpipers and White-rumped Sandpipers. Shorebirds were somewhat more abundant in coastal locations, but were present in small numbers even in the

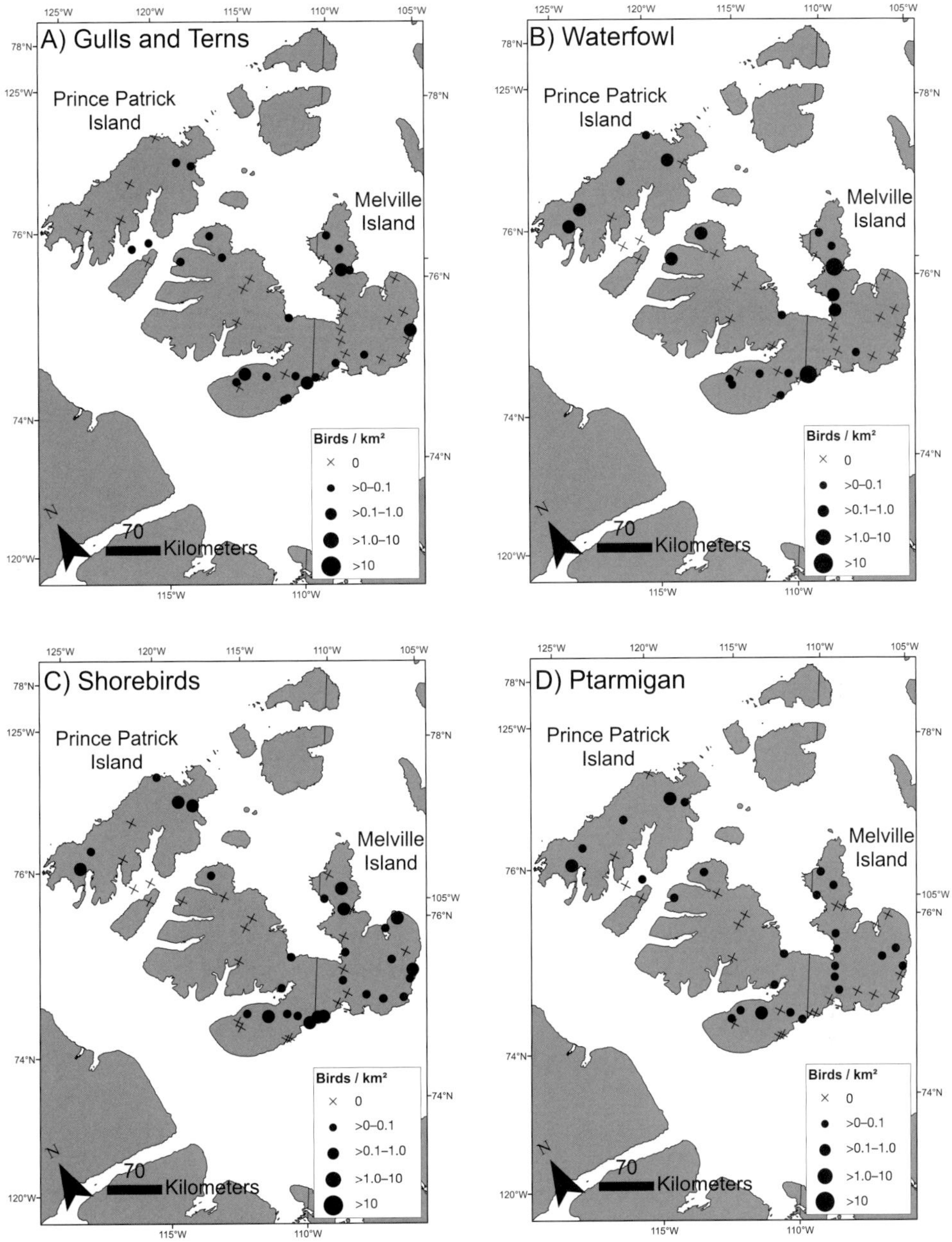

Figure 9.7. Densities of birds recorded during aerial surveys at Melville and Prince Patrick Islands.

center of the island, where suitable wetland habitats are restricted to isolated patches (Fig. 9.5d). Shorebird densities were high, comparable to the Mackenzie Delta and Southampton Island (Fig. 9.5d).

High-arctic Islands Aerial Surveys

The bird communities in the high-arctic islands were markedly different. Whereas waterfowl were the most frequently detected group in the low-arctic islands (Kivalliq, Mackenzie Delta), shorebirds were the most common group in the high-arctic islands. Birds were concentrated mostly along the east coast of Axel Heiberg and the Fosheim Peninsula shorelines (Fig. 9.6). Densities were lower in the western high-arctic islands (Fig. 9.7), with the highest densities along the Dundas and Sabine Peninsulas of Melville Island.

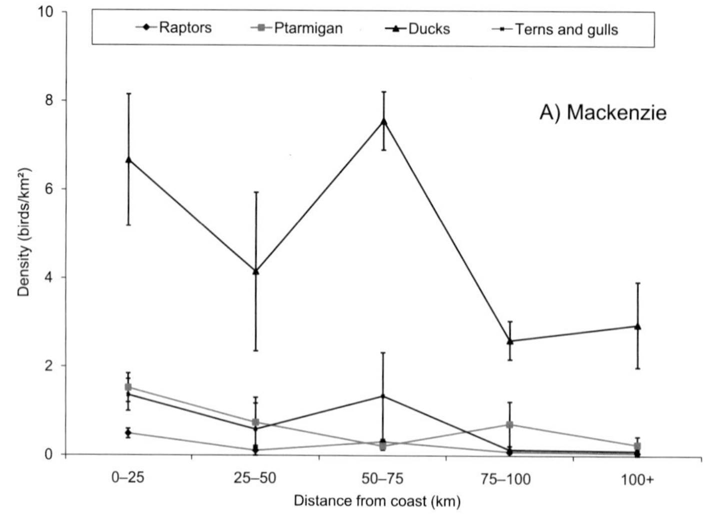

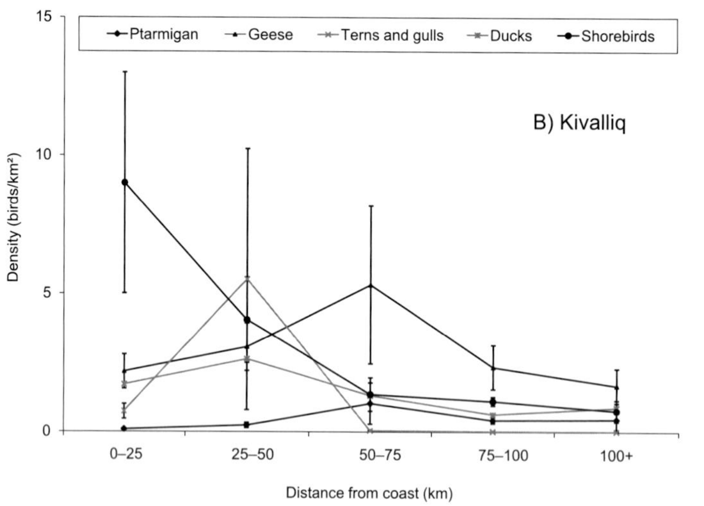

Figure 9.8. Bird densities relative to distance from the coast at (a) Mackenzie delta and (b) Kivalliq. No other regions showed a significant relationship for bird density relative to distance from coast ($P > 0.5$).

In the high-arctic islands, Red Phalaropes replaced Red-necked Phalaropes in the shorebird community (Tables 9.1, 9.2). Red Knots and, at the Fosheim Peninsula of Ellesmere Island, Ruddy Turnstones were the most common shorebird species (Table 9.1). On Ellesmere Island, Red Knots are considered to be *islandica*, which is consistent with our observations of brighter overall coloration and *islandica* rump pattern. The raptors recorded were Gyrfalcons, Rough-legged Hawks, and Snowy Owls.

Habitat Associations

Habitat associations for shorebirds at the Mackenzie Delta are described elsewhere (see Rausch and Johnston, chapter 5, this volume, Pirie et al., chapter 11, this volume). Waterfowl density was positively related to water body size ($t_6 = 4.80$, $P = 0.003$), but this was not true for any other birds during our arctic surveys ($P > 0.30$). At the Mackenzie Delta, the only location where both ptarmigan species were observed, Rock Ptarmigan were associated with dry uplands and Willow Ptarmigan with wet lowlands ($\chi^2 = 10.8$, df = 1, $P = 0.001$). In particular, Rock Ptarmigan outnumbered Willow Ptarmigan along the rocky Tuktoyaktuk Peninsula, while Willow Ptarmigan was the main species present in the Mackenzie Delta proper. No other significant habitat associations were observed, possibly because it was difficult to associate birds with fine-scale habitats due to uncertainty in their precise location, or because shorebirds were often not distinguishable to species, and habitat associations occurred at the species level.

DISCUSSION

Aerial surveys are widely used to census large waterbirds, and are known to provide reliable results (Smith 1995). However, rates of detection for smaller birds such as shorebirds during aerial surveys in the arctic are unknown. Laursen et al. (2008) showed that rates of detection (air vs. ground) for shorebirds in a temperate coastal environment varied widely among species, from greater than 80% to less than 50%. Elliott et al. (2010) demonstrate that detection of shorebirds in a boreal environment was lower still, and in many cases nil. Our results here suggest that rates of detection for shorebirds during aerial surveys in the arctic may also be low; densities of shorebirds from ground surveys in the areas where we conducted aerial surveys were typically many times larger than those recorded from the air (see chapters 3–7 and 14, this volume). For example, the Mackenzie Delta averaged 2.2 shorebirds/km^2 in the aerial surveys, while ground surveys counted 38–51 shorebirds/km^2. Further study of detection rates is warranted, but preliminary results suggest that aerial surveys might provide unreliable absolute estimates of density for shorebirds. However, the aerial surveys could nonetheless make an important contribution to the Arctic PRISM monitoring program.

Aerial surveys can provide supplementary information on range of species to complement the ground-based surveys. For example, we recorded small numbers of medium-sized shorebirds across the interior of Melville and Prince Patrick Islands. These were likely Purple Sandpipers, as they were the most common medium-sized shorebird on ground surveys away from large wetlands. Aerial surveys confirmed their extent across both of these islands and expanded the known range by several hundred kilometers.

Similarly, the broad patterns of distribution and relative abundance revealed by aerial surveys can complement the ground-based surveys to aid in stratification during future rounds of surveys. The spatial coverage of the ground plots is necessarily small; spatial coverage of aerial surveys is orders of magnitude larger, and could be used to confirm habitat associations or regional patterns in distribution and abundance. For example, relative densities across latitudes, and relative densities in coastal versus inland areas (see below) could be used to optimize the allocation of survey effort. However, variability in detection rates remains unknown and could bias even relative abundances. For example, surveys were carried out with slightly different methods at Coats Island versus Southampton Island. Coats Island had over twice the density of shorebirds on ground surveys as Southampton Island (113 vs. 54), yet aerial surveys suggested the reverse (Figs. 9.4–9.5). More study of detection rates is needed in order to maximize the value of aerial surveys to the Arctic PRISM program, but preliminary results do suggest that aerial surveys could potentially play an important role in future surveys.

In contrast to smaller birds, our estimated densities for larger waterbirds are likely reliable (Smith 1995) and represent the first of their kind for many of the inland areas that we surveyed. Although we tended to survey coastal and wetland areas more than interior and non-wetlands areas (because we flew between survey plots that were biased toward coastal or wetlands habitats), we are confident that accurate estimates of large waterbird densities across large portions of the arctic could be attained using aerial surveys such as these. A more formal sampling plan, potentially including covariates such as distance from coast or wetland abundance, could be devised to estimate population sizes over large areas.

As for shorebirds, our surveys also revealed important new insights into the distribution of waterbirds and even mammals. For example, we documented a significant number of breeding Snow Geese on the north side of Coats Island, where no colony existed previously (Gaston and Ouellet 1997), and our observation of a grizzly bear on Melville Island is the most northerly record for the species.

The distribution and relative abundance of shorebirds and waterbirds in coastal versus inland locations is particularly noteworthy, as most previous surveys have focused heavily on coastal areas. In a rare example of inland aerial surveys at Victoria Island and the Queen Maud Gulf, species including raptors, jaegers, and loons occurred with equal abundance at the coast and more than 25 km inland (Cornish and Dickson 1996, Groves et al. 2009). However, some waterfowl such as Common Eiders were essentially restricted to the coast (Raven and Dickson 2009). We found waterbirds and shorebirds to be concentrated near the coast at some locations, but found substantial and often equal densities in inland habitats. This finding agrees with ground-based PRISM surveys in the same regions, where large numbers of shorebirds are often recorded in suitable habitats inland (see also Morrison 1997, where many shorebirds, such as White-rumped Sandpiper, were common at inland wetlands). Coastal areas comprise only a small fraction of any landmass; even if densities in inland habitats are moderate, the majority of a population may still be found away from the coast. Furthermore, some of the species that we surveyed, such as jaegers, Arctic Terns, Glaucous Gulls, and ptarmigan, have a global breeding range primarily in the arctic, a winter range largely within the arctic (ptarmigan) or in remote oceans (jaegers, terns, gulls), and are therefore poorly surveyed by other monitoring schemes. All of these species are readily seen from the air, but occur too sparsely to be efficiently counted from the ground. Thus, by conducting aerial surveys in both coastal and inland areas we are adding significantly to knowledge of their global population size.

We demonstrated that the density of birds recorded on the surveys did not vary over the helicopter speeds used (60–100 km/h), although the reliability of species identification may have varied. Because of the expense of helicopter time, a faster flight speed is cheaper over a fixed flight distance, suggesting that 100 km/h is the most efficient flight speed for surveys. We were unable to examine the effect of transect width, flight altitude, and recording method (i.e., voice vs. computerized data management, observer locations in the aircraft, etc.) and suggest the effect of varying these methods be examined in the future. For example, in Alaska, data recorders are used to track GPS locations and bird sightings in real time, improving the link between sightings and locations (J. Bart, pers. comm.). Until this is possible, we suggest that the current methodology be used so that results can be compared between sites and years. Aside from investigating and improving field methodology, we suggest that future research should focus on estimating detection rates for birds observed during aerial surveys. In particular, even for large birds, detection probabilities for aerial surveys are known to vary according to survey date, species, group size, observer, and observer position (Rumble and Flake 1982, Pollock and Kendall 1987, Gabor et al. 1995, Naugle et al. 2000, Conroy et al. 2008). Species- and observer-specific detection probabilities for a given year could be established using double sampling (i.e., with simultaneous ground surveys) or other methods (Pirie and Johnston, chapter 10, this volume). The eventual adoption of dual observer or double sampling calibration methods would improve the ability of these aerial surveys to assess population trends (Bart and Earnst 2002, Collins 2007, Conroy et al. 2008).

In summary, aerial surveys carried out when traveling between PRISM plots can add significantly to the Arctic PRISM monitoring program. They provide qualitative information on the distribution and relative abundance of shorebirds that can help to refine future surveys, and also provide additional distributional data for other birds and mammals. They could potentially contribute to population estimates for larger birds (waterfowl, jaegers, terns, gulls, and ptarmigan), some of which are otherwise poorly surveyed. Because the helicopter must travel between ground plots during the course of PRISM surveys, the incremental financial costs of traveling at a slower speed and conducting aerial surveys are modest. We propose that further work be carried out to better understand rates of detection, but in the interim, we propose that observers continue to conduct aerial surveys between PRISM plots at 100 km/h ground speed, 25 m altitude, recording all birds seen within 200 m of either side of the aircraft. With continued refinement, what is currently a worthwhile "add-on" could eventually form a key monitoring tool for shorebirds, waterbirds, and larger landbirds.

ACKNOWLEDGMENTS

We thank our many assistants: J. Bart, V. Charlwood, S. Earnst, C. Francis, V. Johnston, M. Kotierk, C. Parker, L. Pirie, S. Mackenzie, J.-L. Martin, K. Monaghan, J. Rausch, and A. Skok. We also thank the excellent pilots at Canadian Helicopters, Helicopter Transport Services, and Universal Helicopters Newfoundland Limited for their expert maneuvering. G. Morrison, J. Bart, and an anonymous reviewer provided excellent comments that improved the manuscript. We thank the Polar Continental Shelf Project and Canadian Wildlife Service for logistical and financial support. This report is PCSP Contribution No. 01209.

ONLINE CONTENT

Abstracts are available in French and Spanish from www.ucpress.edu/go/sab.
Une traduction du résumé est disponible en français.
Una traducción del resumen está disponible en español.

CHAPTER TEN

Survey Methods for Whimbrel

Lisa Pirie and Victoria Johnston

Abstract. Whimbrel breed at low densities, are patchily distributed, and have proven difficult to survey with traditional PRISM methods. We tested a double sampling methodology that used aerial surveys as the "rapid" component to determine whether this method was more effective at surveying Whimbrel in the Mackenzie Delta. Detection rates for aerial surveys were low, detecting on average 23% of the Whimbrel that were actually on the ground within the surveyed area, but rates were relatively consistent among trials. Contrary to our assumptions, Whimbrel did not regularly flush when a helicopter passed. In contrast, detection rates using PRISM methods were high (mean: 2.75), and varied among three sites and four years. The high mean value suggests that Whimbrel approached observers from areas outside of plots, while the high variability suggests that this behavior may vary among plots, habitats, or years. The comparatively consistent detection rates during aerial surveys suggest that this method may hold promise for surveying this species, but stratification of aerial survey blocks by habitat might improve the accuracy of any region-wide extrapolations based on aerial surveys.

Key Words: aerial survey, arctic, detection rate, Mackenzie Delta, monitoring, Northwest Territories, population size, PRISM, shorebirds, Whimbrel.

Whimbrel (for scientific names, see Appendix C) is designated as a sensitive species in the Northwest Territories (GNWT 2006) and a species of high conservation priority nationally (Donaldson et al. 2000, Brown et al. 2001, U.S. Shorebird Conservation Plan 2004). Because of its priority status and its ecological importance in the Mackenzie Delta north of the treeline (hereafter Mackenzie Delta), the species was also chosen by proponents of the Mackenzie Gas Project (MGP) as a Valued Ecosystem Component (any component of the environment that is considered important to those involved in an environmental assessment process) in the MGP environmental impact statement (AMEC Americas Ltd., unpubl. report).

The western North American subspecies of Whimbrel (*N. p. rufiventris*) occurs in discontinuous

Pirie, L., and V. Johnston. 2012. Survey methods for Whimbrel. Pp. 177–184 *in* J. Bart and V. Johnston (editors). Arctic shorebirds in North America: a decade of monitoring. Studies in Avian Biology (no. 44), University of California Press, Berkeley, CA.

Figure 10.1. Whimbrel (*rufiventris* subspecies) breeding range in North America.

breeding populations throughout boreal, subarctic, and low-arctic portions of Alaska east to the Melville Hills in Nunavut, Canada (Fig. 10.1; Skeel and Mallory 1996, Morrison et al 2006). It has a wide-ranging but disjunct distribution, breeding in low densities throughout much of its range (Skeel and Mallory 1996). Previous studies in the Mackenzie Delta identified two distinctly different habitats used by Whimbrel for breeding: wet-sedge low-centered polygons (LCP) on Taglu and Fish Islands in the Outer Mackenzie Delta (hereafter Outer Delta; Fig. 10.2), and much drier sedge/heath/low shrub habitat complexes throughout the Mackenzie Delta (Dickson and Smith 1991, Gratto-Trevor 1996, Ashenhurst 2004, Pirie et al. 2009).

The Arctic Program for Regional and International Shorebird Monitoring (PRISM) conducted surveys in PRISM Region 12, including the Mackenzie Delta, from 2005 to 2007. From the outset of the PRISM program, we knew that acceptable survey precision might not be achieved for some species because of limited ranges or particular habitat requirements. We conjectured that Whimbrel might be one of these species because of its relatively small population and disjunct breeding range. Attaining timely and accurate monitoring information for Whimbrel was critical because of the species' key role in the MGP environmental assessment. Therefore, we decided to investigate the utility of a single-species survey for Whimbrel using a double sampling design with aerial surveys as the rapid method (Bart et al., chapter 2, this volume), and to compare our results to those obtained through regular PRISM surveys (Rausch and Johnston, chapter 5, this volume) in the Outer Delta and the Middle Mackenzie Delta (hereafter Middle Delta; Fig. 10.2).

Studies specific to shorebirds in the Mackenzie Delta region are few, and have only been conducted in the past 20 years. In the late 1980s, Dickson (Dickson and Smith 1991) conducted research in a small region of the Outer Delta to determine if Landsat imagery could be used to accurately identify suitable shorebird habitat. In 1991 and 1992, Gratto-Trevor (1996) repeated Dickson's surveys and extended the study area to include most of the Outer Delta. In 2004, PRISM

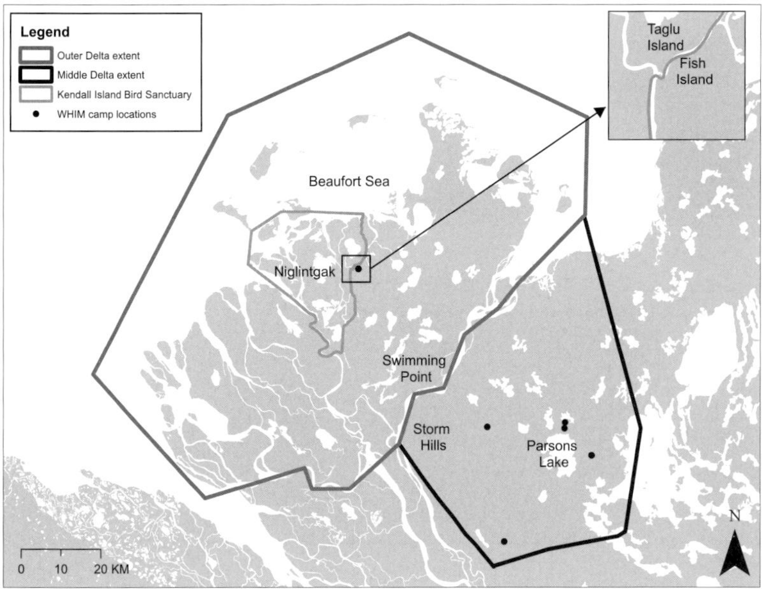

Figure 10.2. Map showing the location of the Mackenzie Delta study area, the Kendall Island Bird Sanctuary boundary, the extents of the Outer and Middle Mackenzie Delta, and the locations of the Whimbrel study plots.

reconnaissance surveys were conducted throughout the Mackenzie Delta. From 2005 to the present, PRISM surveys have been conducted annually in the Mackenzie Delta area. In 2006 and 2007, Pirie (2008) conducted studies focused on Whimbrel in the Outer Delta and described Whimbrel abundance and distribution, nest success, and habitat selection at three scales—landscape, territory, and nest-site—the latter using a predictive habitat model based on satellite imagery.

METHODS

Study Area and Plot Selection

Our study area included plots scattered across the Outer and Middle Delta of the Mackenzie River. The Outer Delta encompasses 857,000 ha of land bordering Mackenzie Bay and the Beaufort Sea coast, extending from Shallow Bay in the west to Kugmallit Bay in the east. The Outer Delta is composed of low-lying alluvial islands that are dominated by saturated wetlands (Kemper 2006). Herbaceous lowland habitats undergo annual spring flooding after ice breakup and snowmelt. Other lowland habitats such as willow shrub flood less frequently. The Outer Delta includes the Kendall Island Bird Sanctuary (KIBS). Wet-sedge low-centered polygon habitat, characterized by polygon ridges with shallow ponds and scattered raised mounds, is typical Whimbrel habitat in this area.

The Middle Delta is the region directly south of the Outer Mackenzie Delta that extends south to the treeline, encompassing approximately 727,000 ha (Fig. 10.2). The landscape in this area is dominated by rolling hills with scattered wetland patches in valleys and depressions. Whimbrel are found in dry tussock upland tundra with nearby adjacent wetlands exhibiting high- and low-centered polygon structure or within the wetlands themselves.

Ground Survey Methods

From 9 to 17 June 2007, intensive ground surveys were conducted at four plots south of Swimming Point (Table 10.1). Because observations were limited to 24 person-hours per plot, limited nest searching was undertaken at these plots. In 2008, a study plot at Parsons Lake was surveyed over multiple days (4–20 June) because it was so large (3,300 ha). Approximately 390 person-hours of survey time was spent at Parsons Lake, and intensive nest searching was undertaken. Selection of these plots was non-random, based on the

TABLE 10.1

Number of Whimbrel observed during aerial and ground surveys of Whimbrel in intensive plots in 2007 and 2008.

The detection ratio is the number of birds estimated (by air) divided by the actual number of birds observed on the ground.

Plot	Plot size (ha)	Aerial survey dates	No. Whimbrel observed on the ground	No. Whimbrel observed from the air	Detection ratio
Fly Camp 1	168	9 June 2007	8	2	0.25
Fly Camp 2	66	13/14 June 2007	8	2	0.25
Fly Camp 3	70	15 June 2007	10	4	0.40
R. Wiacek Plot	150	17 June 2007	6	0	0
Parsons Lake (Survey 1)	3,300	8 June 2008	16	4	0.25
Parsons Lake (Survey 2)	3,300	8 June 2008	16	3	0.19
Fish Island	350	10 June 2008	12	3	0.25
Overall average detection rate					0.23

observation of at least several pairs of Whimbrel at sites in 2005–2007.

A study area comprised of all of the wet-sedge LCP habitat on Fish Island in the Outer Delta was also intensively surveyed in 2008 (approximately 350 person-hours of survey time; previous years' work had shown that Whimbrel were restricted to this habitat on Fish Island). We determined the actual number of breeding territories present in the plots. In all of the plots described above, territories were located by observing Whimbrel that were foraging, exhibiting breeding behaviors, or responding to predators. Approximate territory boundaries were delineated on a map. When a territory was located, observers followed territory holders until they were able to pinpoint the location of the nest. Individual bird movements were recorded with a GPS and the territory was revisited until the nest was located. It took an average of 1–3 visits to locate a nest. These surveys formed the intensive component for our Whimbrel-specific survey methodology. Aerial surveys of these five plots and of Fish Island yielded the rapid count, and comparison of aerial results with ground based results yielded the detection ratio for aerial surveys.

From 2006 to 2008, 119 rapid plots and nine intensive plots (five in 2006) were surveyed throughout the Mackenzie Delta as part of regular PRISM surveys. These results are presented in Tables 10.2 and 10.3 for comparison to the aerial survey results, and are described more fully in Rausch and Johnston (chapter 5, this volume).

Aerial Survey Methods

Aerial surveys of Whimbrel plots in the Middle Delta were conducted in both 2007 and 2008. An aerial survey of the single plot at Fish Island was conducted in 2008. Surveys were conducted using a Bell 206L helicopter with bubble windows. Plots were divided into 200-m swaths with a north–south orientation and were navigated using a Global Positioning System (GPS). Entire plots were surveyed by air, except the Parsons Lake and Fish Island plots, where only subsets of the area were surveyed (for logistical reasons), with complete coverage within those subsets. Initially, surveys were conducted with two observers and were flown at speeds of 80 to 90 km/hr, approximately 15 m above the ground. But in 2007, we also experimented with various combinations of survey height and aerial transect width to determine the method that yielded the highest count of Whimbrel. These trials were conducted on the four intensive Whimbrel survey plots.

Ground distance zones of 0–100 and 101–200 m from the observer were marked on the helicopter

TABLE 10.2

Number of Whimbrel observed during rapid and intensive PRISM surveys of nine plots from 2006 to 2008.

The detection ratio is the number of birds estimated during rapid surveys (see chapter 2, this volume for details) divided by the actual number of birds observed in the plots during intensive surveys.

Plot	Year	Survey dates	No. Whimbrel actually present in plot estimated from intensive surveys	No. Whimbrel observed during rapid surveys	Detection ratio
Taglu A	2006	12 and 18 June	1	1.5	
Taglu B	2006	12/18 June	1	2	
Taglu C	2006	12/18 June	0	0.5	3.25
Taglu D	2006	12/18 June	0	1.5	
Taglu E	2006	12/18 June	0	1	
Taglu A	2007	12/18 June	0	3	
Taglu B	2007	12/18 June	1	0.5	
Taglu C	2007	12/18 June	0	0	
Taglu E	2007	12/18 June	0	0	
Fish F	2007	12/18 June	0	1	4.00
Fish G	2007	12/18 June	0	1.5	
Fish H	2007	12/18 June	0	0	
Fish I	2007	12 June	1	2	
Taglu A	2008	7/16 June	0	0	
Taglu B	2008	7/16 June	0	1	
Taglu C	2008	7/16 June	0	0	
Taglu E	2008	7/16 June	0	0	
Fish F	2008	7/16 June	0	0	1.00
Fish G	2008	7/16 June	1	1	
Fish H	2008	7/16 June	0	0	
Fish I	2008	7/16 June	1	0	
Overall average detection rate					2.75

TABLE 10.3

Comparison of estimates of density of Whimbrel for the Mackenzie Delta derived from aerial surveys and PRISM surveys.

Aerial surveys		PRISM surveys		
No. of survey plots	Density, individuals/km^2 (± SE)	No. of survey plots	Density, individuals/km^2 (± SE)	
	All habitats		Wetland	Vegetated upland
23	0.60 (0.17)	68	1.25 (0.40)	1.62 (0.56)

windows. When Whimbrel were observed, observers recorded the zone in which the observation occurred. The majority (eight of nine) of Whimbrel observations in 2007 were made at distances greater than 100 m, so we concluded that a 200-m transect width was more appropriate for the aerial surveys and maintained that width for 2008. We also concluded that a survey height of 15 m was too low for Whimbrel because observers had difficulty detecting Whimbrel before they exited the 200-m observation zone, and increased height to 30 m in 2008.

In all aerial survey instances, each observer recorded their observations and the time using a hand-held voice recorder and marked the location of Whimbrel observations on their GPS. Whimbrel locations were mapped to ensure observations were not double counted on a subsequent survey swath.

In both years, ground surveyors were stationed near three known Whimbrel nests, where they watched and recorded the reactions of Whimbrel pairs to the passage of the helicopter. Both members of the pair attend the nest, and at this stage of the breeding season (early to mid-incubation) the non-attending mate is within 0–100 m of the nest, alert to any potential predator activities. When a disturbance is present, both mates react, flying up from the ground and calling, in a probable attempt to distract the potential predator from identifying the nest. Helicopters passed 0–250 m from each nest. Observations made by ground surveyors were compared with aerial observations to determine (1) if birds flushed when the helicopter went by (= passed within 250 m), and (2) if flushed birds were seen by aerial surveyors.

Detection rates were calculated by dividing the number of territories estimated from the aerial surveys into the number of territories estimated from territory mapping in the same ground plots.

In 2008, aerial surveys were conducted on the one intensive survey plot (Fish Island) and on 22 aerial survey–only blocks (each composed of six 5 km × 200 m transects spaced 600 m apart) in the Middle and Outer Delta. We surveyed additional aerial blocks to obtain a larger sample. Surveys were conducted using the same methodology as for aerial surveys of Whimbrel plots in 2008. There were no ground surveys associated with these aerial surveys; therefore, we corrected the observed densities using the "aerial detection ratio" calculated from data collected in 2008.

RESULTS

Whimbrel Detection Rates

PRISM Ground-based Surveys

The average detection rate (±SE) of Whimbrel based on three years of PRISM surveys in the Mackenzie Delta was 2.75 ± 1.41 (Table 10.1). This rate being greater than 1.0 indicates that Whimbrel were overestimated during rapid surveys of plots; indeed they were often observed in plots that did not actually contain Whimbrel territories (Table 10.1).

Aerial Surveys

Aerial surveys detected an average (±SE) of 23 ± 4% of the Whimbrel with mapped territories in the plots (Table 10.1). Although the range was 0–0.40, the detection ratio was relatively consistent; four of seven plots had detection ratios of 0.25 and one had a ratio of 0.19.

In two overpasses of four nests, Whimbrel were only observed flushing from two nests. The incubating bird of one nest flushed approximately 300 m ahead of the helicopter and resettled after the helicopter passed. The non-incubating Whimbrel of the same nest was observed flushing during the second helicopter pass, while the incubating bird did not. After the bird flushed, it attacked a nearby Long-tailed Jaeger before settling down in the nest vicinity. Both of these flushes were observed by aerial surveyors. Birds that were present but not seen from the air either did not flush or were simply not noticed by aerial observers. A second previously unobserved nest was observed during the second pass; the incubating Whimbrel flushed from the nest approximately 200 m ahead of the helicopter. Its return to the nest was not recorded.

Whimbrel Densities

Aerial Transect Surveys

Whimbrel density encountered in the aerial survey blocks (corrected using the "aerial detection ratio" of 0.23) was 0.60 ± 0.17 (Table 10.3).

PRISM Surveys

Whimbrel densities from the PRISM ground-based surveys, corrected using the detection ratio of 2.75, were determined for wetlands, vegetated uplands, and dry habitats (Table 10.3). Density was 24% lower in wetlands than in vegetated upland habitats. Whimbrel were not observed in the dry habitat strata, resulting in a density of zero.

DISCUSSION

Comparison of Aerial Surveys and Ground-based PRISM Results

The aerial survey detection rate was low compared to the detection rate based on ground surveys. Whimbrel did not engage in "mobbing" activities during aerial surveys, but have been observed to do so during ground surveys. On the ground, Whimbrel responded aggressively when intruders were in their nesting territories, attracting neighboring Whimbrel who assisted in deterring the potential predators and were included in our counts, resulting in an overestimation of their numbers during rapid ground surveys.

The PRISM detection rate for Whimbrel in the Mackenzie Delta was based on intensive surveys from the Outer Delta region only, and only from plots in the wet-sedge LCP habitat. Therefore, this detection rate may not be suitable for estimating Whimbrel densities in the mixed vegetated upland/wetland habitat mosaic, which is the habitat type that we surveyed in the Middle Delta. Indeed, the Taglu and Fish Island sites, where PRISM intensive surveys were conducted, have exceptionally high concentrations of Whimbrel compared to other wetland habitats in the Outer Delta. Taglu and Fish Island sites appear to be unique from other wetland habitats in that they are more structurally complex with less encroachment of willows (Pirie 2009, Pirie et al. 2009). Pirie et al. (2009) found that wet-sedge LCP habitat outside of Fish and Taglu Islands was either too dry, had poor structural complexity, or had dense shrubs. As a result, the PRISM classification of wetland habitats may be too general for accurately extrapolating Whimbrel densities to get population estimates. Despite plots being situated in only a single habitat type (wet-sedge LCP), we found that the ground-based detection rates varied four-fold among years. In contrast, our detection rates during aerial surveys were low but more consistent among plots and years. For five of seven plots, we found no variation in aerial detection rate despite considerable variation in densities, habitat type, and geographical location. Indeed, in 2008, when aerial survey methods were identical among three surveys in two plots, detection rates varied little (19–25%).

The mainly treeless landscape of the Mackenzie Delta should provide good conditions for detecting Whimbrel during aerial surveys, as should the species' large size. Our aerial surveys detected only 23% of the Whimbrel that were estimated to be resident in a given plot. We had assumed that this gregarious and aggressive species would flush readily, but in our observations, approximately half of the monitored birds did not flush at all when the survey helicopter passed over. Other birds were simply missed because of visual limitations such as the window configuration in the helicopter. So although the detection rate was relatively consistent, we were surprised that it was not closer to 1.

The frequency of air traffic in the Mackenzie Delta may be one reason that Whimbrel often did not flush during passage of the helicopter. Oil and gas activities and other wildlife research in the area contribute to a level of air traffic that is high relative to many arctic areas. Birds may be less likely to flush from their nests when a helicopter flies over if they have experienced such activity previously and have "learned" that it does not pose a danger. One would need to replicate aerial surveys with "no traffic" control locations to test this hypothesis.

Evaluation of Aerial/Ground Double Sampling Design for Whimbrel

Aerial surveys are commonly used to assess waterfowl populations, but there is little precedent for using these methods for shorebirds. In 2000 and 2006, aerial methods were tested to survey boreal shorebirds in the Mackenzie Valley (Elliott et al. 2010), and detection rates were found to be extremely variable; aerial surveys were deemed unsuitable for estimating populations of shorebirds in the boreal regions of the Mackenzie Valley. Aerial surveys associated with the Arctic PRISM program have proven useful for providing qualitative information on the abundance of shorebirds (Elliott and Smith, chapter 9, this volume); however, the utility of aerial surveys as the rapid component of a double sampling survey design had not previously been evaluated for arctic shorebirds.

Our trials show that in the mixed vegetated upland/wetland habitat mosaic that is typical of the Middle Delta, aerial surveys could be an appropriate "rapid" component of a double sampling methodology for Whimbrel. Detection rates were low but relatively consistent for our small sample of plots. Both incubating and non-incubating mates have been observed to flush. During ground visits to nests it was observed that one bird usually flushes first and assesses the situation before alerting the other bird (L. Pirie, pers. obs.). The second bird may not flush in a helicopter pass because the event is short and the "danger" passes quickly. Aerial survey/double sampling trials in other habitats with varying breeding densities need to be undertaken prior to widespread use of an aerial survey double sampling method for Whimbrel in the Mackenzie Delta.

If aerial surveys were to be used at a larger scale, it would be necessary to incorporate stratification among habitat types. Density estimates for PRISM surveys were calculated from habitat-stratified plots, while aerial survey estimates were not. When selecting PRISM survey plots we strive to achieve a 6:3:1 ratio of wetlands, vegetated uplands, and dry habitat. Our aerial plots, which were not selected based on habitat, had an overall habitat ratio of 1:1:1. However, because of inaccuracies in locating birds during the aerial surveys, densities could not be partitioned among habitat types. Because of the large variation in Whimbrels use of the different habitat strata, habitat-specific estimates of density would be preferable. Stratification of aerial survey plots among habitat types or more accurate recording of bird locations could produce habitat-specific estimates of density.

Aerial surveys merit further scrutiny as the rapid component of a double sampling scheme for low-density species with a restricted breeding range, but it is premature to judge whether the methodology used here is a suitable alternative to traditional PRISM methods. Aerial survey methods can greatly reduce field survey time, but they are more costly to undertake than standard PRISM surveys. A more thorough investigation of variation in detection rates, with more years and sites included, will provide better evidence to determine whether this additional cost is justified.

ACKNOWLEDGMENTS

Many people helped with various components of the Whimbrel work and we thank them all. Special thanks to the project-specific field assistants V. Charlwood, K. Hansen-Craik, and K. Sittler. This report is PCSP Contribution No. 01009.

ONLINE CONTENT

Abstracts are available in French and Spanish from www.ucpress.edu/go/sab.
Une traduction du résumé est disponible en français.
Una traducción del resumen está disponible en español.

CHAPTER ELEVEN

Tier 2 Surveys

Lisa Pirie, Victoria Johnston, and Paul A. Smith

Abstract. The Arctic Program for Regional and International Shorebird Monitoring (Arctic PRISM) has been designed to monitor shorebirds across their breeding range. Tier 2 of Arctic PRISM provides region- and site-specific population trend and demographic information. When fully implemented, Tier 2 of Arctic PRISM will provide a network of sites where shorebird numbers and demographic parameters are monitored following standardized protocols. As results from our two pilot sites have demonstrated, a wealth of additional information such as shorebird habitat characteristics, abundance trends of predators, annual nest success, and variation in climate parameters and their impacts can also be obtained. Annual research and monitoring at Tier 2 sites will aid interpretation of the periodic, range-wide Tier 1 surveys by increasing our understanding of year-to-year fluctuations in breeding density. Perhaps most importantly, a network of sites monitoring shorebird reproductive success, return rates, and other aspects of demography may provide insights into the cause of the widespread decline of North American shorebirds.

Key Words: arctic, demographics, East Bay, Kendall Island Migratory Bird Sanctuary, Mackenzie Delta, monitoring, Northwest Territories, Nunavut, PRISM, shorebirds, Southampton Island.

In the three-tiered approach to Arctic PRISM, Tier 2 meets the objective of "regular monitoring of populations at permanent sites." Tier 2 sites are non-randomly selected in areas of known importance to shorebirds and are regularly at relatively high intensity of use—that is, for the entire breeding season for several consecutive years. PRISM is developing a systems plan which will ensure that a representative sample of arctic shorebird breeding sites is contained within the Tier 2 network. These sites should meet most of the following criteria: (1) easy accessibility, (2) established long-term research programs and facilities, (3) high-quality shorebird habitat, and (4) preferably located in or near protected areas such as migratory bird sanctuaries or National Wildlife Areas/Refuges.

Pirie, L., V. Johnston, and P. A. Smith. 2012. Tier 2 surveys. Pp. 185–194 *in* J. Bart and V. Johnston (editors). Arctic shorebirds in North America: a decade of monitoring. Studies in Avian Biology (no. 44), University of California Press, Berkeley, CA.

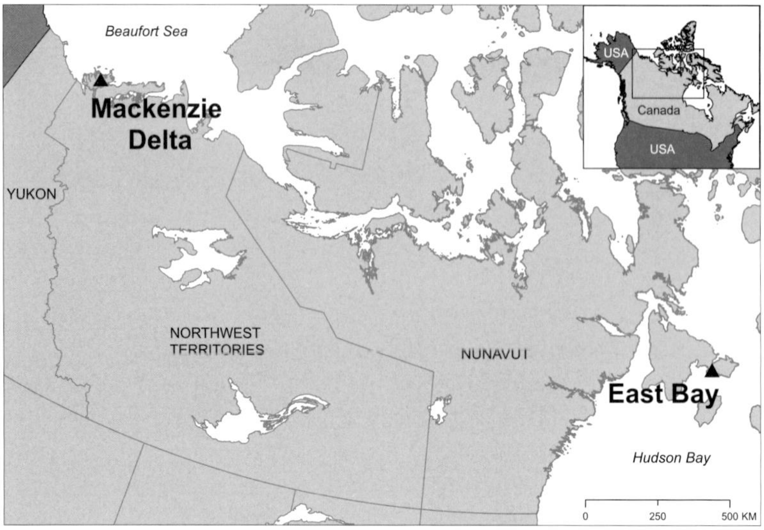

Figure 11.1. Location of the Mackenzie Delta and East Bay Tier 2 sites.

The Tier 1 (double sampling) surveys are carried out regionally over long time intervals (Bart et al., chapter 2, this volume), while the Tier 2 sites contribute localized annual population estimates and trend data. This understanding of fluctuations in local populations from year to year provides detail and context for the Tier 1 survey results. Importantly, these sites also collect (1) demographic data to provide measures of productivity and/or survivorship on arctic breeding grounds, and (2) data on a whole suite of ecological variables, which will help to place avian trends in the context of wider ecological trends at the site. We expect that Tier 2 sites will provide logistical support for priority research projects identified by the Shorebird Research Group of the Americas. They will also be available as a base for research on other fauna and flora. In this paper, we present two case studies to illustrate how Tier 2 fits within the overall scheme of Arctic PRISM.

CASE STUDIES: MACKENZIE DELTA AND EAST BAY

In Canada there are currently two Tier 2 sites. The first site, "Mackenzie Delta," is located in PRISM Region 12 in the outer Mackenzie Delta, straddling the border of the Kendall Island Bird Sanctuary (KIBS) in the Northwest Territories (Fig. 11.1). The Mackenzie Delta, a subdivision of the arctic Coastal Plain region, is a mosaic of low-lying alluvial islands that are dominated by wetlands (Kemper 2006). Annual flooding occurs in herbaceous habitats and less frequently in shrubby habitats. The Tier 2 camp is established here in the first week of June and operates until mid-July, when most shorebird nests have hatched.

Shorebird research at this site began in 1985 in response to increasing interest in hydrocarbon development. Studies conducted from 1985 to 1993 focused on identifying shorebird nesting and staging habitat and using Landsat Thematic Mapper imagery to develop predictive habitat maps (Dickson et al. 1989; Gratto-Trevor 1994, 1996). Additional research efforts included estimating shorebird densities and population sizes (within the study site), and collecting data on breeding phenology, philopatry, and site fidelity (Dickson et al. 1989; Gratto-Trevor 1994, 1996). Gratto-Trevor (1994) also looked at flock size and distribution of fall migrants, seasonal changes in invertebrate prey, and prey distribution. In 1986, a study of preferences and nest success of Whimbrel and Semipalmated Plovers (for scientific names, see Appendix C) was undertaken on Fish Island (Tarves 1987). Intensive survey plots (for Tier 1 surveys of PRISM Region 12) were established in 2005, and these plots have been surveyed annually beginning from June to mid-July, up to and including 2008 (Rausch and Johnston, chapter 5, this volume).

The second site, "East Bay," is in the East Bay Migratory Sanctuary on Southampton Island, Nunavut (Fig. 11.1). This site is situated in an area of coastal lowland tundra interspersed with raised beach deposits. Complete habitat details and descriptions of the shorebird community at this site appear in Smith (2003) and Smith et al. (2007a). Although situated at latitude normally classified as low arctic, the climate and flora of East Bay are mid-arctic in character, influenced by the cold waters of the Foxe Basin (Edlund 1990). Shorebird research has been conducted annually from the beginning of June until the end of July since 1999. A wide variety of research topics have been addressed, including breeding densities, nest survival, timing of breeding, breeding behavior, habitat selection, endocrinology, and toxicology (Perkins 2004, Smith et al. 2007b, Smith 2009). Plot surveys following the PRISM Tier 1 protocol began at this site in 2004.

TABLE 11.1

Density of breeding shorebirds observed in the intensive plots at Tier 2 study sites in the Mackenzie Delta.

	No. of breeding shorebird pairs/ha	
Year	Niglintgak	Taglu/Fish islands
2005	0.26	—
2006	0.93	0.35[a]
2007	—	0.19
2008	—	0.18

—: plots were not surveyed these years.
[a] Surveys only conducted on Taglu Island.

TIER 2 SITE MONITORING PARAMETERS

The following ecological parameters are monitored intensively at our current Tier 2 sites:

1. Shorebird abundance;
2. Shorebird reproduction (nesting success, onset and phenology of nesting);
3. Climate/weather (water levels, weather patterns, snow cover);
4. Habitat (nest habitat, plant community, arthropod availability);
5. Avian and mammalian predator abundance; and
6. Additional monitoring/research (bird checklists, external monitoring programs, student research, shorebird/lemming connection).

Shorebird Abundance

At each Tier 2 site, we conduct a series of intensive surveys on 12-ha plots following the "intensive survey" methodology of PRISM Tier 1 (see Bart et al., chapter 2, this volume, for methods). The data collected during these surveys allows us to determine shorebird densities within the plots (Table 11.1) as well as year to year variation when the same plots are resurveyed. These data are useful as a baseline in environmental assessment processes. At the Mackenzie Delta site, a proposed oil and gas production facility may have a future negative impact on shorebird densities. If the production facility is constructed, we will compare post-construction plot densities to the baseline densities. Comparison of pre- and post-construction plot densities will allow us to better understand how the development is affecting shorebirds and what mitigation is possible.

In 1991–1992, Gratto-Trevor (1994, 1996) conducted breeding shorebird surveys of thirteen 200 m × 200 m plots on Niglintgak Island in KIBS. In 2005, we resurveyed these plots three times throughout the field season (once in mid-June, and twice in early July) following methods described in Gratto-Trevor (1994). The number of shorebirds observed during the three survey times within the 2005 breeding season was used to determine the detectability of shorebirds at different stages during the nesting cycle. We found no statistical difference in the numbers of shorebirds observed during the three visits ($P > 0.05$), suggesting no variability in detectability. The density and species composition data compiled from the two July surveys in 2005 were compared to the July 1991/1992 surveys (Table 11.2). Shorebird populations appeared to have increased since the earlier surveys. However, other PRISM studies (Smith et al., chapter 6, this volume) as well as repeated surveys of a fixed 1.5 × 1.5 km plot at East Bay have indicated dramatic inter-annual fluctuations in shorebird densities (Fig. 11.2; Smith et al., chapter 6, this volume), which suggests that Gratto-Trevor's plots should be surveyed at least two consecutive years before we can determine whether shorebird populations have indeed increased at these sites since the early 1990s. We are currently investigating the cause of the

TABLE 11.2

Numbers of shorebirds observed in Gratto-Trevor's (1994) survey plots in the Mackenzie Delta in 1991/92, versus our observations in 2005.

Species	1991/1992	2005[a]	% change
American Golden-Plover	0	1	+
Semipalmated Plover	1	4	+300
Whimbrel	3	5	+67
Hudsonian Godwit	5	8	+60
Pectoral Sandpiper	2	18	+800
Semipalmated Sandpiper	1	1	0
Stilt Sandpiper	5	6	+20
Long-billed Dowitcher	1	2	+100
Wilson's Snipe	11	5	−54
Red-necked Phalarope	36	41	+14

[a] Total number of birds recorded in each plot, averaged for the two July surveys.

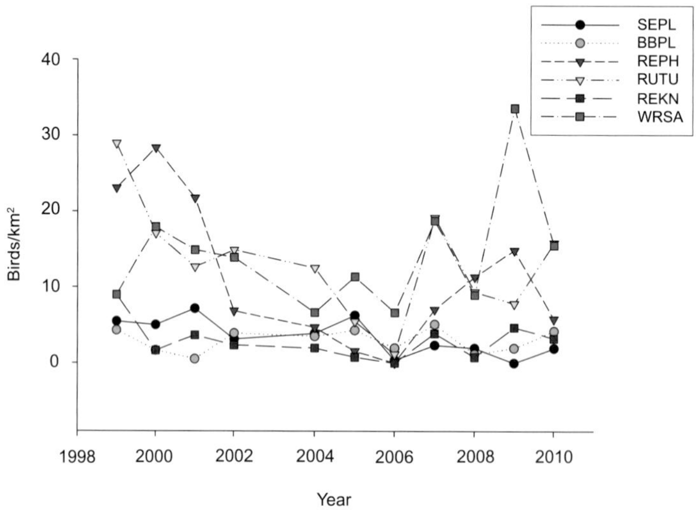

Figure 11.2. Fluctuations in shorebird breeding densities at East Bay from 1999 to 2010 (P. A. Smith, unpubl. data). Species are displayed with AOU 4-letter codes (see Table 11.1).

inter-annual fluctuations in densities at East Bay, but the presence of the variability itself highlights the need for long-term sites such as these.

Reproductive Parameters

Nest Success

Shorebird nest monitoring is conducted regularly at Tier 2 sites. Nests are typically visited once every three days to determine nest fate and hatch date. We monitored nests at the Mackenzie Delta site from 2006 to 2008; hatch success varied from 29% at Niglintgak in 2006 to 56% at Taglu and Fish Islands in 2008. The lower nest success at Niglintgak in 2006 is attributable to the loss of many Red-necked Phalarope nests due to flooding (in the same year at Taglu, which did not flood as dramatically as Niglintgak, hatch success was 48%). In 2007, hatch success on Taglu and Fish Islands was lower, at 31%.

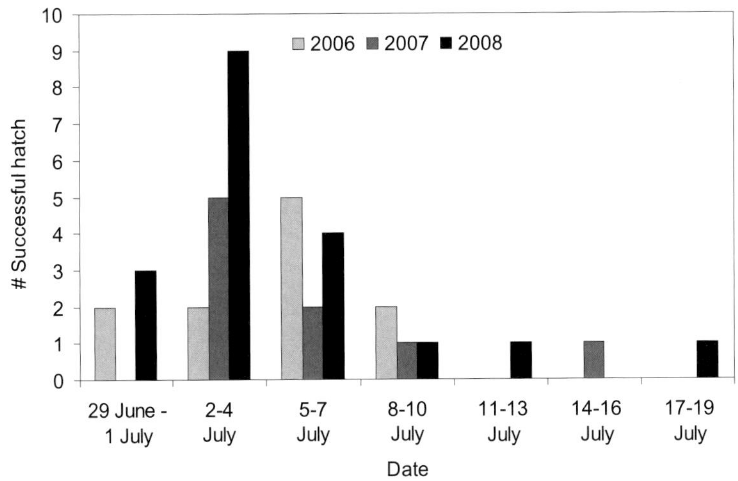

Figure 11.3. Number of successful shorebird nest attempts from three consecutive breeding seasons at the Mackenzie Delta Tier 2 site.

At East Bay from 2000 to 2007, we monitored 899 nests of 12 species. Program MARK was used to assess patterns in daily nest survival. Most of the inter-specific variation could be explained by grouping species by incubation strategy, with uniparental species (only one incubating parent) showing consistently lower nest survival. Over the eight study years, the average survival of nests from lay to hatch was 10% for uniparental species as compared to 42% for biparental species (shared incubation between two parents). Strong temporal patterns suggest that nest survival is highest early and late in the breeding season and depressed mid-season (Smith 2009, Smith and Wilson 2010).

A Whimbrel-specific study in the Mackenzie Delta showed large differences in hatch success between nests situated on two types of land formations. Nests placed on raised mounds showed significantly higher nest success than nests placed on low-centered polygon ridges (Pirie 2008). While further studies are necessary, it is hypothesized that nests placed on ridges are at a greater risk of fox predation because the fox can travel a network of dry ground to reach the nest. Nests on raised mounds may be less likely to be depredated by foxes because they are surrounded by water.

Phenology

The length of the Tier 2 field seasons allows us to monitor variation in timing of breeding for shorebirds. In the Mackenzie Delta, shorebird nest initiation is limited by water levels. In late May and early June, the frozen river channels break up and water floods low-lying habitat in the outer Delta. The extent of flooding is determined by the depth of the snow pack in the catchment basin and the rate of melt, temperature, and ice cover on the rivers; flooding may range from minimal to extreme (Gratto-Trevor 1994). If breakup is gradual and there are no ice jams, the Delta experiences little or no flooding. This may provide suitable nesting habitat immediately for shorebirds arriving on the breeding grounds, whereas in years when flooding occurs, nest initiation may be delayed until water levels sufficiently recede, thus potentially affecting nest success. The three years of nest data for these sites have shown relative synchrony of breeding, with peak nest hatch occurring between 2 and 7 July (Fig. 11.3).

At East Bay, we also studied the timing of shorebirds' arrival to the breeding grounds and their dates of nest initiation. Although weather varied markedly over the years of this study (2000–2007), timing of arrival varied by less than one week, and was not related to local conditions such as temperature, wind, or snowmelt. Timing of breeding was best predicted by the date of 50% snowmelt, with later snowmelt resulting in delayed breeding. Synchrony of breeding was not related to local weather conditions, but was significantly greater in late breeding years. Our results suggest a relatively fixed date for the termination of nest initiation, after which nesting is no longer profitable (Smith 2009, Smith et al. 2010).

Climate/Weather Parameters

Climate change is expected to have dramatic effects on arctic landscapes through shifts in habitat types, melting of permafrost, and changes in weather patterns (ACIA 2005). All current scenarios predict a rise in sea level (Markham 1996, Watkinson et al. 2004), which may alter coastal areas used by shorebirds through inundation and erosion. Increased variability in weather patterns may result in more frequent and intense storms as well as more severe droughts and flooding (Markham 1996, ACIA 2005). Warmer temperatures may lead to the drying up of shallow ponds (Smith et al. 2005, Riordan et al. 2006), a preferred habitat for aquatic shorebird prey. The mismatch hypothesis (where there is a mismatch in the timing of maximum food availability for raising chicks and the timing of parents feeding the chicks) has been proposed as a potential result of climate change. The mismatch between the hatch of arthropods and the hatch of shorebird chicks may have negative consequences for chick growth and survival (Gratto-Trevor 1997). Because changes in climate could have a wide variety of negative effects on breeding shorebirds and their habitats, Tier 2 sites are used to monitor local weather and habitat conditions.

Trends in Water Levels, Weather, and Snow Cover

At Tier 2 sites we monitor weather patterns throughout the breeding season and document major weather events and their impacts on shorebirds. Weather is recorded daily at each site using hand-held weather meters or *in situ* weather stations (see Rausch and Johnston, chapter 5, this volume, Smith et al., chapter 6, this volume, for methods and results).

On river deltas, like the Mackenzie Delta Tier 2 site, water levels in wetland habitats fluctuate on a yearly and seasonal basis. During the breeding season, water levels may further fluctuate depending on temperatures, wind, or tides. Because these habitats are low-lying, small changes in water levels have a large effect on the landscape. Changes in water levels have been shown to affect shorebird prey abundance and distribution, availability of suitable nesting habitat, and even rates of predation (Danks 1992; Gratto-Trevor 1994, 1997; Beveridge 2007). At the Mackenzie Delta, we set up graduated stakes in selected ponds and areas of standing water within each of our study plots. These stakes are then monitored every two days to record water depth. Thus far, we have collected water depth data from Taglu Island (2006–2008) and Fish Island (2007–2008). The analysis of water depth trends at these two areas over the years will provide insight into rates of shorebird nest success. Lower water levels in the Delta may make it more difficult for nest predators to locate shorebird nests because there is a larger area of dry ground to search. When water levels are high there are fewer areas of dry ground, possibly making it easier for predators to locate nests. In 2008, for example, water levels were much lower, and nest success higher, than the previous two years. Alternatively, high water levels may also make ponds unsuitable for shorebird prey that prefer shallow pools of water, while if ponds dry up too early in the breeding season there may not be adequate prey available for brooding chicks. In future years, as we collect and interpret water level data at the Mackenzie Delta site, we will be able to quantify the relationships between water levels, predation risk, prey availability, and shorebird reproductive success.

Extreme weather events are rare in the Mackenzie Delta but can have a profound, immediate effect on water levels. The Mackenzie Delta site experienced a major windstorm on 3 July 2006. Strong offshore winds (speeds up to 60 km/hr) forced a 20–100 cm rise in coastal water levels, which flooded study plots at Niglintgak and destroyed 45% of Red-necked Phalarope nests in lowlands ($n = 58$; Beveridge 2007). While the majority of shorebird species in the Delta nest on areas of raised ground in the wetlands, Red-necked Phalaropes often nest in clumps of sedge only a few centimeters above the water, so they are particularly susceptible to flooding events (L. Pirie, pers. obs.).

Water levels are more stable in the coastal wetlands of East Bay, but harsh weather events are common, and snow often persists until mid-June. Across years, the date of 50% snow cover has varied markedly, from 1 June (in 2001) to 19 June (in 2000). We monitor timing of snowmelt, and have found that it plays an important role in determining the timing of shorebird breeding (Smith et al. 2010; Smith et al., chapter 6, this volume). Snow can fall throughout the breeding season, and storms with high winds and cold rain are not uncommon. The storms can lead to temporary nest abandonment (Smith 2009), or could have negative consequences for the survival of newly hatched chicks.

Habitat

Changes in climate may induce subtle (or not so subtle) changes in shorebird breeding habitat. In delta areas, the frequency and duration of annual flooding, rate of sediment deposition, and erosion play an important role in determining the composition of vegetation (Gratto-Trevor 1997). Changes in vegetation (e.g., increased cover by shrubs) can reduce the amount of suitable nesting habitat for shorebirds (Pirie 2008, Pirie et al. 2009). Another agent of habitat change is the increasing arctic white goose population. The majority of arctic migratory bird sanctuaries were first established to protect the Snow Goose, yet overgrazing by these same geese now threatens the tundra grassland habitats within the sanctuaries. Dramatic increases in goose abundance, and consequent increases in grazing pressure on tundra grasslands, has led to a reduction in the sward height of the grass as well as an increase in "grubbing," where geese dig up the grasses' rhizomes and degrade the habitat (Kerbes et al. 1990, Abraham and Jefferies 1997, Kotanen and Jefferies 1997). Evidence suggests that these heavily grazed habitats are less suitable for shorebirds (J. Hines, unpubl. data). A study that measured shorebird densities in the Lesser Snow Goose colony at the Banks Island No. 1 Bird Sanctuary found that shorebird densities increased as one moved away from the goose colony; densities continued to increase up to 10 km away from the colony edge, where they stabilized (Hines et al. 2010). Sites that are even more heavily colonized by geese, such as the McConnell River Bird Sanctuary and the Dewey-Soper Bird Sanctuary, also have lower shorebird densities within areas of heavy goose grazing pressure than those outside of the colonies (J. Rausch, pers. comm.).

Nest Habitat

Nest habitat data is collected for each shorebird nest found in the intensive plots at Tier 2 sites. Nest habitat measurements are made at the nest and 1 m away in each cardinal direction and include vegetation height, percent cover, dominant vegetation type and the percentage ground cover of each, ground moisture, distance to water, and surface roughness. These data can be used to determine shorebird habitat preferences across the arctic, correlate habitat characteristics with shorebird densities, and may be compared with future habitat data when/if habitat changes occur. Ultimately, we will be able to determine whether shorebirds are adapting to the changes created by climate change or human activities or if they simply no longer use that location.

In the Mackenzie Delta, Whimbrel nests situated on hummocks exhibited markedly different vegetation characteristics than nests on ridges (Pirie 2008). At East Bay, an analysis of nest habitat data revealed species-specific habitat preferences and the consequences of these nest habitat choices in terms of reproductive success (Smith et al. 2007a, 2007b). For example, some shorebird species were found to select nest sites with taller surrounding vegetation than random sites, offering a higher degree of concealment. As grazing pressure from geese reduces the height of the vegetation, the availability of suitable nest sites for these species may be reduced.

Periodic Plant Collection (Floristic Changes)

At each Tier 2 site, we made a representative plant collection. Samples of all plant species, including shrubs, herbs, and graminoids, were collected, pressed, identified, and stored. Future plant collections will be made to determine whether there have been any shifts in plant species composition for each area in response to climate change or as a result of human-induced impacts such as the proposed Mackenzie Gas Project.

Arthropod Monitoring

Many shorebird species lay their eggs so that hatch occurs during the peak of arthropod (particularly mosquito and midge) emergence (Holmes and Pitelka 1968). Flies, mosquitoes, and chironomids form the bulk of a shorebird chick's diet, and if chick hatch occurs at a time of inadequate prey abundance, chick survival may be greatly reduced (Meltofte et al. 2007b). This "mismatch hypothesis," proposed by Visser et al. (2004), is a leading theory for effects of climate change on wildlife. At the Mackenzie Delta, we have been monitoring timing of arthropod emergence since 2005, using a variety of qualitative and quantitative means; so far, no ideal method of collecting arthropods has been found. Future plans are to adopt the methodology used at the East Bay site. At East Bay, "malaise traps" are used to monitor seasonal variation in arthropod abundance. Phenology of insect emergence clearly varies among years in relation to weather, and a predictive model has been constructed to hindcast the seasonal abundance of arthropods in past years. Work to determine the effect on shorebirds of mismatches in insect emergence and chick hatch is ongoing (E. Bolduc, unpubl. data).

Other Wildlife and Predator Abundance

Jaegers, arctic foxes, and gulls are common nest and juvenile shorebird predators in the arctic, and their predation of birds and their eggs can have a significant impact on nest success (Smith et al. 2007a, Smith 2009). Increased human presence results in higher numbers of predators who are attracted by supplemental food (garbage) or by the creation of nesting or denning habitat and predator perches (Johnson et al. 2007). At Tier 2 sites, we monitor the numbers of all potential predators and their activities in the shorebird breeding grounds. We also observe if predator activity increases during certain weather conditions and how shorebirds react to predator presence. We can also learn more about predator behavior in response to our research and monitoring projects—do predators take cues regarding nest location from our presence? We also observe if predator activity increases during certain weather conditions and how shorebirds react to predator presence. At the Mackenzie Delta, predator studies have thus far been limited to counting numbers of predators, but this method of counts per observer hour has been demonstrated to be a useful index (Hochachka et al. 2000).

At East Bay, we modeled the effect of predator numbers on the survival of shorebird nests, and contrasted their effect with that of lemmings and weather. Among years, variation in the encounter rate of predators was an extremely strong predictor of nest survival. In contrast, we found little effect of weather or abundance of lemmings, a supposed primary prey of tundra predators, on the survival of shorebird nests. Changes in arctic fox abundance explained the variation in nest survival of biparental species, whereas changes in abundance of jaegers explained variation in nest survival of uniparental species. Our results demonstrate that there is a correlation between predator abundance and variation in shorebird nest survival, but that the strength of the correlation may be influenced by both the behavior of the predators and the behavior of the nesting shorebirds (Smith 2009).

Bird Checklist Surveys

At Tier 2 sites, we conduct several bird checklist surveys per week for Tier 3 of the Arctic PRISM program (see Armer et al., chapter 12, this volume). Checklist surveys are used to detect trends in species distribution and, in some cases, in long-term trends in species population size. We also keep daily and seasonal records of all the bird and mammal species observed at each study site. These daily indices of abundance have proven useful to determine timing of shorebird arrival (see Smith et al. 2010), inter-annual changes in predator abundance, and a variety of other broad patterns of abundance within and among years.

Contributions to External Monitoring Programs

Both Tier 2 sites regularly contribute to the PlantWatch and IceWatch programs, joint ventures between the Canadian Nature Federation and Environment Canada's Ecological Monitoring and Assessment Network Coordinating Office (NatureWatch 2009). These programs evolved to help identify changes that may be occurring at a large spatial scale in the environment. PlantWatch has selected a set of plant species that are used across the arctic and sub arctic as indicators for environmental change. Observers in our field camps locate those species and record the date of the first bloom and mid-bloom of the flowering phase, as well as the date of leaf-out. For IceWatch, we select lakes within our study area to monitor spring "breakup." Observers recorded the percentage of ice on each lake and the day that the ice completely disappears from the lake.

RESEARCH

Tier 2 sites provide a logistic base for shorebird and non-shorebird research external to PRISM. The creation of these sites allows students and other researchers to conduct research at established study sites, which greatly simplifies project logistics.

Student Research

A number of graduate projects have focused on shorebird ecology at our Tier 2 sites. In the Mackenzie Delta, species-specific studies have been conducted on Red-necked Phalarope (Beveridge 2007; Walpole et al. 2008a, 2008b) and Whimbrel (Pirie 2008, Pirie et al. 2009). Both projects focused primarily on bird habitat use. East Bay has had, and continues to have, a wide range of shorebirds studies investigating topics such as Ruddy Turnstone behavioral endocrinology and breeding biology (Perkins 2004, Perkins et al. 2007), shorebird nest success (Smith et al. 2007a, Smith 2009, Smith and Wilson 2010), Red Phalarope nesting association with Sabine's Gulls (Smith et al. 2007b), shorebird toxicology (A. Hargreaves, unpubl. data), Semipalmated Plover foraging behavior (E. Davies, unpubl. data, P. Woodard, unpubl. data), shorebird nest defense (D. Turner, unpubl. data), nest camouflage (J. Gobin, unpubl. data), and sexual selection and parasitism (D. Edwards, unpubl. data). Non-shorebird projects have also been conducted, including research on the reproductive biology of Sabine's Gulls (Stenhouse et al. 2001) and arctic tadpole shrimp (L. Lachowsky, unpubl. data).

Lemming–Shorebird Connection

Lemmings are a major source of food for many predatory birds and foxes in arctic habitats (Pitelka et al. 1955, Angerbjörn et al. 1999). Due to the cyclical nature of lemming populations, predator pressure on shorebird eggs and young increases during years when lemmings are scarce (Underhill et al. 1989, Blomqvist et al. 2002). In the Mackenzie Delta, we record all of the lemmings we see during the course of our daily activities and calculate the number of lemmings seen per observation hour to determine the stage of the lemming cycle in a given year. In addition to recording lemming numbers/observer, we conduct a snap-trapping study at the East Bay site, following standard protocols (Krebs et al. 2008). This information is used to compare indices of lemming abundance to the proportion of successful shorebird nesting attempts. Contrary to our expectations, there was no consistent relationship between the abundance of lemmings and the survival of shorebird nests, and we found no support for an effect of lemming abundance in any of our nest survival models. In fact, we saw substantial fluctuations in shorebird nest survival even in four consecutive years where lemmings were extremely scarce (Smith 2009).

THE WAY AHEAD

It will take a number of years to fully develop the Arctic PRISM Tier 2 network in Canada. The first step is to undertake a feasibility study that identifies and prioritizes additional Tier 2 site locations. A number of candidate sites have been suggested, but a formal SWOT (strengths, weaknesses, opportunities, and threats) exercise should be undertaken to optimize the site establishment process. A reasonable goal is to have one additional priority Tier 2 site established in the next five years.

The other priority activities for Arctic PRISM Tier 2 development is to continue full participation in the Arctic Shorebird Demographic Network and publish all Tier 2 data collected to date.

ACKNOWLEDGMENTS

The number of people who have contributed to the East Bay and Mackenzie Delta Tier 2 sites are far too numerous to mention individually; they know who they are and we thank them for all their hard work. This report is PCSP Contribution No. 01109.

ONLINE CONTENT

Abstracts are available in French and Spanish from www.ucpress.edu/go/sab.
Une traduction du résumé est disponible en français.
Una traducción del resumen está disponible en español.

CHAPTER TWELVE

Arctic PRISM Tier 3

PROGRESS NOTES FROM THE NORTHWEST TERRITORIES–NUNAVUT BIRD CHECKLIST SURVEY

Lindsay A. Armer, Craig S. Machtans, and Brian T. Collins

Abstract. Checklist data comprise the third tier of sampling for the Arctic PRISM program. Data provided to date to the Northwest Territories–Nunavut Bird Checklist Survey has led to the revision of breeding ranges for 20 shorebird species in the Canadian arctic. Checklist data are showing some utility to detect long-term population trends in both arctic shorebirds and landbirds. Analyses in progress will lead to improved data consistency via targeting specific geographic locations and habitats for annual checklist data collection.

Key Words: arctic, birds, checklist data, distribution, Northwest Territories, Nunavut, PRISM.

The Northwest Territories–Nunavut Bird Checklist Survey (NNBCS) program was established in 1995 by the Canadian Wildlife Service as an economical approach to monitoring the distribution, abundance, and breeding status of all bird species in the Northwest Territories (NWT) and Nunavut. At its inception, there existed only scattered baseline data on bird species in either territory. Since then, the NNBCS has amassed over 125,000 records on 12,600 checklists from 6,800 unique locations, including historical (pre-1995) data sets (Canadian Wildlife Service Prairie and Northern Region, unpubl. data; Fig. 12.1). The NNBCS follows typical protocols for a checklist data collection program (Droege et al. 1998). Volunteers provide information on geographic location, general habitat, count type (e.g., traveling vs. stationary), area or distance surveyed, survey duration, number of each species observed, and breeding code for each species.

Arctic PRISM includes three tiers (Skagen et al. 2003). Tiers 1 and 2 are shorebird-specific surveys to estimate species abundance over time and geographic region. Tier 3 is the collection of checklist data through the existing structure of the NNBCS. The objectives of Tier 3 are threefold: (1) to detect variations in breeding distribution and phenology, (2) to provide general trends in shorebird abundance and distribution over time, and (3) to detect annual variations in shorebird distribution (Skagen et al. 2003). This note summarizes the NNBCS's progress on each of these objectives.

Armer, L. A., C. S. Machtans, and B. T. Collins. 2012. Arctic PRISM Tier 3: progress notes from the Northwest Territories–Nunavut Bird Checklist Survey. Pp. 195–200 *in* J. Bart and V. Johnston (editors). Arctic shorebirds in North America: a decade of monitoring. Studies in Avian Biology (no. 44), University of California Press, Berkeley, CA.

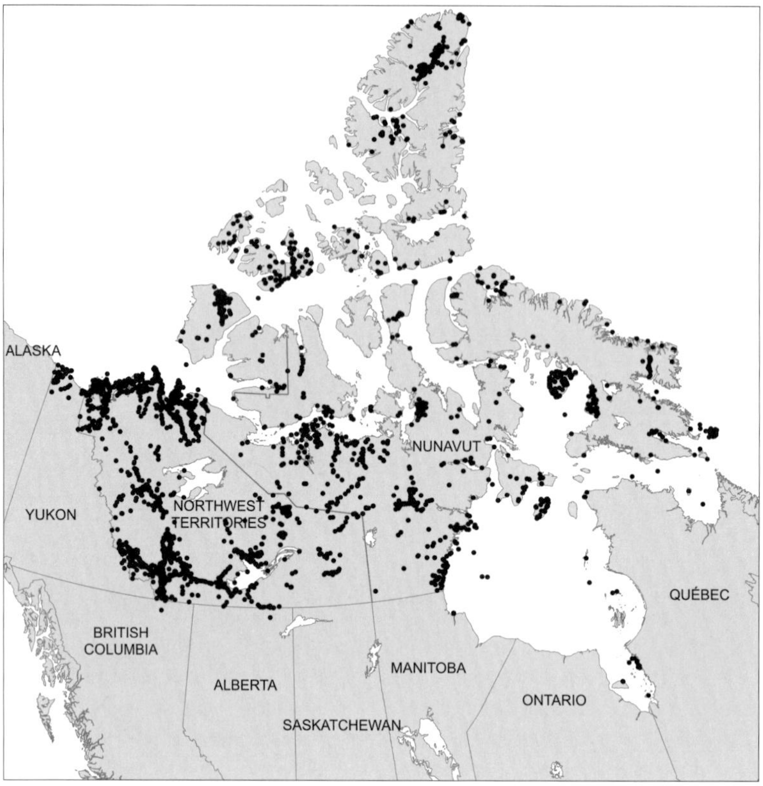

Figure 12.1. Distribution of observations in the Northwest Territories–Nunavut Bird Checklist Survey database as of 2009.

The NNBCS has made significant progress in identifying breeding range and phenology of arctic shorebirds. The NNBCS provided revised digital ranges in 2009 to update NatureServe.org's Digital Distribution Maps of the Birds of the Western Hemisphere by directly editing their GIS files (ver. 3.0, Ridgely et al. 2007). The NNBCS collects data for 21 shorebird species targeted by PRISM, and records from checklist data led to the revision of breeding ranges for 20 of these species (only the range for Red-necked Phalarope [for scientific names, see Appendix C] did not change). While some changes were minimal (e.g., addition of an arctic island not previously included in the range), some were more drastic. Stilt Sandpiper was reported as breeding in a thin band along the Beaufort Sea from Alaska to Cape Parry, NWT; along the southern shore of Victoria Island; and in the south near Churchill, Manitoba, by the Birds of North America species account (Klima and Jehl 1998). The *National Geographic Field Guide to the Birds of North America* (5th ed.; Dunn and Alderfer 2006) reported an expanded range that included a large section of mainland southern Nunavut. Observations provided from Arctic PRISM and NNBCS extended the known breeding range south from the arctic coast to include Snap and Daring Lakes, NWT, and from the Hudson Bay coast at Arviat, Nunavut, westward inland to Bernier Lake, Nunavut (Fig. 12.2). A confirmed breeding observation submitted to the NNBCS in 2008 (M. Gebauer, pers. comm.) further extended the probable breeding range to include the Kivalliq region of Nunavut, indicated by the dashed line and "?" in Figure 12.2. Godfrey (1986) reported potential breeding in this region, but until now it had not been confirmed.

With many regions yet to be surveyed across Nunavut and the Northwest Territories, we expect that the NNBCS will continue to revise and update breeding and migratory ranges for shorebirds. These revisions will continually be provided to NatureServe.org for global updates so that the information is readily accessible to researchers and the general public. All revised shorebird ranges are currently available through the authors until the NatureServe maps are updated with the NNBCS data.

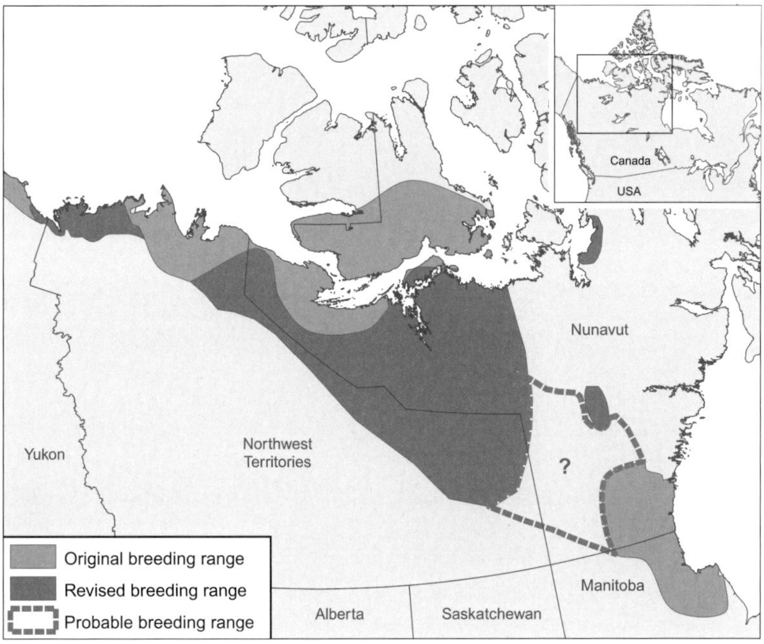

Figure 12.2. Revised breeding range of Stilt Sandpiper in the Northwest Territories and Nunavut, Canada.

Some progress has also been made in using NNBCS data to detect general population trends for arctic shorebirds and landbirds. Research from a similar checklist program—Études des Populations d'Oiseaux du Québec—has shown that general population trends can be detected from checklist data (Droege et al. 1998). Droege et al. (1998) found that even when observer or geographic bias associated with checklist data is considered, "large and abrupt changes" in population numbers are likely to be detected.

Trend and power analyses of all data in the NNBCS database were completed in early 2009. Presence/absence data at each site were used for the analysis, and checklist observations were grouped into hexagonal cells to provide an increased revisiting rate over time. Hexagons were sized as 1, 0.5, or 0.1 degrees latitude in width at a 60th parallel reference (equivalent to 56 km, 28 km, and 5.6 km widths) in order to assess the extent that cell size would influence the result. The proportion of sites in a cell where the species was present was analyzed using a logistic regression with factors for cell and year. Annual indices for the whole Bird Conservation Region (BCR 3 in this case) were calculated, and the population trajectory was then determined by using a simple linear regression on logistically transformed annual indices. Further details on the analysis technique are available from the authors.

When summarized on the 1-degree latitude/longitude scale, 42% of degree blocks with data only have one year of observations, and 95% of the blocks with data have less than 12 years of surveys. Essentially, this points to a weakness that was noted in a preliminary analysis of the checklist database in 2000—that the program should encourage observers to revisit sites periodically over time to track changes in abundance or distribution. Even with that caveat, the data are now sufficient to generate some meaningful results.

For the period 1987–2007 and all bird species within BCR 3, 12 of 60 species (20%) showed statistically significant trends (positive or negative) when NNBCS data were aggregated into hexagons with a width of 1 degree. For smaller spatial aggregations of data (0.5 and 0.1 degree hexagon widths), the proportion of species with statistically significant trends was 9 of 57 (16%) and 13 of 65 (20%), respectively. Table 12.1 shows the five statistically significant trends for arctic shorebirds calculated with current data from the NNBCS using the analytical technique described above. Like any other trend data, at least one other corroborating source is necessary to validate the trends.

TABLE 12.1

Significant results from presence/absence analysis with different cell sizes, Bird Conservation Region 3 (arctic) from NNBCS data.

	Hexagonal cell width (degrees latitude)[a]					
	1.0		0.5		0.1	
Species	Slope	*P*-value	Slope	*P*-value	Slope	*P*-value
Positive trends						
Black-bellied Plover	0.071	0.296	0.075	0.212	0.075*	0.020
Dunlin	0.109*	0.008	0.088*	0.013	0.093*	0.032
Semipalmated Sandpiper	0.072*	0.043	0.072*	<0.001	0.058	0.305
Negative trends						
Baird's Sandpiper	−0.122*	0.003	—		—	
Least Sandpiper	−0.028	0.423	−0.039	0.396	−0.082*	0.010

[a] Refers to the width of a hexagonal cell, in degrees, at 60° latitude within which checklist observations were grouped.
*Trends with $P < 0.05$.

The slopes shown in Table 12.1 are based on a logistic regression through the derived annual indices of detection probabilities. The change in distribution is presented as slopes in order to facilitate comparison of the estimated change across the different cell sizes. Due to the curvilinear shape of a logistic regression, the magnitude of the change in detection probability depends on both the slope and initial detection proportion.

The analysis presented in Table 12.1 is based on the presence/absence of the species in the analysis cells. Hence, the resulting analysis describes changes in the distribution of the species rather than changes in abundance *per se*. Some preliminary work to analyze the observed number of individuals, rather than presence/absence, was done. However, there was large variability in the number of individuals recorded, making this an ineffective approach to detect changes in abundance with the current data set. Checklist projects are most effective at monitoring changes in population distribution rather than population abundance. The analysis done here provides a more quantitative measure of distributional change, which would complement a monitoring protocol measuring population abundance. Comparing and contrasting the two results will give some insight into the manner in which changes in abundance are occurring.

The results could be further quantified to report the proportion of times the species is detected in a cell in the initial and final years. This would give a more readily understandable measure of the magnitude of the change. However, there is no way to adjust the results for detectability, and proportion of times detected should not be interpreted as an estimate of the underlying occupancy rate.

Some preliminary work on calculating the power of the NNBCS to detect a given level of distributional change was completed. The magnitude of change required to be detected was set at a decline of one-half the initial proportion over a period of 20 years (analogous to the target for PRISM Tier 1; Bart et al., chapter 2, this volume). Due to the curvilinear shape of the logistic regression, this gave different required slopes depending on the initial proportion. Power for a monitoring program depends on the number of sites, the number of years of monitoring, and the regularity by which the same cell is monitored over time. Predicting the regularity of visits is difficult because the survey is increasing in prominence, which suggests the number of sites visited each year could still be increasing; but the sampling is opportunistic, which suggests that regular monitoring of the same sites will be haphazard.

Table 12.2 highlights the results of the preliminary power analysis. The most common shorebird species could be adequately monitored for changes in cell occupancy (distributional changes that may be correlated with population declines) with little additional effort if the power target

TABLE 12.2

Required relative sample size (R) for 80% power to detect a decline of 50% with 21 years of data (hexagon width = 1.0) in Bird Conservation Region 3 (arctic) from NNBCS data.

R = 1.0 with the current data.

Species	Initial proportion	Required slope	Observed SE	Estimated parameters for 21 years	
				SE	*R*
American Golden-Plover	0.243	−0.042	0.039	0.019	1.64
Semipalmated Plover	0.151	−0.039	0.038	0.019	1.81
Semipalmated Sandpiper	0.203	−0.041	0.041	0.020	1.91
Black-bellied Plover	0.097	−0.037	0.041	0.020	2.34
Pectoral Sandpiper	0.109	−0.038	0.050	0.024	2.34
Red Knot	0.024	−0.035	0.050	0.024	3.28
Ruddy Turnstone	0.136	−0.038	0.055	0.027	3.86
Least Sandpiper	0.065	−0.036	0.058	0.028	4.74
Red Phalarope	0.123	−0.038	0.064	0.031	5.36
Baird's Sandpiper	0.128	−0.038	0.066	0.032	5.64
Red-necked Phalarope	0.104	−0.037	0.089	0.043	10.74
Dunlin	0.116	−0.038	0.094	0.046	11.73
Stilt Sandpiper	0.062	−0.036	0.160	0.078	36.83

was as specified. Conversely, there is little hope of detecting changes in several species even with dramatic increases in effort in the checklist program. Reaching the specified power for a selected subset of species could likely be achieved by focusing existing effort within the NNBCS, rather than merely trying to increase overall effort.

Data for non-shorebird species has been a significant contribution of Arctic PRISM to the NNBCS. Arctic PRISM collects breeding observations from landbird species that occur on and off of survey plots. These "incidental" observations from Arctic PRISM represent an important contribution to monitoring the breeding phenology and occurrence for otherwise poorly monitored species such as Snow Bunting, Horned Lark, Lapland Longspur, Common Redpoll, and Hoary Redpoll.

In the future, the NNBCS will continually provide PRISM with updated distribution information and estimated population trends of arctic shorebirds and landbirds. Improvement is still required to improve the NNBCS's ability to detect annual distribution of shorebirds. For example, there is a need to designate specific locations for consistent annual reporting, as suggested by Dunn et al. (2001). Although the NNBCS receives consistent annual checklist data from two or three locations in arctic Canada, it is unclear if they are in the most suitable habitat or geographic locations to allow for annual checklist surveys. Clearly, more than three locations will be required to detect annual changes in distribution over such a vast geographic region. Existing, long-term research locations that have not yet contributed data to the database would be desirable targets for expanding the database. More significant and structured involvement from the National Parks in BCR 3 would also be a useful and low-cost way to increase effort. Finally, financial assistance for observers to supply data from specific locations would also be a way to increase the effectiveness of the NNBCS. These are common challenges for volunteer-based programs in areas lacking proper roads, but ones that the NNBCS and PRISM could address over the coming years.

ACKNOWLEDGMENTS

Thanks to the hundreds of volunteers and researchers who have and continue to contribute data to the NWT-NU Bird Survey Checklist, this makes this analysis possible. Thanks to S. Kennedy for help with figure preparation. Thanks also to the PRISM team for envisioning a role for citizen science in long-term monitoring. Funding and administration of the NWT-NU Bird Checklist Survey is provided by the Canadian Wildlife Service, Environment Canada.

ONLINE CONTENT

Abstracts are available in French and Spanish from www.ucpress.edu/go/sab.
Une traduction du résumé est disponible en français.
Una traducción del resumen está disponible en español.

CHAPTER THIRTEEN

Design of Future Surveys

Jonathan Bart and Paul A. Smith

Abstract. This brief chapter addresses two related issues: how effort should be allocated to different parts of the sampling plan and, given optimal allocation, how large a sample will be required to achieve the PRISM accuracy target. Simulations based on data collected to date showed that two plots per cluster on rapid surveys, two intensive camps per field crew-year, 2–4 intensive plots per intensive camp, and 2–3 rapid surveys per intensive plot is the most efficient allocation of resources. Using this design, we investigated how crew-years should be allocated to each region in order to meet the PRISM accuracy target most efficiently. The analysis indicated that 40–50 crew-years would achieve the accuracy target for 18–24 of the 26 species breeding widely in the arctic. This analysis was based on assuming that two rounds of surveys were conducted and that a 50% decline occurred between them. We discuss the complexity of making these estimates and why they should be viewed as first approximations.

Key Words: Alaska, arctic, Canada, monitoring, PRISM, sampling effort, shorebirds, survey effort.

In this chapter, results from the first decade of Arctic PRISM are used to address two related issues: how effort should be allocated to different parts of the sampling plan and, with the optimal allocation, how large a sample will be required to achieve the PRISM accuracy target described in Johnston and Bart (chapter 1, this volume). We assume the reader is familiar with Bart et al. (chapter 2, this volume, Methods).

OPTIMAL ALLOCATION OF EFFORT

From Bart et al. (chapter 2, this volume), the squared CV of the estimated population size may be written

$$\left[CV(\hat{Y})\right]^2 = \sum_{h=1}^{L} W_h^2 \left(\frac{V(\hat{\bar{X}}_h)}{\bar{X}_h^2} + \frac{V(\hat{R}_h)}{R_h^2} \right) \qquad (1)$$

where $W_h = Y_h/Y$ is the fraction of the population in stratum h. Substituting the estimators for

Bart, J., and P. A. Smith. 2012. Design of future surveys. Pp. 201–210 *in* J. Bart and V. Johnston (editors). Arctic shorebirds in North America: a decade of monitoring. Studies in Avian Biology (no. 41), University of California Press, Berkeley, CA.

$V(\hat{\bar{X}}_h)$ and for $V(\hat{R}_h)$ from Bart et al. (chapter 2, this volume) yields

$$[CV(\hat{Y})]^2 \cong \sum_{h=1}^{L} W_h^2 \Bigg[\frac{1}{\bar{X}_h^2} \Bigg(V(\bar{x}_h) + \sum_{a=1}^{E} \beta_{ha}^2 V(\bar{u}_{ha}) - 2\sum_{a<}^{E}\sum_{d}^{E} \beta_{ha}\beta_{hd} Cov(\bar{u}_{ha}, \bar{u}_{hd}) \Bigg) \Bigg] + W_h^2 \left(\frac{V(\bar{x}'_h)}{\bar{X}_h^2} + \frac{V(\bar{y}_h)}{\bar{Y}_h^2} - \frac{2Cov(\bar{x}'_h, \bar{y}_h)}{\bar{X}_h \bar{Y}_h} \right). \quad (2)$$

Expression (2) can be rearranged to isolate the sample sizes and the population variances and covariances (see Appendix 13.1). The result may be expressed as

$$[C\hat{V}(\hat{Y})]^2 = \sum_{h=1}^{H} W_h^2 \Bigg\{ \frac{1}{c_h}\left[(1-f_h) g_{h1} + \frac{1}{n_h}\left(g_{h2} + \frac{g_{h3}}{m_h} \right) \right] + \frac{1}{c'_h}\left[(1-f'_h) g_{h4} + \frac{1}{n'_h}\left(g_{h5} + \frac{1}{m'_h}\left(g_{h6} + \frac{g_{h7}}{o'_h} \right) \right) \right] \Bigg\} \quad (3)$$

where the g-values are population variances, the *f*-values are finite population corrections, and the other terms are sample sizes. The prime (′) indicates intensive plots. The c_h, n_h, m_h, and o_h are the numbers of crew-years, clusters, plots per cluster, and surveys/plot, respectively, in stratum h. We assume that second- and third-stage sample sizes are the same within strata and that second- and third-stage finite population corrections are negligible.

We expressed the estimated proportion of the species' population in stratum h (w_h), as

$$W_h = \frac{\hat{Y}_h}{\Sigma \hat{Y}_h} = \frac{A_h \hat{\bar{Y}}_h}{\Sigma A_h \hat{\bar{Y}}_h}. \quad (4)$$

In well-surveyed regions, we had direct estimates of A_h. For other regions, we made tentative estimates of how strata would be defined and how large each would be. For most areas, we assumed the strata would be wetlands and uplands. We used the Circumpolar Arctic Vegetation Map to identify wetlands and to identify areas unsuitable for shorebirds (CAVM Team 2003). The CAVM only portrayed large lakes. To estimate the area in smaller lakes—which would be considered outside the stratum—we used a Geographical Information System (GIS) layer covering Canada that reported the fraction of each 1 km^2 covered by water as depicted on 1:50,000 topographic maps (R. Fernandes, pers. data). For the area of water in Alaskan strata, we used the landcover map by Muller et al. (1999). We let $A_h = A^*_h F_h$, where A^*_h was the size of stratum h as delineated on the vegetation map, and F_h was the estimated proportion of the stratum not covered by deep water.

It may be seen from expression (3) that the CV depends critically on the population variances (g-values). We had far too little data to estimate the g-values for every species and stratum. Most of the g-values, however, showed strong relationships with density (Fig. 13.1). To estimate the population variance (g) for each species, we therefore expressed the g-values as functions of density, estimated the parameters in the relationship using all of the data, and then used the models to estimate stratum-specific g-values. We investigated linear, exponential, and polynomial models to estimate the g-values. For most of the g-values, the best model was obtained by fitting a linear curve to ln(g) plotted against $\ln(\bar{y})$ and back-transforming to estimate the model parameters.

A simulation written in TrueBASIC® was then used to identify the most efficient sampling plan for a fixed total number of crew-years (c) (Kemeny and Kurtz 1992). A crew-year is defined as an 8–10 day field season with 3–4 clusters (typically 6–8 plots) covered per day. The program cycled through various values for number of rapid surveys on intensive plots (o'), number of plots per intensive camp (m'), number of intensive plot clusters per crew-year (n'), and number of plots per cluster in rapid surveys (m). For each array of values, the program used the estimated variances and covariances to calculate the overall CV of the estimated population size for each species, and it recorded various summary statistics such as the mean CV and the proportion of species for which the CV was less than a user-supplied value. We evaluated how to allocate resources among m, n', m', and o' based on these results.

We selected inputs for the analysis based on the fieldwork. We used sampling plans with total number of crew-years (c) = 24, 30, and 40, distributed fairly evenly among the regions. These values provide fair, good, and excellent achievement of the PRISM accuracy target. In all analyses, we assumed that intensive plots were completed in each crew-year ($c' = c$) and that the number of clusters per crew-year (n) was equal to 24 (Canada) or 40 (Alaska). We initially set the number of plots per zone in rapid surveys (m) = 2, number of intensive plot clusters per crew-year (n') = 2, number of plots per intensive camp (m') = 4,

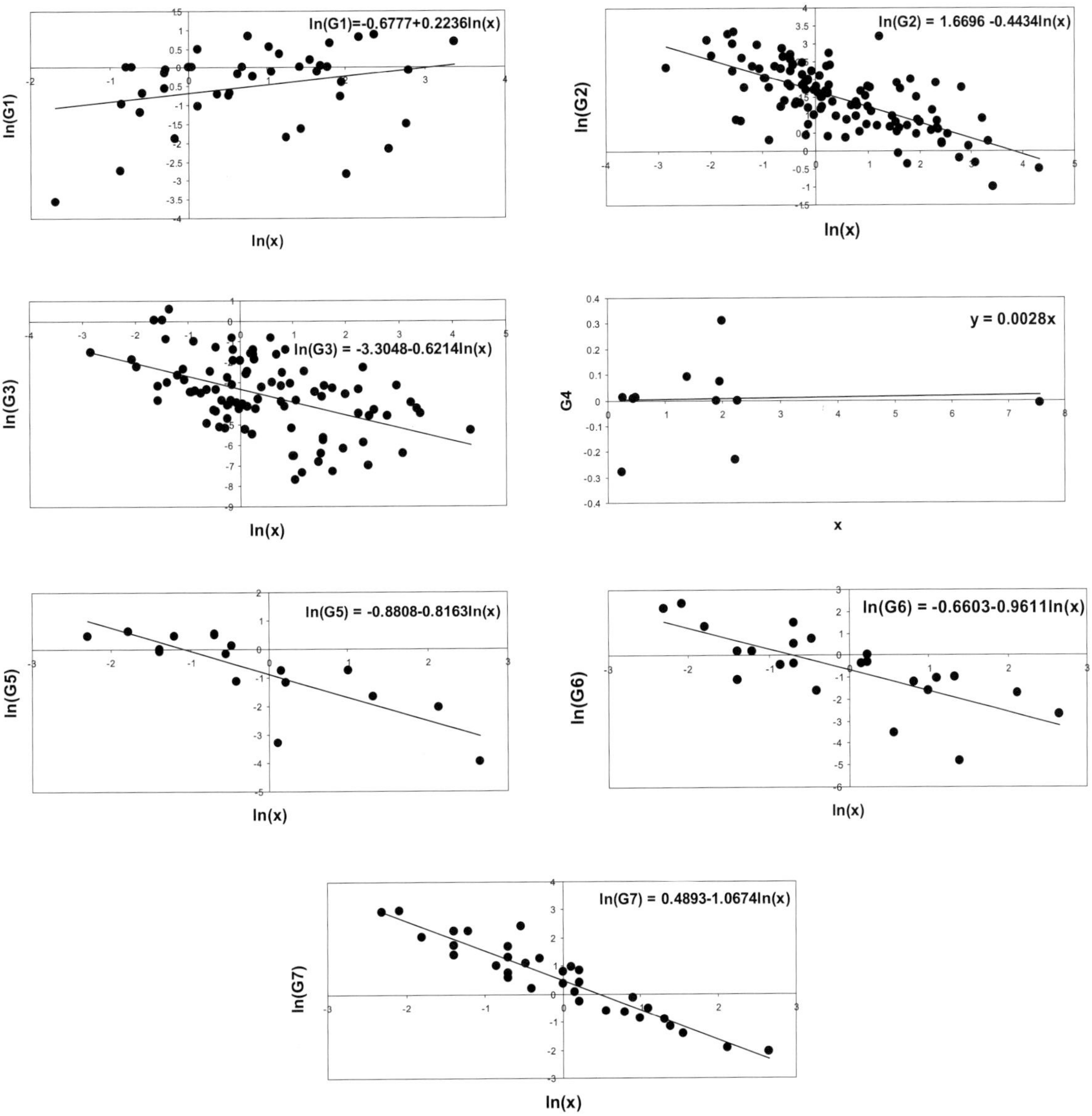

Figure 13.1. Equations used to estimate the seven population variances (g-values).

and number of rapid surveys on intensive plots (o') = 2, and then we varied each of these four quantities, adjusting other variables to maintain constant effort as much as possible.

We varied the number of rapid surveys on intensive plots, o', from one to five, and we decreased n so that the total number of plots surveyed remained constant. Results for $o' = 2$ and $o' = 3$ were similar (Table 13.1). Other values of o' resulted in fewer species adequately monitored and higher average CVs. Conducting more than one survey of each intensive plot reduces the risk of bias due to timing of the rapid surveys on these plots. The analysis suggests that conducting two or three surveys of intensive plots is optimal.

We next evaluated the effect of changing the number of plots per intensive camp (m'). There is no practical way to change other parts of the sampling plan to offset changes in m'. The effect of m' is important to assess, though, because different numbers of people can be brought to intensive camp, and because different intensive survey methods use different numbers of people per plot. The analysis showed that precision,

TABLE 13.1

Precision in relation to sample size for three sampling plans (A, B, and C) in which number of camps (c) = 24, 30, and 40, respectively.

	No. of species adequately monitored			Average CV		
Changes	Plan A	Plan B	Plan C	Plan A	Plan B	Plan C
$o' = 1$; increase n by 4	7	9	13	0.370	0.316	0.264
$o' = 2$ (base condition)	7	10	14	0.353	0.300	0.251
$o' = 3$; n reduced by 4	7	10	14	0.355	0.303	0.255
$o' = 4$; n reduced by 8	6	9	14	0.367	0.314	0.256
$o' = 5$; n reduced by 12	5	7	10	0.399	0.345	0.282
$m' = 2$; no other changes	6	9	12	0.391	0.335	0.282
$m' = 3$; no other changes	7	9	14	0.366	0.313	0.261
$m' = 4$; (base condition)	7	10	14	0.353	0.300	0.251
$m' = 5$; no other changes	10	12	14	0.345	0.293	0.244
$m' = 6$; no other changes	10	12	15	0.339	0.288	0.239
$n' = 1$ (no other changes)	6	9	9	0.428	0.370	0.312
$n' = 2$ (base condition)	7	10	14	0.353	0.300	0.251
$n' = 3$ (no other changes)	10	12	15	0.324	0.273	0.226
$m = 2$ (base condition)	7	10	14	0.353	0.300	0.251
$m = 3$; n reduced by 33%	6	9	12	0.379	0.325	0.273
$m = 4$; n reduced by 50%	6	8	12	0.404	0.349	0.295

NOTES: o' = rapid surveys/intensive plot, m' = intensive plots/camp, n' = intensive camps/rapid camp, m = rapid plots/zone, n = number of clusters/crew-year, c' = number of intensive plots/crew-year. In all plans, the "base condition" was $n = 16$ (Canada) or 36 (Alaska), $m = 2$, $c' = c$, $n' = 2$, $m' = 4$, and $o' = 2^1$. CVs were estimated for the 17 low arctic species and White-rumped Sandpiper.

and the number of adequately monitored species, increased rapidly as m' increased from two to four, but the gains, especially for precision, were smaller for $m' = 5$ or 6 (Table 13.1). This suggests that surveying four intensive plots per camp is worthwhile, regardless of which survey method is used.

We next addressed the costs and benefits of having one or three, rather than two, intensive camps per crew-year (n'). Having only one intensive camp reduced precision substantially (Table 13.1). The additional increase in precision, with three intensive camps, was modest (~10%). We did not attempt to adjust other components of the sampling plan to keep resources constant because the ways of doing so, and their practicality, would vary widely across the study area. We suggest that a plan with two intensive camps per crew-year provides the best balance of cost and effectiveness.

Finally, we addressed the effect of cluster size (i.e., number of plots per cluster) in the rapid surveys, estimating precision with m = 2, 3, and 4 and reducing the number of clusters to maintain the total number of plots. Precision declined as m increased from two to four, showing the advantage of having more locations rather than more plots per location (Table 13.1). We ignored increases in cost due to increased travel time because if clusters are sampled consecutively along a flight line, costs vary little with different values for m.

In summary, the following sample sizes appear to provide an efficient allocation of resources: two to three rapid surveys per intensive plot (o'), two to four intensive plots per intensive camp (m'), two intensive camps per field crew-year (n'), and two plots per cluster on rapid surveys (m). We regard the choice of c, the number of crew-years, as a sample size issue and discuss it in the next section.

SAMPLE SIZE REQUIREMENTS

The PRISM accuracy target is 80% power to detect a 50% decline occurring in no more than 20 years, using a significance level of 15% and acknowledging effects of potential bias (Skagen et al. 2003). In this section, we estimate sample size requirements, using the optimal design identified above, to achieve this target. We begin, however, with a brief explanation of why the "50% and 20 years" parameters were selected.

One of the main reasons for setting the decline at 50% was that detecting smaller declines rarely leads to significant conservation efforts, so it does not make sense to expend scarce resources detecting such declines. For example, in the current Breeding Bird Survey (BBS) estimates of trends during 1980–2007, 41 species have estimated declines exceeding 40% with P-values of <0.005. In other words, we have good evidence for major declines in at least 41 species. Among these, no significant investigations or conservation actions are underway for the species exhibiting declines of less than 50%, whereas significant resources have been expended for at least some of those species with declines of greater than 50% (e.g., Cerulean Warbler, Rusty Blackbird).

Furthermore, declines of 40–50% are often followed by recoveries without active management. For example, during 1966–1979, according to the BBS, 33 species declined more than 40% with P-values <0.01. During 1980–2007, 13 of these species increased, only ten of them showed continuing declines significant at the 5% level, and *all* of these had declines exceeding 50% during the earlier period. Thus, we do not believe that scarce resources should be devoted to identifying declines of less than 50%. We think that a better plan is to spend the funds needed to identify seriously declining species (declines greater than 50%) and then use the resources that could have been used to detect smaller declines, to discover the causes of declines exceeding 50%. This information will generally provide a better basis for predicting whether the declines will continue than the trend information alone.

The second part of the accuracy target is that a 50% decline should be detectable if it occurs in no more than 20 years. A decline occurring even faster is easier to detect because the decline during 20 years is then greater than 50%. The interval length, 20 years, was selected in part because changes at a large scale of greater than 50% in less than 20 years are rare. It does not make sense to design a survey that is capable only of detecting declines that rarely occur. On the other hand, waiting more than 20 years to detect a decline seems unacceptable from an adaptive management perspective. For additional discussion of how the parameters in the accuracy target were selected, see Bart et al. (2004).

We now return to discussing the sample sizes needed to achieve the accuracy target adopted by the shorebird initiative. If the population is declining at a rate equal to 50% in 20 years (about 3.4% per year), but the surveys are less than 20 years apart, then the total decline would be less than 50% and the decline would be harder to detect than if the surveys were 20 years apart. For example, if surveys were conducted 12 years apart, the total decline would be only 34%, and would be much more difficult to detect than if the surveys were 20 years apart. Surveys in Alaska can be conducted in a few years, but it seems likely that managers will not want to wait 20 years for a second survey. They will probably conduct the survey roughly once a decade. This means that within ~20–25 years, three surveys will have been conducted. At present, it is difficult to predict how much value will be added by the second survey or how an analysis using all three surveys will be conducted (regression with three data points is usually not recommended). Thus, the second survey will certainly add considerable value, but we do not know of a reliable way to predict how much it will increase precision. The situation in Canada is more complex because a round of surveys will take up to ten years. It seems likely that if the second round shows declines, then a third round will be undertaken more rapidly to determine if the apparent declines have continued.

In the face of this complexity, we have decided to present the estimated accuracy that would be achieved if only two surveys were conducted and a 50% decline occurred between the surveys. Methods for making these estimates are now presented.

Let the estimated ratio of population sizes, y_2/y_1, equal r and let the ratio of actual sizes be

R (e.g., $R = 0.5$ if population size declines 50% between the two surveys). We will declare the sample estimate significant if $|r - 1|/SE(r) > Z_{0.925} = 1.44$. We assume a large enough sample size to use Z-values and thus treat the $SE(r)$ as known. Since R is well below 1.0, we assume $r < 1$ and use $1 - r$ in place of $|r - 1|$. Power, the probability of rejecting the null hypothesis of no change in population size, is

$$P\left[\frac{1-r}{SE(r)} > 1.44\right] = \left[\frac{1-r}{SE(r)} - \frac{1-R}{SE(r)} > 1.44 - \frac{1-R}{SE(r)}\right]$$

$$= \left[N(0,1) > 1.44 - \frac{1-R}{SE(r)}\right]. \qquad (5)$$

A standard normal variable exceeds -0.84 with probability 0.8. Thus, achieving power of 0.8 requires a $SE(r)$ such that

$$1.44 - \frac{1-R}{SE(r)} = -0.84$$

or

$$SE(r) = \frac{1-R}{2.28} = 0.22 \qquad (6)$$

with $R = 0.5$.

The equation for the value of $SE(r)$ obtained in the surveys is

$$SE(r) = R\left(\frac{V(\hat{Y}_1)}{Y_1^2} + \frac{V(\hat{Y}_2)}{Y_2^2} - \frac{2Cov(\hat{Y}_1, \hat{Y}_2)}{Y_1 Y_2}\right)^{0.5}$$

$$\equiv R(CV_1^2 + CV_2^2 - 2C)^{0.5} \qquad (7)$$

where C is the relative covariance. Two problems arise in using expression (7) to estimate sample size requirements. If a new set of plots were visited in round two, then C would equal 0. It is anticipated that plots will be revisited, and nonzero covariance is therefore expected. The covariance term must thus be estimated. Second, as the population declines, it is difficult to predict how the CV will be affected. Since we lack data sets collected 20 years apart in the arctic, we had to estimate CV_2 and the relative covariance using other data. To do this, we expressed the $SE(r)$ in a different form,

$$SE(r) = CV_1 \alpha R(1 + \beta)^{0.5} \qquad (8)$$

where

$$\alpha = \left(\frac{CV_1^2 + CV_2^2 - 2C}{CV_1^2 + CV_2^2}\right)^{0.5}$$

$$\beta = \left(\frac{CV_2}{CV_1}\right)^2.$$

Thus, α is the ratio of SEs with paired and unpaired plots and β is the ratio of CVs in the two rounds of surveys.

We used data from the Breeding Bird Survey to estimate α and β. We initially used all species ($n = 37$) in which the numbers reported during 1980–2007 declined by at least 25% and the species was recorded on at least 500 routes. Three species were excluded due to various irregularities in the data, leaving us with 34 species for analysis. We used data from 1985 and 2004 and calculated α and β for each species. Values of α showed no trend with R, so we used the average value, which was 0.73. Values for β showed a fairly strong relationship with R. We used the value of an exponential trend line, fit to the observations, at $R = 0.5$, which was $\beta = 1.1$. Inserting these values for α and β, 0.5 for R, and 0.22 for the target SE and solving for CV_1 yields 0.42. Note that this target assumes the same plots are visited in round one and two. If different plots are visited, then the target CV drops to 0.31.

The PRISM target is for range-wide estimates. It thus has to be adjusted for species that occur widely outside the arctic. It can be shown (see Appendix 13.1) that dividing the accuracy target (0.42) by the square root of the proportion of the range within the arctic yields the CV_1 which must be achieved, assuming that a survey of equal precision is carried out across the rest of the range. If such a survey is not conducted outside the arctic, then the needed sample size within the arctic would be larger, but potential bias would also be larger because trends might not be the same inside and outside the arctic.

Sample size was estimated by selecting the total number of crew years and then assigning one crew-year to each region. The rest of the crew-years were then randomly assigned and CV_1 was calculated using density and the population variances (g-values). The number of species for which the species-specific target CV was met (i.e., "covered species") was then calculated. This procedure was repeated 50,000 times and the allocation of crew-years to regions with the largest number of covered species was reported.

The results (Table 13.2) showed that as the number of crew-years increased from 30 to 50, the number of adequately covered species (out of 26) increased from 18 to 24. The species most difficult to cover were Buff-breasted Sandpiper, Purple Sandpiper, and Hudsonian Godwit (for scientific

TABLE 13.2

Number of species for which the accuracy target is met with different sample sizes.

A. Division of effort between Canada and Alaska and overall coverage.

Number of crew-years	Total	30	35	40	45	50
	Canada	19	22	27	28	34
	Alaska	11	13	13	17	16
No. of species for which the accuracy target is met		18	22	22	23	24

B. Predicted CVs for individual species. Shaded areas indicate that the accuracy target is not met.[a]

Species	Prop. of range in arctic[b]	Target CV_1	Number of crew-years 30	35	40	45	50
Black-bellied Plover	0.83	0.46	0.29	0.25	0.24	0.23	0.20
American Golden-Plover	0.59	0.55	0.28	0.30	0.29	0.27	0.21
Pacific Golden-Plover	0.66	0.52	0.51	0.49	0.38	0.34	0.34
Semipalmated Plover	0.39	0.67	0.52	0.61	0.61	0.41	0.40
Whimbrel	0.30	0.77	0.43	0.45	0.34	0.35	0.29
Hudsonian Godwit	0.26	0.82	1.57	2.15	1.53	2.17	1.27
Bar-tailed Godwit	0.85	0.46	0.37	0.39	0.39	0.35	0.36
Ruddy Turnstone	0.95	0.43	0.51	0.29	0.29	0.29	0.28
Black Turnstone	0.72	0.49	0.46	0.46	0.46	0.46	0.46
Rock Sandpiper	0.74	0.49	0.41	0.37	0.41	0.38	0.38
Purple Sandpiper	0.99	0.42	2.34	1.22	1.21	1.22	1.20
Red Knot	0.99	0.42	0.52	0.34	0.34	0.33	0.33
Sanderling	0.99	0.42	0.40	0.34	0.34	0.34	0.26
Dunlin	0.80	0.47	0.36	0.34	0.34	0.33	0.33
Semipalmated Sandpiper	0.72	0.49	0.25	0.27	0.26	0.23	0.21
Western Sandpiper	0.64	0.53	0.47	0.47	0.47	0.47	0.47
Least Sandpiper	0.21	0.92	0.54	0.44	0.43	0.41	0.39
White-rumped Sandpiper	0.98	0.42	0.59	0.33	0.33	0.33	0.31
Baird's Sandpiper	0.78	0.48	0.37	0.34	0.34	0.33	0.24
Pectoral Sandpiper	0.95	0.43	0.29	0.34	0.32	0.28	0.23
Buff-breasted Sandpiper	1.00	0.42	0.49	0.56	0.56	0.52	0.38
Long-billed Dowitcher	0.79	0.47	0.39	0.52	0.52	0.32	0.38
Stilt Sandpiper	0.68	0.51	0.39	0.46	0.41	0.40	0.30
Wilson's Snipe	0.06	1.71	0.33	0.35	0.29	0.33	0.26
Red-necked Phalarope	0.43	0.64	0.30	0.33	0.30	0.32	0.28
Red Phalarope	0.89	0.45	0.54	0.31	0.31	0.31	0.29

[a] Sampling error affects cell values, so decisions about whether to shade a cell were made by inspecting all cells in the row (e.g., see first cell for Long-billed Dowitchers).

[b] Calculated using digital range maps from Ridgely et al. (2007) and arctic as defined by CAVM Team (2003)

names, see Appendix C). With 50 crew years, the accuracy target for Buff-breasted Sandpipers was just met. For Purple Sandpipers, the target was far below the estimated $CV(r)$; however, the range for this species is poorly known. For example, in the simulation we included no birds in Region 10 because our only formal surveys there did not detect it. Most of the region, however, lies within its known range and we detected two birds per plot on reconnaissance surveys on Melville and Prince Patrick Islands even though these areas are thought to be at the edge—or outside—of the range. Thus, Purple Sandpipers may actually be covered by the survey, though we cannot guarantee this. Hudsonian Godwits were rarely recorded on our surveys, and results with different numbers of crew-years were inconsistent (Table 13.2). It is difficult to predict how well or poorly this species will be covered by Arctic PRISM surveys. The numbers of crew-years for Canada varied from 19 to 34; the number for Alaska varied from 11 to 17. This analysis thus indicated that expending 40–50 crew-years per round of survey is likely to achieve the accuracy target for most though not all of the target species.

In conclusion, the estimation process in this chapter required numerous simplifications and assumptions. We lacked data from many areas and had little data for some species, so we had to estimate densities and then estimate variance components from these estimated densities. In general, estimating variance components is more difficult than estimating means (because of the squared term in variances). Thus, we cannot expect that the analyses in this chapter yielded highly precise estimates of future power to detect trends. On the other hand, there is no reason to think the estimates of precision are biased, and such uncertainty is nearly always a part of predicting the power of future surveys. Furthermore, the analysis does provide clear guidelines on how to allocate effort among stages of the sampling plan and between Alaska and Canada, and these predictions are probably much more accurate than the predicted absolute levels of precision. The analysis thus provides useful guidelines that can be used immediately for design of the surveys and plausible—though unproven—predictions about the total effort that will be needed. Fortunately, the precision actually obtained by the surveys can be checked against the predictions in this chapter as the survey proceeds and effort can be adjusted up or down as needed.

APPENDIX 13.1

Derivation of Population Variances

The population variances (g-values) were derived by writing the equations for the actual variances and covariances, deriving estimators for the terms these equations contained, and then collecting terms that described each stage of the sampling plan. The variances and covariances are

$$V(\bar{x}_h) = \frac{(1-f_h)S^2(\bar{x}_{hk})}{c_h} + \frac{\overline{S^2(\bar{x}_{hki})}}{c_h n_h} + \frac{\overline{S^2(x_{hkij})}}{c_h n_h m_h} \quad \text{(A1)}$$

$$V(\bar{x}'_h) = \frac{(1-f'_h)S^2(\bar{x}_{hk})}{c'_h} + \frac{\overline{S^2(\bar{x}'_{hki})}}{c'_h n'_h} + \frac{\overline{S^2(x'_{hkij})}}{c'_h n'_h m'_h} + \frac{\overline{S^2(x'_{hkijl})}}{c'_h n'_h m'_h o'_h} \quad \text{(A2)}$$

$$Cov(\bar{u}_{ah}, \bar{u}_{dh}) = \frac{(1-f_1)Cov(\bar{u}_{hak}, \bar{u}_{hdk})}{c_h} + \frac{\overline{Cov(\bar{u}_{haki}, \bar{u}_{hdki})}}{c_h n_h} + \frac{\overline{Cov(\bar{u}_{hakij}, \bar{u}_{hdkij})}}{c_h n_h m_h} \quad \text{(A3)}$$

The variance formulas are from Theorem 10.3 in Cochran (1977), assuming that the second- and third-stage finite population corrections (fpc) are negligible. The covariance formula may be derived using the methods used to prove Theorem 10.3. $V(\bar{u}_{ha})$ and $V(\bar{y}_h)$ are similar in form to $V(\bar{x}_h)$; $Cov(\bar{x}'_{ha}, \bar{y}_h)$ is similar to $Cov(\bar{u}_{ha}, \bar{u}_{hd})$.

Unbiased estimators of the variances and covariances in expressions (A1)–(A3) are

$$\hat{S}^2(\bar{x}_{hk}) = s^2(\bar{x}_{hk}) - \overline{s_2^2(\bar{x}_{hki})/n_h} \quad \text{(A4)}$$

$$\overline{\hat{S}^2(\bar{x}_{hki})} = \overline{s^2(\bar{x}_{hki})} - \overline{s^2(x_{hkij})/m_h} \quad \text{(A5)}$$

$$\overline{\hat{S}^2(x_{hkij})} = \overline{s^2(x_{hkij})} \quad \text{(A6)}$$

$$\overline{\hat{S}^2(\bar{x}'_{hkij})} = \overline{s^2(\bar{x}'_{hkij})} - \overline{s^2(x'_{hkijl})/m'_h} \quad \text{(A7)}$$

$$\overline{\hat{S}^2(x'_{hkijl})} = \overline{s^2(x'_{hkijl})} \quad \text{(A8)}$$

$$\widehat{Cov}(\bar{u}_{hak}, \bar{u}_{hdk}) = cov(\bar{u}_{hak}, \bar{u}_{hdk}) - \overline{cov(\bar{u}_{haki}, \bar{u}_{hdki})/n_h} \quad \text{(A9)}$$

$$\overline{\widehat{Cov}(\bar{u}_{haki}, \bar{u}_{hdki})} = \overline{cov(\bar{u}_{haki}, \bar{u}_{hdki})} - \overline{cov(u_{hakij}, u_{hdkij})/m_h} \quad \text{(A10)}$$

$$\overline{\widehat{Cov}(u_{hakij}, u_{hdkij})} = \overline{cov(u_{hakij}, u_{hdkij})} \quad \text{(A11)}$$

These expressions may be derived from expression (10.37) in Cochran (1977). For other formulas for intensive plots, add ′ to the sample size variable (e.g., n'_h rather than n_h).

Substituting the standard formulas for the terms above into expressions (A1)–(A3) and collecting terms yields

$$g_{h1} = \frac{1}{\bar{x}_h^2}\left[\hat{S}^2(\bar{x}_{hi}) + \sum_{a=1}^{E} b_{ha}^2 \hat{S}^2(\bar{u}_{hai}) - 2\sum_{a<}^{E}\sum_{d}^{E} b_{ha}b_{hd}\hat{Cov}(\bar{u}_{hai}, \bar{u}_{hdi})\right] \quad \text{(A12)}$$

$$g_{h2} = \frac{f_1}{\bar{x}_h^2}\left[\overline{\hat{S}^2(\bar{x}_{hij})} + \sum_{a=1}^{E} b_{ha}^2 \overline{\hat{S}^2(\bar{u}_{haij})} - 2\sum_{a<}^{E}\sum_{d}^{E} b_{ha}b_{hd}\overline{\hat{Cov}(\bar{u}_{haij}, \bar{u}_{hdij})}\right] \quad \text{(A13)}$$

$$g_{h3} = \frac{1}{\bar{x}_h^2}\left[\overline{\hat{S}^2(x_{hkij})} + \sum_{a=1}^{E} b_{ha}^2 \overline{\hat{S}^2(u_{hakij})} - 2\sum_{a}^{E}\sum_{<d}^{E} b_{ha}b_{hd}\overline{\hat{Cov}(u_{hakij}, u_{hdkij})}\right] \quad \text{(A14)}$$

$$g_{h4} = \frac{\hat{S}^2(\bar{x}'_{hi})}{\bar{x}'^2_h} + \frac{\hat{S}^2(\bar{y}'_{hi})}{\bar{y}_h^2} - \frac{2\hat{Cov}(\bar{x}'_{hi}, \bar{y}_{hi})}{\bar{x}'_h\bar{y}_h} \quad \text{(A15)}$$

$$g_{h5} = \frac{\overline{\hat{S}^2(\bar{x}'_{hij})}}{\bar{x}'^2_h} + \frac{\overline{\hat{S}^2(\bar{y}'_{hij})}}{\bar{y}_h^2} - \frac{\overline{2\hat{Cov}(\bar{x}'_{hij}, \bar{y}_{hij})}}{\bar{x}'_h\bar{y}_h} \quad \text{(A18)}$$

$$g_{h6} = \frac{\overline{\hat{S}^2(\bar{x}'_{hkij})}}{\bar{x}'^2_h} + \frac{\overline{\hat{S}^2(\bar{y}'_{hkij})}}{\bar{y}_h^2} - \frac{\overline{2\hat{Cov}(\bar{x}'_{hkij}, \bar{y}_{hkij})}}{\bar{x}'_h\bar{y}_h} \quad \text{(A19)}$$

$$g_{h7} = \frac{\overline{\hat{S}^2(x'_{hkijl})}}{\bar{x}'^2_h} + \frac{\overline{\hat{V}(y'_{hkij})}}{\bar{y}_h^2} \quad \text{(A20)}$$

The term g_{h7} is structurally different from the others because the estimated variance of the estimated number on a given plot, y_{hkij}, is either assumed to be 0.0 or is obtained from a modified Lincoln–Peterson method rather than from repeated sampling. No covariance term is included in g_{h7} because the intensive and rapid surveys of a given plot are independent.

Target $CV(r)$ for Species Breeding Outside the Arctic

Let $\hat{Y}_a$ denote the estimated the size of population in the arctic and $\hat{Y}_b$ denote the estimated size of the remaining population. Then we may write

$$CV(\hat{Y}_a + \hat{Y}_b) = \frac{\sqrt{V(\hat{Y}_a + \hat{Y}_b)}}{\hat{Y}_a + \hat{Y}_b}$$

$$= \sqrt{CV(\hat{Y}_a)^2 f_a^2 + CV(\hat{Y}_b) f_b^2} \quad \text{(A21)}$$

where $f_a = \dfrac{Y_a}{Y_a + Y_b}$ is the proportion of the population in the arctic. Let the accuracy target for the range-wide estimate be T, and assume that $CV(\hat{Y}_a) = T/\sqrt{f_a}$, $CV(\hat{Y}_b) = T/\sqrt{f_b}$. These surveys are "equally precise" in the sense that they have equal precision for a species whose range is 50% in the arctic and 50% outside the arctic. Substituting these values into (A21) yields $CV(\hat{Y}_a + \hat{Y}_b) = T$. This completes the proof. We lacked estimated population sizes, especially for outside the arctic, so we estimated f_a and f_b as the fraction of the species' ranges inside and outside the arctic.

ONLINE CONTENT

Abstracts are available in French and Spanish from www.ucpress.edu/go/sab.

Une traduction du résumé est disponible en français.

Una traducción del resumen está disponible en español.

PART FOUR

Synthesis

CHAPTER FOURTEEN

Summary

Jonathan Bart and Paul A. Smith

Abstract. This chapter summarizes results in previous chapters by providing estimated densities and population sizes, in the areas we have surveyed, for Alaska, Canada, and both regions combined. A total of 1,554 rapid plots, covering 232 km^2, and 83 intensive plots were surveyed during the study. The monograph presents more than 600 density estimates and more than 200 estimated population sizes. Densities for all shorebirds combined were highest in the Yukon–Kuskokwim Delta and next highest on the Alaska Peninsula and the NPRA. They were lower in eastern Alaska and most of Canada but were high in the Foxe Basin, including Southampton, Coats, Prince Charles, and Air Force Islands. Although densities were highest in wetlands, wetlands only contained about 45% of the shorebirds in our study area; uplands, in which densities were generally low, contained an estimated 27% of the total population. Uplands therefore should not be ignored in monitoring shorebird populations. We argue that the field and analytic methods are now well developed and can be relied on to produce rigorous estimates of density, habitat relationships, population size, and trends in population size. Comparison of our results with Morrison et al.'s (2006) estimates of population size indicate general agreement at the rank-order level, but where we could make detailed evaluations of their estimates, the Morrison et al. (2006) estimates are generally lower. All of the data, programs, and other results are available, free of charge, at http://greatbasin.wr.usgs.gov/CBM/default.asp?PageID=1.

Key Words: Alaska, arctic, Canada, density, monitoring, population size, PRISM, shorebirds.

Previous chapters explained the history and goals of the Arctic PRISM program (Johnston and Bart, chapter 1, this volume), described the general methods (Bart et al., chapter 2, this volume), and reported results from numerous regional surveys (chapters 3–8, this volume). This chapter presents combined estimates for all regions surveyed to date and compares the estimated densities and population sizes to previous estimates. We also identify major revisions of range maps, particularly those in the Birds of North America (BNA) species accounts, which our surveys indicate may be warranted.

Bart, J., and P. A. Smith. 2012. Summary. Pp. 213–238 *in* J. Bart and V. Johnston (editors). Arctic shorebirds in North America: a decade of monitoring. Studies in Avian Biology (no. 44), University of California Press, Berkeley, CA.

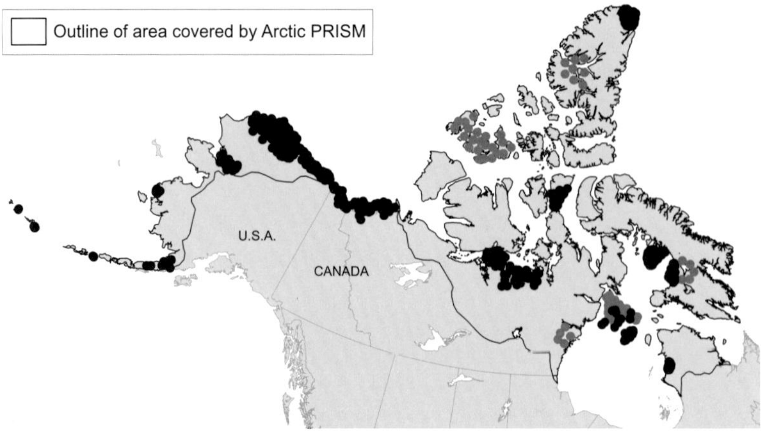

Figure 14.1. Areas that have been surveyed (black dots) and in which reconnaissance work has been conducted (gray dots).

As noted in Johnston and Bart (chapter 1, this volume), accurate estimates of density and population size are critical for managing shorebirds. For example, areas of critical habitat, such as those designated under the Western Hemisphere Shorebird Reserve Network, the Important Bird Areas program, and the Ramsar Convention, are all evaluated on the basis of proportions of species' populations which they contain. The shorebird conservation plan for the United States seeks to restore populations to numerical targets which are based on estimates of population size (Brown et al. 2001). Estimated densities and population sizes are also used to identify species at risk and causes of population decline, help design and refine management and conservation programs, and document success or the need for continued effort.

Despite their importance, accurate estimates of density and population size have proven difficult to obtain for many shorebird species. Some species disperse widely during migration, and varying lengths of stay makes enumeration difficult even for species that aggregate at stop-over sites. Estimates provided in Morrison et al. (2006) represent the best information currently available, and although the authors have devoted substantial effort to refining the estimates, the inherent difficulties in estimating population size at stop-over locations and wintering areas mean that 47 of 75 (63%) of the taxa described have population estimates that are considered correct to only an order of magnitude.

In this chapter, we present estimated densities and population sizes throughout the areas we have surveyed, evaluate the methods, compare our results to those of Morrison et al. (2006), and discuss the next steps in extending these surveys to the rest of the arctic in Alaska and Canada.

METHODS

We estimated population sizes for surveyed areas in Alaska, in Canada, and for both areas. Methods for the surveys are presented in Bart et al. (chapter 2, this volume). As noted there, the analytic approach ("combined estimates") does not permit us simply to sum the regional estimates to get the estimated totals for Alaska or Canada. A separate analysis must be carried out including all regions to be combined. Thus, the estimates in this chapter are the best ones for all surveyed regions in Alaska and all surveyed regions in Canada. The detection ratios were different in Alaska and Canada (because in Alaska surveyors estimated which birds were breeding on the surveyed plots, whereas in Canada all birds were counted), so we could not conduct a single analysis combining all regions. Instead, we added the results from Alaska and Canada to obtain the totals.

RESULTS

Areas Surveyed

Between 1997 and 2007, we completed rapid surveys on 1,554 plots, covering 232 km^2, in 26 regions widely distributed across the North American arctic (Fig. 14.1, Table 14.1;

TABLE 14.1

Region sizes, survey effort, and surveyed areas.

Region number	Name	Area in region (km²)	*n* rapid plots surveyed	Surveyed area (km²)
1.1	Québec	19,003	98	10.38
3.1	Prince Charles Island/Air Force Island	9,051	49	7.06
3.2	West Baffin	7,765	50	7.06
4.1	Southampton Island	24,877	24	3.02
4.2	Coats	3,170	29	3.48
8.1	Queen Maud Gulf—west	20,593	28	3.50
8.2	Queen Maud Gulf—east	23,310	40	4.49
8.3	Kent Peninsula	11,574	52	7.37
9.1	Creswell Bay	139	56	6.30
9.2	Somerset Island	4,220	25	4.37
10.1	Alert	135	61	7.53
12.1	Yukon NS	9,156	60	7.09
12.2	Mackenzie Delta—west	6,913	32	3.63
12.3	Mackenzie Delta—east	4,384	26	2.94
12.4	Mackenzie Delta—Mackenzie Gas Project stratum	301	53	5.89
13.1	Arctic National Wildlife Refuge	4,793	179	23.26
13.2	Central North Slope	10,826	42	5.99
13.3	Foothills-east	2,727	24	3.64
14.1	National Petroleum Reserve—Alaska NE	1,690	236	63.08
14.2	National Petroleum Reserve—Alaska coast	32,800	121	18.80
14.3	Foothills—west	20,527	14	2.23
14.4	Colville River	20	21	3.29
16.1	Selawik NWR	15,170	42	6.75
17.1	Yukon Delta—central	853	78	7.40
18.1	Alaska Peninsula	24,493	54	6.07
19.1	Alaska Maritimes	9,243	60	7.04
Total	All	267,731	1,554	232.00

chapters 3–7, this volume). Intensive surveys were completed on 83 plots. Surveyors recorded 806 pairs of shorebirds. Nests were found for 74% of these birds. Detailed reconnaissance surveys have been conducted in four additional areas on which rapid surveyors recorded 8,766 indicated pairs of shorebirds of 34 species (Bart et al., chapter 8, this volume).

Detection Ratios

The combined detection ratio (and CV) for Alaska was 0.81 (0.08) based on 567 indicated pairs. In Canada, the combined detection ratio was 1.27 (0.17) based on 242 pairs in intensive plots. The reason for the difference in rates is that in Alaska, surveyors attempted to exclude

TABLE 14.2
Species-specific detection ratios.

Species	Alaska			Canada		
	n of pairs	Detection ratio	SE	*n* of pairs	Detection ratio	SE
Black-bellied Plover	14	0.76	0.13	1	0.50	0.74
American Golden-Plover	9	0.88	0.26	4	2.63	1.05
Semipalmated Plover	1	1.25	0.38	0	—	—
Whimbrel	0	—	—	3	3.67	0.64
Hudsonian Godwit	0	—	—	1	7.00	10.34
Bar-tailed Godwit	6	0.54	0.21	0	—	—
Ruddy Turnstone	9	0.93	0.14	0	—	—
Rock Sandpiper	11	0.36	0.00	0	—	—
Dunlin	29	0.97	0.16	28	1.36	0.32
Semipalmated Sandpiper	111	0.75	0.13	43	1.19	0.26
Western Sandpiper	129	0.71	0.04	0	—	—
White-rumped Sandpiper	0	—	—	32	1.20	0.31
Pectoral Sandpiper	96	0.84	0.05	32	0.72	0.10
Buff-breasted Sandpiper	5	0.60	0.03	1	0.50	—
Long-billed Dowitcher	3	2.19	1.18	0	—	—
Stilt Sandpiper	4	0.61	0.11	12	0.57	0.26
Wilson's Snipe	1	2.33	1.26	0	—	—
Red-necked Phalarope	94	0.91	0.12	59	0.97	0.17
Red Phalarope	43	0.77	0.16	26	2.25	0.88

birds not nesting on the plot, whereas in Canada surveyors did not exclude birds (except for obvious transients). In Canada, two species, Whimbrel and Hudsonian Godwit (for scientific names, see Appendix C), had high ratios; however, only a few were present, so they have little influence on the combined ratio (it is 1.20 when they are excluded).

We investigated how detection ratios varied among species by calculating species-specific ratios for Alaska and Canada (Table 14.2). In Alaska, the detection ratios were variable among species with small sample sizes but were quite similar among species with large sample sizes (Fig. 14.2). The pattern was similar for Canada but less clear due to the smaller sample sizes. We discuss the most appropriate detection ratios to use for these and a few other species in the species accounts. See "Detection Ratios" in Bart et al. (chapter 2, this volume) for a more detailed discussion of when to use species-specific ratios.

Species Accounts

Black-bellied Plover

Black-bellied Plovers were widespread but generally uncommon. Breeding individuals were observed in 13 of the 26 surveyed subregions. Rapid surveyors recorded 212 indicated pairs. Intensive plots contained only 15 pairs, 14 of which were in Alaska. The detection ratio in Alaska was close to the combined ratio (Table 14.2).

Densities varied widely among regions, but were typically lower in Canadian regions than in Alaska (Table 14.3), especially in wetlands and moist areas. The highest densities were in wetlands on the NPRA coast region and on Prince Charles and Air Force

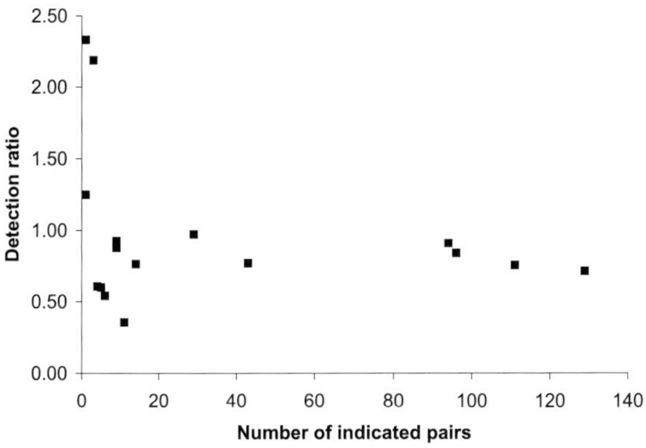

Figure 14.2. Detection ratios from Alaska plotted in relation to sample size.

TABLE 14.3
Estimated densities of Black-bellied Plovers (birds/km²).

Region	Wetlands	CV	Moist areas	CV	Uplands	CV
Prince Charles Island/Air Force Island	1.21	0.61	11.23	0.17	3.08	0.76
West Baffin	0.46	0.98	0.55	1.04		
Southampton Island			1.58	0.99		
Coats Island	2.19	0.45				
Creswell Bay					3.73	0.39
Somerset Island					4.53	0.72
Mackenzie Delta—east	1.26	1.03				
Central North Slope	5.06	0.59	1.67	0.69		
Foothills—east			2.64	0.99		
National Petroleum Reserve—Alaska NE	6.40	0.19	4.38	0.55		
National Petroleum Reserve—Alaska coast	11.05	0.21	3.20	0.42	1.33	1.08
Yukon Delta—central	2.78	0.80	7.16	0.59		
Alaska Peninsula			5.81	1.44	1.75	0.66

Islands, and both of these estimates had small CVs. Densities were much more similar among habitat types than for most other species. The estimated breeding densities were consistent with the statement by Paulson (1995:17) that density was "typically 1 to a few pairs/km^2 on breeding grounds."

The estimated population size (and CV) in surveyed areas was 262,733 (0.25) in Alaska and 51,560 (0.48) in Canada. The total for both areas was 314,293 (0.23). The Morrison estimate for this species is 200,000. Large areas exist that have not yet been sampled and within which density is probably at least 2 birds/km^2, so we suspect the population size is considerably more than 300,000.

We encountered Black-bellied Plovers at three sites outside the breeding range depicted in the BNA (Paulson 1995). These sites were the Alaska Peninsula (three pairs), Somerset Island (four probable nests), and Prince Charles Island (seven pairs).

American Golden-Plover

American Golden-Plovers, one of the most widespread shorebirds in the arctic, were recorded in 14 of 15 surveyed regions in Canada and seven of 11 surveyed regions in Alaska. Rapid surveyors recorded 222 indicated pairs. Only 12 pairs were present on intensive plots, nine in Alaska and

TABLE 14.4
Estimated densities of American Golden-Plovers (birds/km²).

Region	Wetlands	CV	Moist areas	CV	Uplands	CV
Québec	0.12	1.02				
Prince Charles Island/Air Force Island	0.55	1.01			2.12	0.59
West Baffin	1.22	0.84				
Southampton Island	2.97	0.83	4.21	0.64		
Coats Island	2.60	0.49			6.64	0.17
Queen Maud Gulf—west					0.71	1.13
Queen Maud Gulf—east			2.42	0.93	1.32	1.08
Kent Peninsula	0.18	1.13	0.99	0.93	4.13	0.44
Creswell Bay			2.97	0.49	4.35	0.49
Somerset Island			0.49	1.08	0.05	1.35
Yukon North Slope	1.06	0.99	0.69	0.99	3.50	0.68
Mackenzie Delta—west	0.56	1.06	3.32	0.58		
Mackenzie Delta—east	2.03	1.05				
Mackenzie Delta—Mackenzie Gas Project stratum	0.46	0.70	1.22	0.73		
Arctic National Wildlife Refuge	2.46	0.31	4.30	0.26	0.41	0.78
Central North Slope	1.06	1.04	3.37	0.36		
Foothills—east			1.82	0.94		
National Petroleum Reserve—Alaska NE	2.88	0.23	6.80	0.40		
National Petroleum Reserve—Alaska coast	2.50	0.32	8.80	0.32		
Foothills—west					5.50	0.79
Colville River					1.78	0.79

three in Canada. The detection ratio in Alaska was 0.88 (Table 14.2), quite close to the combined ratio (0.81), so there was no justification for using the species-specific ratio.

Estimated densities were generally lowest in wetlands (Table 14.4). In Alaska, they tended to be highest in moist areas; in Canada there may have been a tendency for higher densities in uplands, but the pattern was weak. Our findings were consistent with the statement by Johnson and Connors (1996a) that this species prefers to nest in "relatively dry, open tundra," or "moist habitats." Five of six estimated densities, summarized by Johnson and Connors (1996a), were >5 birds/km^2, whereas densities that high were uncommon in our surveys. It is difficult to know how representative of large areas are the studies of Johnson and Connors (1996a).

The estimated population size (and CV) in surveyed areas was 282,249 (0.30) in Alaska and 208,570 (0.26) in Canada. The total for both areas was 490,819 (0.21). The Morrison estimate for this species is 200,000. Large areas exist in the arctic that have not yet been surveyed and within which density is probably at least 1 bird/km^2. About 40% of the range is outside the arctic. The population is thus probably well above 500,000.

The BNA range map is consistent with our results, although it shows Coats Island as outside the range, whereas we found birds there (Johnson and Connors 1996a).

TABLE 14.5

Estimated densities of Semipalmated Plovers (birds/km²).

Region	Wetlands	CV	Moist areas	CV	Uplands	CV
Québec	0.92	0.41				
Kent Peninsula					0.43	0.98
Arctic National Wildlife Refuge	0.36	0.73	0.12	1.00	0.25	1.05
Foothills—east	20.48	0.49	0.25	1.16		
National Petroleum Reserve—Alaska NE	0.04	1.01				
Colville River					3.89	0.50
Alaska Peninsula	0.52	1.14	4.05	0.57		

Pacific Golden-Plover

This species was recorded only in western Alaska. See McCaffery et al. (chapter 3, this volume) for a summary of its distribution, abundance, and habitat relationships.

Semipalmated Plover

Semipalmated Plovers were encountered on rapid surveys in five regions in Alaska but only two in Canada. Rapid surveyors recorded 20 indicated pairs. Only one pair was present on intensive plots.

Estimated densities were variable but generally low (Table 14.5). High estimates in the Alaska Peninsula and Foothills East regions were each based on a single bird and are therefore unreliable. A high estimate from the Colville River was based on three birds but probably still should not be considered reliable. Semipalmated Plovers breeding in the arctic are usually associated with coastal habitats (Smith et al. 2007a) or well-drained gravel or shale areas inland (Nol and Blanken 1999), and our randomly selected plots were seldom in these restricted habitats. No birds were seen in the Mackenzie Delta on plots because their preferred habitat in the delta (man-made gravel pads) did not occur within our plots. Gratto-Trevor (1996) identified breeding plovers in earlier surveys in the Mackenzie Delta.

The estimated population size (and CV) in surveyed areas was 30,599 (0.76) in Alaska and 18,020 (0.39) in Canada. The total for both areas was 48,619 (0.50). The high CV reflects how poorly this species was sampled in our study. The Morrison estimate is 150,000. The estimated arctic population, based on the Morrison estimate multiplied by the fraction of the species' range within the arctic, is 58,500. A large area of this species' range in the arctic has not yet been sampled, so we suspect the arctic population is larger than suggested by the Morrison estimate, but we cannot be certain based on our surveys. We cannot evaluate the range map in the BNA because the species is so sparsely distributed. Large areas (e.g., on the North Slope) where the species is depicted as being present actually have few if any birds due to lack of suitable habitat.

Greater Yellowlegs

This species was recorded only in western Alaska. See McCaffery et al. (chapter 3, this volume) for a summary of its distribution, abundance, and habitat relationships.

Whimbrel

Whimbrels were encountered on rapid surveys in six regions in Alaska and in four regions in Canada. Rapid surveyors recorded 66 indicated pairs, but none were present on intensive plots.

The estimated densities for this species were highly variable (Table 14.6), with most estimates less than two birds/km² but much higher estimates from Selawik and from one of the three subregions in the Mackenzie region. Both of the high estimates were based on numerous birds. They included several plots with two or three pairs. The species is known for nesting in loose

TABLE 14.6
Estimated densities of Whimbrels (birds/km²).

Region	Wetlands	CV	Moist areas	CV	Uplands	CV
Yukon North Slope	1.42	0.52	2.68	1.00		
Mackenzie Delta—west	1.78	0.77				
Mackenzie Delta—east	0.69	0.89				
Mackenzie Delta—Mackenzie Gas Project stratum	5.37	0.53	6.76	0.60		
Arctic National Wildlife Refuge			0.54	0.78		
Foothills—east			0.25	1.16	0.73	0.96
National Petroleum Reserve—Alaska NE	0.12	0.60	0.11	0.73		
National Petroleum Reserve—Alaska coast			1.11	0.76		
Colville River					0.40	0.97
Selawik National Wildlife Refuge	5.44	0.36			5.13	0.85

colonies (Skeel and Mallory 1996), so these numbers are not unreasonable. On the other hand, the species is also known for mobbing behavior (Pirie and Johnston, chapter 10, this volume), especially late in the season, which is when the surveys at Selawik were conducted. It thus may have been difficult to determine how many pairs were nesting on the plot. If we assume that records of multiple birds were due to overcounting and reduce all the records of two or three pairs at Selawik and Mackenzie to one, then the estimated densities would decline by an average of 38%, in which case the estimated densities would still be substantially higher in these areas than elsewhere. It thus seems likely that the pattern in the densities is accurate even if the high values are positively biased.

The estimated population size (and CV) in surveyed areas was 71,098 (0.41) in Alaska and 19,683 (0.62) in Canada. The total for both areas was 90,781 (0.34). The Morrison estimate for this species is 66,000. Much of the range in the arctic, especially in western Alaska and west of Hudson Bay, remains to be surveyed, and only 30% of the species' breeding range is within the arctic. Notwithstanding the uncertainties about our estimated densities, discussed above, it seems likely that the total population is well in excess of 100,000.

The range map in the BNA appears consistent with our results (Skeel and Mallory 1996). In particular, we did not find the species east of the Mackenzie Delta despite having surveyed numerous areas in that part of Canada.

Hudsonian Godwit

We encountered this species very rarely. Three individuals, judged not to be breeding on the plots, were recorded on the Alaska Peninsula; a pair and an individual were recorded at Selawik; and six individuals were recorded on the Mackenzie Delta. These records are consistent with past records (Elphick and Klima 2002) but do not permit us to make meaningful estimates of density or population size. We expect that a special survey effort will be needed if this species is to be surveyed on its breeding grounds. See Rausch and Johnston (chapter 5, this volume) for additional discussion of the records in the Mackenzie Delta.

Bar-tailed Godwit

Bar-tailed Godwits were recorded mainly from the Yukon–Kuskokwim Delta and the NPRA, though single birds were recorded at Selawik and on the central North Slope of Alaska. The species-specific detection rate, based on a fairly

TABLE 14.7

Estimated densities of Bar-tailed Godwits (birds/km²).

Region	Wetlands	CV	Moist areas	CV	Uplands
Central North Slope			0.40	1.01	
National Petroleum Reserve—Alaska NE	2.26	0.27	1.59	0.68	
National Petroleum Reserve—Alaska coast	1.07	0.66	0.71	0.58	3.19
Selawik National Wildlife Refuge	0.60	0.98			
Yukon Delta—central	2.79	0.56	4.69	0.52	

large sample of 35 indicated pairs, was 0.54 with a SE of 0.21 (Table 14.2). The upper bound of the 95% confidence interval is 0.95, so the estimate is not close to being significantly different from the overall rate.

Densities in the Yukon Delta and NPRA in wetlands and moist areas were generally 1–4 birds/km² (Table 14.7). This species was rarely recorded in uplands. Overall density was especially high in the Yukon–Kuskokwim Delta because this area lacked uplands (i.e., dry, often tussocky, areas) in contrast to the NPRA, where uplands were common.

The estimated population size (and CV) was 81,208 (0.67). The Morrison estimate was 76,500. All but 7,400 of the estimated birds were on the North Slope. We suspect, however, that most of the actual population is on the Yukon–Kuskokwim Delta. We have no information about how many birds are present between Selawik and Point Hope. The actual population is thus probably much larger than our estimate. The range map in the BNA is fairly accurate, though some minor adjustments might be made on the North Slope (Bart et al., chapter 4, this volume).

Marbled Godwit

This species was recorded only in western Alaska. See McCaffery et al. (chapter 3, this volume) for a summary of its distribution, abundance, and habitat relationships.

Ruddy Turnstone

Ruddy Turnstones were recorded in four regions in Canada and three regions in Alaska. Rapid surveyors recorded 85 indicated pairs but most (71) of them were in three sites: NPRA and Prince Charles/Air Force Islands (presumably *A. i. morinella*) and Alert (presumably *A. i. interpres*). Only nine pairs were present on intensive plots, all in Alaska, so a reliable species-specific detection ratio could not be calculated.

Densities were quite variable (Table 14.8), reflecting the small sample sizes in most areas and the fact that this species often breeds in loose colonies (Perkins et al. 2007). The species was sparsely distributed across western Alaska and the North Slope, was not found on the Canadian mainland, and was present at most of the surveyed sites in the Canadian Archipelago. The five density estimates based on eight or more birds varied from 1.5 to 6.4 birds/km², and this range probably describes the densities in high-quality habitat. The high density in moist areas in Prince Charles and Air Force Islands is based on just two birds and thus should not be considered a reliable estimate. Overall, there was no clear pattern across habitats.

The estimated population size (and CV) in surveyed areas was 3,859 (0.38) in Alaska and 45,163 (0.42) in Canada. The total for both areas was 49,022 (0.38). The Morrison estimate was 245,000. The large CV reflects how difficult this species is to survey. Most of this species' core breeding range (the Canadian Arctic Archipelago) has not yet been surveyed, so the total population size is presumably much larger than our estimate. For example, we estimated 450 birds in our 135-km² study area at Alert, and on central Ellesmere and Axel Heiberg Islands we recorded 68 birds on just 11 plots covering ~1.5 km² of good shorebird habitat. Our results provide no reason to suggest a revision of the Morrison estimate.

TABLE 14.8

Estimated densities of Ruddy Turnstones (birds/km²).

Region	Wetlands	CV	Moist areas	CV	Uplands	CV
Prince Charles Island/Air Force Island	6.37	0.52	22.46	0.17	4.90	0.64
Coats Island	2.00	0.70				
Creswell Bay					1.24	0.73
Alert	3.34	0.34				
Arctic National Wildlife Refuge	0.40	0.57	0.60	0.82		
National Petroleum Reserve—Alaska NE	1.09	0.57	1.54	0.76		
Yukon Delta central	0.58	0.74	2.36	0.24		

Black Turnstone

This species was recorded only in western Alaska. See McCaffery et al. (chapter 3, this volume) for a summary of its distribution, abundance, and habitat relationships.

Rock Sandpiper

This species was recorded only in western Alaska. See McCaffery et al. (chapter 3, this volume) for a summary of its distribution, abundance, and habitat relationships.

Purple Sandpiper

Purple Sandpipers are restricted to the low and mid-arctic regions of Canada and are uncommon throughout this area (Payne and Pierce 2002). Only three birds were recorded on our regular surveys in Canada, and two of these were on Prince Charles Island, for which there are no previous breeding records. We did not find this species on Coats, Southampton, West Baffin, or Somerset Islands, though all of these are included in the range as depicted by Payne and Pierce (2002). On the reconnaissance survey of Melville and Prince Patrick Islands, we recorded 76 birds on 35 plots, widely distributed across both islands. Payne and Pierce (2002) were uncertain whether Prince Patrick Island was in the species' range and excluded Melville Island from the range. Our results suggest that the core breeding range for this species may be in the Queen Elizabeth Islands. The Morrison estimate for this species was 25,000. Our results provide no basis for re-evaluating their estimate.

Red Knot

We recorded single birds on West Baffin and Somerset Islands and two indicated pairs on Prince Charles and Air Force islands. The single bird on Baffin was previously banded in Argentina, suggesting that these birds were *C. c. rufa*. Another banded bird (banded in Delaware Bay) that was sighted off-plot on Southampton Island was also of the *rufa* subspecies. The main concentration (23 indicated pairs) occurred at Alert (*C. c. islandica*). The estimated density (and CV) was 4.80 (0.30). During reconnaissance surveys of Melville and Prince Patrick Islands, we recorded 77 birds (presumably *C. c. islandica*) on 22 plots widely distributed across both islands. On central Ellesmere and eastern Axel Heiberg Islands, another 31 birds were recorded on 11 ground survey plots and 139 birds on 3,500 km of aerial transect surveys. These results are consistent with the known range of *C. c. islandica* (Harrington 2001). We have no basis for estimating population size of either subspecies.

Sanderling

Sanderling were recorded on regular surveys only on Somerset Island (seven indicated pairs) and at Alert (11 indicated pairs). During reconnaissance surveys of Melville and Prince Patrick Islands, we recorded 24 birds on five plots.

TABLE 14.9

Estimated densities of Sanderlings (birds/km²).

Region	Wetlands	CV	Moist areas	CV	Uplands	CV
Somerset Island			2.27	0.65	9.97	1.08
Alert	2.30	0.42				

TABLE 14.10

Estimated densities of Dunlins (birds/km²).

Region	Wetlands	CV	Moist areas	CV	Uplands	CV
Québec	0.43	0.59				
Prince Charles Island/Air Force Island	3.95	0.70			1.18	0.79
West Baffin			3.40	0.84		
Southampton Island	12.53	0.46	7.38	0.46		
Coats Island	12.94	0.30				
Queen Maud Gulf—west	4.26	0.59	4.03	0.58	4.12	3.09
Queen Maud Gulf—east	7.47	0.30			2.00	1.90
Kent Peninsula	9.21	0.39	2.97	0.93	0.37	0.37
Arctic National Wildlife Refuge	2.89	0.41	0.04	1.02		
Central North Slope	8.27	0.43	3.92	0.48		
National Petroleum Reserve—Alaska NE	16.63	0.24	14.45	0.34		
National Petroleum Reserve—Alaska coast	32.05	0.17	5.12	0.39	4.87	3.48
Selawik National Wildlife Refuge	2.99	0.70				
Yukon Delta—central	198.99	0.19	8.91	0.42		
Alaska Peninsula	65.84	0.56			31.81	18.40

The few densities obtained for this species (Table 14.9) have high CVs and so cannot be considered reliable, but they do suggest that densities of two or more birds/km² may not be uncommon. We do not have sufficient data to estimate population size.

The BNA range map shows the species extending south to Southampton Island, where we did not find it, and occurring only along the coasts of Melville and Prince Patrick Islands (MacWhirter et al. 2002). Three of our locations for the species were in the interior of Melville Island, suggesting that the species is not restricted to the coast of these islands. Overall, our surveys indicate that the species may be most common in the Queen Elizabeth Islands.

Dunlin

Dunlin were widespread and abundant across our study area. Rapid surveyors recorded 1,065 indicated pairs, 955 of them in Alaska. Intensive plots had 57 pairs. The detection ratio for Alaska was 0.97 with a 95% CI of 0.66 to 1.29. The ratio in Canada (1.36) was close to the combined ratio (1.27). There was thus no justification for using either of the species-specific ratios.

Estimated densities were highest in wetlands in the Yukon Delta, where they averaged an amazing 199 birds/km² (Table 14.10). The next highest estimated densities were 66 in the Alaska Peninsula (wetlands), 32 in wetlands in the NPRA coast region, and 32 in uplands on the Alaska

TABLE 14.11

Estimated densities of Semipalmated Sandpipers (birds/km²).

Region	Wetlands	CV	Moist areas	CV	Uplands	CV
Québec	3.59	0.26				
Southampton Island	5.83	0.79	13.44	0.32		
Coats Island	60.81	0.25	7.94	0.32	6.64	0.17
Queen Maud Gulf—west	3.57	0.70	4.52	0.77	5.24	0.68
Queen Maud Gulf—east	4.69	0.67	4.84	0.93	11.73	0.28
Kent Peninsula	10.37	0.63	1.25	1.02	1.75	0.76
Yukon North Slope	2.61	0.70	0.30	1.06		
Mackenzie Delta—west	3.15	0.68	12.98	0.54	3.00	1.04
Mackenzie Delta—east	11.94	0.37			15.96	0.47
Mackenzie Delta—Mackenzie Gas Project stratum	3.58	0.61	1.23	1.03		
Arctic National Wildlife Refuge	15.32	0.22	8.15	0.29	0.18	1.06
Central North Slope	29.06	0.38	18.79	0.25		
Foothills—east			6.22	0.96		
National Petroleum Reserve—Alaska NE	49.73	0.14	29.25	0.21		
National Petroleum Reserve—Alaska coast	68.48	0.17	24.64	0.28	4.81	0.81
Selawik National Wildlife Refuge	18.29	0.58				
Yukon Delta—central	69.74	0.20	2.95	1.28		

Peninsula. Other densities were much lower. Overall, density was highest (by far) in wetlands and lowest in uplands, and higher in western Alaska than in northern Alaska or Canada.

The estimated population size (and CV) in surveyed areas was 1,698,017 (0.40) in Alaska and 448,720 (0.26) in Canada. The total for both areas was 2,146,737 (0.32). The Morrison estimates were 1,300,000 for Alaska and 225,000 for Canada. Given the high density in our Yukon Delta study area, and that only 1% of this region has been surveyed, the Alaska population is probably well above 2,000,000 birds. Substantial parts of the species' range in Canada remain to be surveyed. We therefore suspect both of the Morrison estimates are low.

The BNA range map is largely consistent with our results; however, we observed six indicated pairs in the West Baffin region, 11 indicated pairs on Prince Charles and Air Force Islands, and three indicated pairs in Northern Québec, suggesting that the breeding range for *C. a. hudsonia* may extend farther north and east than previously suspected (Warnock and Gill 1996).

Semipalmated Sandpiper

Semipalmated Sandpipers were encountered in 17 regions and were consistently absent only in the Queen Elizabeth Islands. On rapid surveys, we recorded 1,774 indicated pairs; 137 pairs occurred on the intensive plots. The detection ratios were 0.75 for Alaska and 1.19 for Canada. Each value was close to the corresponding combined estimate.

Estimated densities were much higher in Alaska than in Canada (Table 14.11), except for a remarkable density of 61 birds/km² on Coats

TABLE 14.12

Estimated densities of Western Sandpipers (birds/km^2).

Region	Wetlands	CV	Moist areas	CV	Uplands	CV
Arctic National Wildlife Refuge	0.14	0.98				
National Petroleum Reserve—Alaska coast	6.26	0.48	5.68	0.53	15.36	0.42
Selawik National Wildlife Refuge	4.10	0.60				
Yukon Delta—central	14.42	0.60	172.10	0.17		
Alaska Peninsula	1.55	0.76			1.75	0.66

Island. The species was found in all three habitats, but densities were consistently highest in wetlands and lowest in uplands. This is consistent with the variety of breeding habitat preferences, ranging from "wet sedge" to "variously drained upland tundra" described in Gratto-Trevor (1992).

The estimated population size (and CV) in surveyed areas was 1,447,560 (0.15) in Alaska and 809,697 (0.21) in Canada. The total for both areas was 2,257,257 (0.12). The Morrison estimate is 2,000,000. Large areas in the range remain to be surveyed both in Alaska (Yukon Delta) and Canada (e.g., the mainland west of Hudson Bay), and the species' range extends well outside the arctic. We therefore suspect the Morrison estimate is low. Morrison et al. (2006) also suggest that 75% of the breeding population may migrate through the Bay of Fundy, on the Atlantic coast of Canada. Our results suggest that this is an overestimate and a significant proportion might migrate through western Canada and the United States.

Western Sandpiper

Rapid surveyors recorded 224 indicated pairs of Western Sandpipers, all but five of them on the Yukon–Kuskokwim Delta, Selawik NWR, or in the coastal region of the NPRA. The species-specific detection rate, based on 129 indicated pairs, was 0.71 with a confidence interval of 0.63–0.80 (Table 14.2). The detection rate for this species thus may have been slightly lower than the combined rate of 0.81. Using a detection rate of 0.71 instead of 0.81 would increase densities and population sizes by 14%. Precision would be slightly higher because the CV of the detection ratio for Western Sandpipers was 0.06, whereas the CV of the combined estimate was 0.08. In this section we use the combined estimate of the detection ratio, which we regard as conservative with respect to both population size and measures of precision.

The highest densities, by far, occurred in moist areas of the Yukon–Kuskokwim Delta, where the estimated density was an astonishing 172 birds/km^2 (Table 14.12). The estimated densities in uplands of the coastal NPRA and in wetlands on the Yukon Delta study area both exceeded 14 birds/km^2, but values from all other regions and habitats were markedly lower. As noted in McCaffery et al. (chapter 3, this volume), although our estimate is remarkably high, it is consistent with past estimates in the Yukon–Kuskokwim Delta.

The estimated population size (and CV) in the surveyed areas was 474,347 (0.26). The North Slope estimate, where our surveys have been fairly complete, was about 322,000. Because we have only surveyed about 1% of the Yukon–Kuskokwim Delta, where the densities are so high, we do not have a good basis for estimating population size for this species or comparing our estimate with the Morrison estimate of population size. See McCaffery et al. (chapter 3, this volume) for additional discussion of this species.

The range map in the BNA shows the range not extending south of our Yukon Delta study area and occurring on the North Slope in only a few scattered locations (Wilson 1994). More recent maps such as National Geographic Society (2006) show the range extending south to the base of the Alaska Peninsula and over more of the North Slope, which is consistent with our results.

TABLE 14.13

Estimated densities of White-rumped Sandpipers (birds/km^2).

Region	Wetlands	CV	Moist areas	CV	Uplands	CV
Prince Charles Island/Air Force Island	47.30	0.43	112.31	0.17	58.48	0.63
West Baffin	18.83	0.33	22.11	0.38		
Southampton Island	10.53	0.54				
Coats Island	4.39	0.64	5.27	0.43		
Queen Maud Gulf—west					0.90	1.00
Kent Peninsula	1.11	1.05			0.27	1.03
Creswell Bay	12.57	0.42	17.85	0.27	15.53	0.26
Somerset Island	13.42	0.17	0.50	1.03		
National Petroleum Reserve—Alaska NE	0.55	0.90	0.14	1.01		
National Petroleum Reserve—Alaska coast	0.25	1.01				

Least Sandpiper

Least Sandpipers occur widely across the boreal zone of North America but only penetrate into the tundra in a few places and for short distances. Our records of this species reflect that distribution. They were recorded in eight regions, all bordering the boreal zone, but were common only on the Alaska Peninsula, where our plots frequently contained coniferous trees and where 36 of the 58 indicated pairs occurred. In the Alaska Peninsula, the estimated densities were 18.2, 0.0, and 9.6 birds/km^2 in wetlands, moist areas, and uplands, respectively. In other areas, estimated densities did not exceed 2.5 birds/km^2. The estimated population size in the areas we surveyed was 367,143 (0.27). Since nearly 80% of the species' range is outside the arctic, and substantial areas within the arctic breeding range have not yet been surveyed, we suspect the population size is larger than the Morrison estimate of 700,000.

White-rumped Sandpiper

White-rumped Sandpipers occurred in ten regions, almost exclusively in Canada, though five indicated pairs were recorded in the NPRA, which is consistent with past reports (Parmelee 1992b). The detection ratio in Canada was 1.20, which is close to the combined estimate.

The estimated densities were much higher on Prince Charles and Air Force Islands than elsewhere (Table 14.13). They did not show a strong preference for wetlands, as most other shorebirds do.

The estimated population size (and CV) in surveyed areas was 633,236 (0.28). Prince Charles, Air Force, and Southampton Islands were estimated to have nearly 555,000 birds. Large areas within the range of this species have not yet been surveyed, suggesting that the Morrison et al. (2006) estimate of 1,120,000 may be low. On the other hand, we do not at present know whether any high-density areas remain to be surveyed. If all other areas have low densities, then the Morrison estimate could be close to the real population size.

Baird's Sandpiper

Baird's Sandpipers occurred mainly in the Canadian Archipelago, though a few birds were recorded on the North Slope. The species was never common; only 30 indicated pairs were recorded.

In Alaska, most estimated densities were well below 1 bird/km^2, though the estimate was 14.36 in wetlands in the Arctic NWR (Table 14.14). The CV for this estimate was 1.0, however, indicating that the estimate was not reliable. Estimated densities were all less than 4 birds/km^2 in Canada.

TABLE 14.14
Estimated densities of Baird's Sandpipers (birds/km²).

Region	Wetlands	CV	Moist areas	CV	Uplands	CV
Queen Maud Gulf—west					1.41	1.13
Kent Peninsula	2.52	0.52				
Creswell Bay			1.78	0.60	3.73	0.44
Somerset Island			3.65	0.63	2.49	1.08
Yukon North Slope	0.29	1.03	0.69	0.99		
Arctic National Wildlife Refuge	14.36	1.00	0.25	1.01		
Central North Slope			0.55	1.02		
National Petroleum Reserve—Alaska NE	0.02	1.02				
National Petroleum Reserve—Alaska coast			0.32	1.00		

The estimated population size (and CV) in surveyed areas was 22,797 (0.69) in Alaska and 25,249 (0.50) in Canada. The total for both areas was 48,046 (0.42). The Morrison et al. (2006) estimate for this species was 234,000. Given how much of the range remains to be surveyed, our results seem consistent with the Morrison estimate.

Pectoral Sandpiper

Pectoral Sandpipers were one of the most widely and frequently recorded species in both Alaska and Canada. They were recorded in 17 of the 26 regions. They were absent only from the southernmost regions (Aleutian Islands, Alaska Peninsula, and Québec) and the northernmost regions (Somerset Island north of Creswell Bay, and Alert). The detection ratio for Alaska was 0.84, nearly identical to the combined rate. In Canada, the ratio was only 0.72 with a small SE of 0.10, indicating that the ratio may have been lower than the combined ratio. But the ratio from Alaska, based on a much larger sample, suggests that the low rate in Canada may have been caused by sampling error rather than by a true difference in detection ratios. We used the combined rate in the analyses below, in part because it produces conservative estimates, but we recognize that if the ratio in Canada was really 0.72, then our estimated densities and population sizes for Canada have substantial negative bias.

Estimated densities were highest from the North Slope of Alaska to the Queen Maud Gulf Migratory Bird Sanctuary (Table 14.15). The maximum values were 30–40 birds/km² in Alaska and 15–20 birds/km² in Canada. This species occurred in all three habitats, but estimated densities were generally highest in wetlands and lowest in uplands.

The estimated population size (and CV) in surveyed areas was 1,146,956 (0.18) in Alaska and 432,767 (0.29) in Canada. The total for both areas was 1,579,723 (0.15). The Morrison estimate was only 475,000, less than a third of our estimate. Our estimate had a small CV, and large areas, particularly in mainland Canada where densities were high, remain to be surveyed. It thus seems probable that the Morrison estimate is substantially too small.

Buff-breasted Sandpiper

Buff-breasted Sandpipers were recorded on the North Slope of Alaska and at Creswell Bay on Somerset Island. Only 30 indicated pairs were recorded, making this one of the least-often sighted species. Only six indicated pairs occurred on intensive plots, so reliable estimates of the species-specific detection rate could not be calculated.

This species was recorded only in wetlands and moist areas (Table 14.16). Seven of the estimates were between 0.9 and 3.0 birds/km². In regions where birds were found in both wetlands

TABLE 14.15

Estimated densities of Pectoral Sandpipers (birds/km²).

Region	Wetlands	CV	Moist areas	CV	Uplands	CV
Prince Charles Island/Air Force Island	0.71	0.70			0.06	1.11
Southampton Island			1.93	0.98		
Queen Maud Gulf—west	4.03	0.57	1.50	1.09	4.44	0.92
Queen Maud Gulf—east	14.02	0.41	18.00	0.58	1.32	1.08
Kent Peninsula	16.24	0.30	2.22	0.69	2.75	0.49
Creswell Bay	2.79	0.67	2.38	0.47		
Yukon North Slope	10.36	0.41	2.15	0.78		
Mackenzie Delta—west	5.16	0.65	1.66	0.96		
Mackenzie Delta—east	16.41	0.33			6.71	0.65
Mackenzie Delta—Mackenzie Gas Project stratum	2.73	0.92	1.84	1.03	3.20	1.12
Arctic National Wildlife Refuge	17.70	0.22	7.79	0.26	2.36	0.63
Central North Slope	22.60	0.27	19.09	0.27		
Foothills—east			8.86	0.67	1.57	0.99
National Petroleum Refuge—Alaska NE	35.69	0.13	32.69	0.25		
National Petroleum Refuge—Alaska coast	37.09	0.18	9.44	0.23	23.32	0.43
Colville River					4.96	1.19
Yukon Delta—central	4.75	0.99	8.60	0.22		

TABLE 14.16

Estimated densities of Buff-breasted Sandpipers (birds/km²).

	Wetlands		Moist areas	
Region	Dens	CV	Dens	CV
Arctic National Wildlife Refuge	1.10	0.72	0.04	1.02
Central North Slope	2.30	1.01	0.91	0.71
National Petroleum Refuge—Alaska NE	1.62	0.66	0.04	1.01
National Petroleum Refuge—Alaska coast	1.35	0.98	0.92	0.71
Creswell Bay			2.97	0.55

and moist areas, density was consistently higher in wetlands.

Due to the small size of the Creswell Bay region, only 281 birds were estimated to be breeding there. In Alaska, the estimated population size (and CV) in surveyed areas was 42,558 (0.44). The Morrison et al. (2006) estimate was 30,000. Most of the range of this species remains to be surveyed, so we suspect the Morrison estimate is low.

Short-billed Dowitcher

This species was recorded only in western Alaska. See McCaffery et al. (chapter 3, this volume) for

TABLE 14.17

Estimated densities of Long-billed Dowitchers (birds/km²).

Region	Wetlands	CV	Moist areas	CV	Uplands	CV
Mackenzie Delta—west	1.26	1.05				
Mackenzie Delta—Mackenzie Gas Project stratum	0.68	0.92				
Québec	0.44	0.59				
Yukon North Slope	5.59	0.51				
Arctic National Wildlife Refuge	1.04	0.71	0.99	0.58	0.43	1.04
Central North Slope	5.23	0.44	5.12	0.44		
Foothills—east			2.64	0.99		
National Petroleum Reserve—Alaska NE	8.92	0.17	6.93	0.34		
National Petroleum Reserve—Alaska coast	29.10	0.19	11.62	0.23	10.59	0.49
Colville River					0.99	1.19
Selawik National Wildlife Refuge	1.20	0.98				
Yukon Delta—central	1.24	0.83	8.17	0.70		
Alaska Peninsula					1.75	0.66

a summary of its distribution, abundance, and habitat relationships.

Long-billed Dowitcher

Long-billed Dowitchers were recorded in 13 regions from the Alaska Peninsula to Québec. Rapid surveyors recorded 402 indicated pairs, 350 of them in NPRA. Unfortunately, only three pairs were present on intensive plots, far too few to justify a species-specific detection ratio.

The species was generally recorded in wetlands or moist areas, though it also occurred in uplands in four regions (Table 14.17). Densities were particularly high in the coastal NPRA (e.g., 29 and 12 birds/km^2 in wetlands and moist areas) and were lower, though still above 5 birds/km^2, in the Yukon–Kuskokwim Delta, central North Slope, and Yukon North Slope areas. Elsewhere, the estimates were close to one bird/km^2.

The estimated population size (and CV) in surveyed areas was 704,387 (0.17), with 99% of the total coming from Alaska, mainly the North Slope. The Morrison estimate was 316,000. Given the small CV for our estimate, and the relatively high density in moist areas of the Yukon–Kuskokwim Delta (most of which remains to be surveyed), we suspect the population size is substantially larger than 700,000.

Stilt Sandpiper

Stilt Sandpipers were regularly recorded from the North Slope of Alaska east to the western part of the Queen Maud Gulf Migratory Bird Sanctuary. Rapid surveyors recorded 171 indicated pairs. Intensive plots contained 16 indicated pairs, with a small sample of four in Alaska and a marginal sample of 12 in Canada. For both areas, the detection ratios were lower than the combined estimates, 0.61 in Alaska and 0.57 (95% CI = 0.06–1.08) in Canada. Due to the small sample sizes, we used the combined ratios but acknowledge that these ratios may have been high. The estimated densities and population sizes would have been approximately twice as large if the species-specific ratios had been used.

Densities were consistently highest in wetlands (up to about 6 birds/km^2) and lowest in uplands

TABLE 14.18

Estimated densities of Stilt Sandpipers (birds/km²).

Region	Wetlands	CV	Moist areas	CV	Uplands	CV
Queen Maud Gulf—west			3.73	0.85		
Kent Peninsula	3.47	0.43			1.86	0.68
Yukon North Slope	3.75	0.49				
Mackenzie Delta—west	1.41	0.65	1.66	0.96		
Mackenzie Delta—east	3.24	0.51			1.89	1.08
Mackenzie Delta—Mackenzie Gas Project stratum	2.40	0.73	0.61	1.03		
Arctic National Wildlife Refuge	4.25	0.36	1.39	0.57		
Central North Slope	3.91	0.64	2.41	0.55		
National Petroleum Reserve—Alaska NE	3.27	0.24	1.19	0.42		
National Petroleum Reserve—Alaska coast	6.34	0.28	2.03	0.51		

(Table 14.18). They were slightly higher on the North Slope than in the Canadian regions. Overall, estimates within habitat types were remarkably consistent.

The estimated population size (and CV) in surveyed areas was 124,180 (0.22) in Alaska and 52,737 (0.47) in Canada. The estimated total was 176,917 (0.21). The Morrison estimate was 557,600. Much of the range has not yet been surveyed, so the actual population size is probably considerably larger than our estimate. We thus have no basis for suggesting any change to the Morrison estimate.

Wilson's Snipe

Wilson's Snipe were recorded in 13 regions, widely distributed across our study area from the Alaska Peninsula to Québec, but not into the Canadian Archipelago. Rapid surveyors recorded 98 indicated pairs. Intensive plots contained only one indicated pair.

Densities were highly variable across the range (Table 14.19) but CVs were generally large, making it impossible to know whether the variation reflected variation in density or sampling error. An estimate of 79 birds/km² from the Colville River presumably reflects sampling error rather than high density, though the habitat there (wet brushy areas) is both unusual in our study area and typical of where the species occurs elsewhere in its range. Unlike most other species, there was no strong tendency for density to vary among habitats.

The estimated population size (and CV) in surveyed areas was 242,045 (0.49) in Alaska and 74,210 (0.31) in Canada. The estimated total was 316,255 (0.38). The Morrison estimate was 120,000. Our study area includes a very small fraction of this species' range, so the total population size is surely much larger than the Morrison estimate.

Red-necked Phalarope

Red-necked Phalaropes were widely distributed across our study area, occurring in 16 regions from the Alaska Peninsula to the eastern part of the Queen Maud Gulf Migratory Bird Sanctuary. They were not recorded in Québec (though our sample there was small) or the Canadian Archipelago. Rapid surveyors recorded 927 indicated pairs. Intensive plots contained 153 pairs, giving us an excellent basis for calculating species-specific detection ratios. These ratios (and CVs) were 0.91 (0.13) in Alaska and 0.97 (0.18) in Canada. Both species-specific rates are lower, but not significantly lower, than the corresponding combined rate, so we used the combined ratios. Use of the species-specific rates would increase estimated densities and population sizes by 12% and 31% in Alaska and Canada, respectively.

TABLE 14.19

Estimated densities of Wilson's Snipe (birds/km²).

Region	Wetlands	CV	Moist areas	CV	Uplands	CV
Québec	0.97	0.48				
Queen Maud Gulf—west	1.88	1.01				
Yukon North Slope	1.89	0.64	0.72	0.99		
Mackenzie Delta—west	10.95	0.44	1.66	0.96	1.57	1.03
Mackenzie Delta—east	1.69	1.00			3.84	0.50
Mackenzie Delta—Mackenzie Gas Project stratum	3.45	0.38	2.80	0.69	9.60	1.12
Foothills—east			3.11	0.96		
National Petroleum Reserve—Alaska NE	0.25	0.58	0.21	0.61		
National Petroleum Reserve—Alaska coast			0.09	0.99		
Colville River			79.07	0.81	7.22	0.26
Selawik National Wildlife Refuge	0.17	1.04				
Yukon Delta—central	2.73	1.11	8.29	0.25		
Alaska Peninsula	8.16	0.63			9.78	0.53

Using the species-specific rates would produce estimates with slightly larger CVs in Alaska and nearly identical CVs in Canada.

Estimated densities were often quite high (Table 14.20); 13 strata had estimates exceeding 10 birds/km², and in the Yukon–Kuskokwim Delta the estimates in both wetlands and moist areas exceeded 55 birds/km². Densities were slightly lower in Canada than in Alaska.

The estimated population size (and CV) in surveyed areas was 692,347 (0.20) in Alaska and 235,520 (0.34) in Canada. The estimated total was 927,867 (0.17). The Morrison estimate was 2,500,000. Our surveyed area excludes many areas where the species is probably common, especially in the Yukon–Kuskokwim Delta, where densities were extremely high, so the population size in the arctic is surely much larger than our estimate. More than half of this species' range is outside the arctic. We thus suspect the Morrison estimate is low.

Red Phalarope

Red Phalaropes were widely distributed across our study area, occurring in 15 regions from the Yukon–Kuskokwim Delta to Baffin Island (but not Québec) and north to Alert. Rapid surveyors recorded 1,387 indicated pairs. Intensive plots contained 69 indicated pairs, giving us a good basis for calculating species-specific detection ratios. These ratios (and CVs) were 0.77 (0.21) in Alaska and 2.25 (0.39) in Canada. The ratio from Alaska is quite close to the combined rate. The ratio from Canada is far from the combined rate, but the CI extends from 0.53 to 3.98, so there is no good basis for using a species-specific detection ratio in Canada.

Estimated densities were often high (Table 14.21); five strata had estimates of 40 or more birds/km² and one had an estimate of 72 birds/km². This species was strongly associated with wetlands, though some regions had high estimates in moist areas and even occasionally in uplands.

Red Phalaropes were abundant along the coast of central Alaska and in the eastern Canadian arctic, and less dense elsewhere. Previous accounts report the Red Phalarope as absent from the Mackenzie Delta (Gratto-Trevor 1996), and we observed none there. Adjacent to this region, they were rare on the Yukon North Slope and the Arctic NWR to the west, as well as the Queen Maud Gulf

TABLE 14.20

Estimated densities of Red-necked Phalaropes (birds/km²).

Region	Wetlands	CV	Moist areas	CV	Uplands	CV
Queen Maud Gulf—west	1.37	0.70	1.50	1.09	3.22	0.73
Queen Maud Gulf—east	10.05	0.92				
Kent Peninsula	3.47	1.00	1.25	1.02		
Yukon North Slope	14.04	0.36	2.92	0.58		
Mackenzie Delta—west	17.33	0.39	1.66	0.96	3.14	1.03
Mackenzie Delta—east	11.35	0.48			14.52	0.60
Mackenzie Delta—Mackenzie Gas Project stratum	16.60	0.45	2.27	0.86	35.11	0.96
Arctic National Wildlife Refuge	15.04	0.26	2.90	0.42	2.49	0.57
Central North Slope	8.14	0.52	3.73	0.91		
Foothills—east			3.11	0.96	1.57	0.99
National Petroleum Reserve—Alaska NE	11.99	0.17	8.16	0.26		
National Petroleum Reserve—Alaska coast	18.49	0.22	9.77	0.32	10.28	0.65
Colville River					2.98	1.19
Selawik National Wildlife Refuge	2.43	0.76			0.24	1.15
Yukon Delta—central	64.03	0.17	55.02	0.21		
Alaska Peninsula	8.66	0.15				

TABLE 14.21

Estimated densities of Red Phalaropes (birds/km²).

Region	Wetlands	CV	Moist areas	CV	Uplands	CV
Prince Charles Island/Air Force Island	71.68	0.30	22.46	0.17	40.32	0.48
West Baffin	42.15	0.28	39.59	0.33		
Southampton Island	21.82	0.28	5.10	0.64		
Coats Island	27.50	0.34				
Queen Maud Gulf—west	2.71	0.93				
Kent Peninsula	7.55	0.65			0.54	1.03
Creswell Bay	44.68	0.30	25.58	0.28	13.67	0.33
Somerset Island	13.42	0.17	1.01	1.03	7.48	1.08
Alert	0.21	1.03				
Yukon North Slope			0.30	1.06		
Arctic National Wildlife Refuge	6.42	0.41	1.45	0.49		
Central North Slope	10.64	0.50	3.83	0.53		
National Petroleum Reserve—Alaska NE	23.88	0.25	20.92	0.34		
National Petroleum Reserve—Alaska coast	40.02	0.18	9.33	0.39		
Yukon Delta—central	4.86	0.67	3.49	0.94		

Migratory Bird Sanctuary and Kent Peninsula to the east. Published range maps do not list the Red Phalarope as breeding on Southampton Island and Coats Island, but we found them to be common there.

The estimated population size (and CV) in surveyed areas was 589,104 (0.17) in Alaska and 1,028,573 (0.21) in Canada (including a remarkable 524,000 birds on Prince Charles and Air Force Islands). The estimated total was 1,617,000 (0.15). The Morrison estimate was 2,500,000. The surveyed area excludes nearly all of the Yukon–Kuskokwim Delta, where the estimated density was nearly 5 birds/km^2 in the small area we did survey. Many areas close to Prince Charles and Air Force Islands, where there were such large numbers, have not been surveyed. The population is thus probably substantially larger than our current estimate.

Results for All Shorebirds Combined

Previous chapters did not report the densities or estimated population sizes of all shorebirds, so we present that information in some detail here (Table 14.22). The wetlands and moist habitat of the Yukon–Kuskokwim Delta had the highest and second highest densities of shorebirds for any area yet surveyed, with an astonishing 414 and 299 shorebirds/km^2, respectively. It is unfortunate that only a small fraction of this highly productive area has been surveyed. We have no way to estimate region-wide densities. Densities were lower, though still considerable, on the Alaska Peninsula and much lower in the Alaska Maritimes region. The observed densities were also low in Selawik; however, we believe that may be due to a late survey there (see McCaffery et al., chapter 3, this volume).

The second highest densities occurred in the arctic Coastal Plain of the North Slope of Alaska. It is noteworthy that the Coastal Plain of the North Slope is much larger than the Yukon–Kuskokwim Delta. It is hard to predict at present which area supports more shorebirds. Further east in Alaska, densities in both wetlands and moist areas declined to less than half the values in the NPRA.

Densities throughout the Canadian arctic were generally much lower. In wetlands in the Yukon Territory, Mackenzie Delta, Kent Peninsula, and Queen Maud Gulf Migratory Bird Sanctuary, densities are half or less the density in the Arctic NWR and less than 25% of the density of shorebirds in coastal wetlands of the NPRA. Densities were even lower in Québec, though we have not yet sampled a large area there.

Although our regions in mainland Canada supported comparatively low densities of shorebirds, densities were much higher in the lower arctic archipelago. Density in wetlands on Prince Charles/Air Force, Coats, Baffin, and Southampton Islands, and at Creswell Bay and Kent Peninsula (effectively an island) were all higher than any of the densities in wetlands in the Yukon North Slope, Mackenzie Delta, or Queen Maud regions of mainland Canada (an average 80 shorebirds/km^2 for the former regions vs. 38 shorebirds/km^2 for the latter). The pattern was similar in moist areas. Further north in the arctic archipelago, density appears to decline again. It was close to zero on northern Somerset Island and 11 shorebirds/km^2 at Alert. Reconnaissance surveys on Prince Patrick, Melville, and Ellesmere Islands indicate low densities there as well.

Fewer conclusions can be reached about population sizes because so many areas remain to be surveyed (Canada) or surveyed completely (western Alaska). The estimates for the Alaska Maritimes, Alaska Peninsula, and Selawik study areas must be regarded with caution, though they do suggest a large population on the Alaska Peninsula and much smaller numbers in the Aleutian Islands. As noted previously, no estimate can be made for the Yukon–Kuskokwim Delta. The NPRA is estimated to have more than 4,000,000 shorebirds. The CVs there are small enough to give this estimate considerable credibility. Another 1,000,000 shorebirds are estimated to occur in Alaska east of the Colville River.

In Canada, the Yukon North Slope and Mackenzie Delta are estimated to have about 500,000 shorebirds. The Queen Maud Gulf Migratory Bird Sanctuary is estimated to have about 900,000 shorebirds. Perhaps the biggest surprise, however, is the high estimates for Southampton and Prince Charles/Air Force Islands. These two fairly small regions are estimated to have a combined total of nearly 2,000,000 shorebirds. Numbers in surveyed areas further north are considerably smaller.

Estimating numbers present, by habitat, in our surveyed areas is also of interest (Table 14.23). While wetlands certainly have the highest densities, they were estimated to include only about 45% of all shorebirds in our surveyed areas. Moist areas and uplands had roughly equal proportions of the remaining 55% of the shorebirds. This finding has important implications for survey

TABLE 14.22

Estimated densities (birds/km²) and population sizes of shorebirds (all species), and their coefficients of variation (CV).

	Wetlands		Moist areas		Uplands		Population size	
Region	Dens	CV	Dens	CV	Dens	CV	Est	CV
Québec	6	0.25					122,895	0.25
Alert	11	0.28					1,434	0.28
Yukon North Slope	42	0.28	10	0.52	4	0.68	107,737	0.31
Mackenzie Delta—west	43	0.25	23	0.40	8	0.66	184,109	0.23
Mackenzie Delta—east	51	0.21			43	0.38	210,674	0.23
Mackenzie Delta—Gas Project stratum	38	0.34	17	0.43	48	0.70	10,115	0.51
Prince Charles Island/Air Force Island	133	0.33	168	0.17	110	0.51	1,121,368	0.32
West Baffin	64	0.21	66	0.32	0		480,479	0.22
Southampton Island	54	0.27	34	0.39	0		740,310	0.17
Coats Island	113	0.23	13	0.17	13	0.17	144,425	0.26
Queen Maud Gulf—west	18	0.26	15	0.45	20	0.54	367,607	0.32
Queen Maud Gulf—east	36	0.50	25	0.44	16	0.26	548,862	0.28
Kent Peninsula	54	0.27	10	0.38	14	0.28	176,903	0.27
Creswell Bay	60	0.27	54	0.22	42	0.21	6,751	0.19
Somerset Island	27	0.17	8	0.32	27	0.99	70,551	0.65
Arctic National Wildlife Refuge	81	0.23	28	0.21	6	0.48	155,548	0.18
Central North Slope	96	0.30	64	0.24	0		820,920	0.22
Foothills—east	20	0.49	29	0.59	4	0.80	47,464	0.53
National Petroleum Reserve—Alaska NE	165	0.13	128	0.16			244,794	0.12
National Petroleum Reserve—Alaska coast	253	0.12	92	0.18	73	0.38	3,897,794	0.15
Foothills—west					5	0.79	112,398	0.79
Colville River			79	0.81	22	0.60	943	0.66
Selawik National Wildlife Refuge	40	0.35			7	0.84	340,033	0.35
Yukon Delta—central	414	0.12	299	0.12			308,280	0.11
Alaska Peninsula	117	0.20	16	0.92	61	0.54	1,785,931	0.32
Alaska Maritime			7	0.58			61,385	0.58

design. It makes sense to concentrate surveys in wetlands, but the other areas should not be sampled too lightly as they in fact contain half of the population. Moist and upland areas have often been ignored in past shorebird surveys, and this may have led to a serious underestimation of population size for some species.

DISCUSSION

This monograph contains nearly 600 estimated shorebird densities (each one for a species, in a region, in a habitat) and more than 200 estimated population sizes (each one for a shorebird species in a region). It also contains numerous estimates

TABLE 14.23

Area and estimated numbers of shorebirds by habitat in the regions surveyed to date.

	Landcover		Shorebirds (millions)	
Habitat	Area (km^2)	Percent	Number	Percent
Wetlands	50,531	0.21	5.3	0.45
Moist areas	79,530	0.33	3.3	0.28
Uplands	109,288	0.46	3.2	0.27
Total	239,350	1.00	11.8	1.00

derived from these, such as the number of all shorebirds in each region and in groups of regions. The fieldwork involved surveying dozens of areas that have seldom if ever been visited by shorebird biologists. All of the fieldwork and analyses were carried out using a single, flexible method that included a sophisticated, Geographical Information System (GIS)–based sampling plan; field methods developed during nearly a decade of work in numerous locations across Alaska and Canada by dozens of skilled biologists; and a special extension of double sampling that recognized the existence of "zones" and other aspects of the design. Below, we summarize some of the major findings from this work.

Regional Patterns

In general, we found shorebirds at higher densities in Alaska than in Canada, with over 400 shorebirds/km^2 in wetlands of the Yukon–Kuskokwin Delta, and densities of over 100 shorebirds/km^2 not uncommon elsewhere in the state. Densities were much lower throughout Canada, but tended to be higher in the lower arctic archipelago than on the mainland. What Canada lacks in density, however, it makes up in size; our GIS analyses suggest that the Canadian arctic contains nearly five times more suitable shorebird habitat than Alaska.

In western Alaska, we conducted brief surveys on four National Wildlife Refuges, aimed primarily at developing methods that could be used for more comprehensive surveys of these areas in the future. In the Yukon Delta, our study area covered only about 1% of the Yukon Delta NWR but was still one of the largest coastal areas in the NWR ever surveyed for all shorebirds using rigorous methods. Some of our best data come from the North Slope of Alaska, where we surveyed more than 600 plots in the NPRA, the central coast, and the Arctic NWR. These surveys produced the first estimates of abundance of shorebirds in northern Alaska and clearly established the preeminence of the NPRA as one of the world's great shorebird regions.

In Canada, we carried out the first systematic ground surveys of the Yukon North Slope since the early 1970s and the first shorebird surveys of the Queen Maud Gulf Migratory Bird Sanctuary. Extensive surveys were also carried out on Southampton, Coats, Prince Charles, and Air Force Islands and showed that these areas, especially the latter three, have surprisingly high shorebird densities, quite possibly the highest in all of arctic Canada. Surveys further north, including reconnaissance surveys in the Queen Elizabeth Islands, clarified the distribution of high-arctic species and revealed how those remote areas can be surveyed despite the extremely low density of shorebirds.

Evaluation of Field and Analytic Methods

Most past studies of arctic shorebirds have suffered from selection bias to an unknown degree and have used estimation methods susceptible to measurement bias of unknown magnitude. As a result, when different studies have produced markedly different estimates, it has usually been difficult to determine how much of the difference was due to sampling error or bias rather than biological differences. The problems of selection and measurement bias are complex, and will probably never be completely solved, especially in large-scale studies. However, in developing Arctic PRISM we have tried to reduce the problems caused by non-random selection of plots and measurement error. In this section, we briefly review and evaluate the methods, focusing on problems that we believe have been solved and those that remain to be solved.

The key to the current methods used in Arctic PRISM (and in several other regional surveys) is obtaining a complete partition of the study area into plots. With large study areas and small plots, this may require that hundreds of thousands of plots must be delineated. This task can only be accomplished using automated, GIS methods. The main difficulty in carrying out the partition in most environments is that natural borders need to be recognized but are irregular at a small scale. If these borders are ignored, many plots are created that are not suitable because they cross major habitat boundaries. For example, a plot may be largely water or may be half wetlands (usually a high-quality habitat) and half unvegetated bedrock.

In the early years of Arctic PRISM we tried a variety of *ad hoc* methods to circumvent these problems without creating a complete sampling frame. These efforts, however, inevitably caused us difficulties in the estimation stage. Eventually, we contracted with Hawthorne Beyer, developer of the widely used "Hawth Tools" for ArcGIS, to produce a comprehensive "Arctic PRISM" extension to GIS which would automate the various steps in producing the sampling frame. The process was complex for the Arctic PRISM surveys because we assigned plots to clusters, zones, and strata (see Bart et al., chapter 2, this volume). The tools available in Arctic PRISM have been migrated to H. Beyer's Geospatial Modelling Environment (GME) program, which is available online free of charge. We anticipate gradually refining the Arctic PRISM components. We encourage anyone designing large-scale surveys to consider using this extension. See "Access to the Data" below for how to obtain a copy of the extension and its user manual. We welcome correspondence from people interested in using the extension.

Once a partition of the study area into plots has been accomplished, the investigator has great freedom in how to proceed. Stratification can be used, or not, and can be based on any variables that are available at the region-wide level. Sampling plans within strata may be simple or complex, and may vary between strata. Plots may be of any size and plot size can vary within a study (e.g., as always happens if plot borders follow natural borders), though extreme variation could reduce precision substantially. Investigators may delete areas they are willing to assume have virtually no birds. The survey method may be constrained by search time or distance between observers, or unconstrained. Surveyors may work alone or with others. Effort can be allocated to rapid and intensive surveys in any way the investigators wish. While training is certainly worthwhile, there is no assumption that detection rates are constant across surveyors, habitats, regions, dates, or weather conditions. Perhaps most critical, there is no assumption that rapid surveyors detect all or a constant fraction of birds breeding on the plot. The only assumptions are that the nominal sampling plan is followed which includes an assumption that rapid surveyors have no knowledge of the birds present on intensive plots, and that the intensive surveys yield unbiased estimates of breeding density.

Although the analytic method is fairly complex, mainly because so many estimates may be needed, a comprehensive program, DS, exists to carry out all the calculations. The program and a detailed user's manual are available free of charge (see "Access to the Data" in this chapter).

Comparison with Previous Estimates

A large group of shorebird experts, led by R. I. Guy Morrison, made the first comprehensive efforts to estimate continent-wide population sizes of all shorebirds breeding in the U.S. and Canada (Morrison et al. 2001b, 2006). A notable feature of their estimates is that each is accompanied by an accuracy rating, a number from 1 to 5, with "1" being the lowest accuracy. For most arctic species, the rating was 1, an "educated guess," or 2, an "estimate based on broad-scale surveys . . . likely to be in the right order of magnitude." A few estimates for arctic species were given a rating of 3, indicating the true value was thought to be within the range (0.5 × estimate, 1.5 × estimate).

We obtained sufficient information for 17 species to provide a preliminary evaluation of the Morrison estimates (Table 14.24). This evaluation is based upon three questions: (1) How well do the Morrison estimates of population size, classified simply as small, medium, large, correspond to our estimates? (2) Do our results support the accuracy ratings assigned by Morrison et al. (2006) to their estimates? (3) Do our results suggest a tendency for the Morrison estimates to be either high or low?

TABLE 14.24

Species for which our study suggests that the estimated population sizes by Morrison et al. (2006) are low.

Species	Morrison estimate	Accuracy rating	Our estimate[a]	Reasons for believing the Morrison estimate is low
Black-bellied Plover	200,000	2	314,293	Our estimate exceeds Morrison estimate; common on Yukon Delta; >50% of range not surveyed
American Golden-Plover	200,000	2	490,819	Much of range in the arctic not surveyed; 40% of range is outside the arctic
Pacific Golden-Plover	50,000	2	47,865	Common in the Yukon Delta; most of range not surveyed
Greater Yellowlegs	100,000	2	59,544	>95% of range outside the arctic
Whimbrel	66,000	2	90,781	Our estimate exceeds the Morrison estimate; much of the range in the arctic not surveyed; 70% of range outside the arctic
Bar-tailed Godwit	90,000	3	81,208	Common in Yukon Delta; much of rest of range not surveyed
Dunlin	1,525,000	2	2,146,737	Our estimate exceeds the Morrison estimate; extremely abundant in the Yukon Delta; most of range in Canada not surveyed
Semipalmated Sandpiper	2,000,000	2	2,257,257	Abundant in the Yukon Delta; most of range in Canada not surveyed
Least Sandpiper	700,000	2	367,143	Much of arctic range remains to be surveyed; 80% of the range outside the arctic
White-rumped Sandpiper	1,120,000	3	633, 236	Most of range not yet surveyed
Pectoral Sandpiper	500,000	2	1,579,723	Our estimate exceeds Morrison estimate; current estimate far above Morrison estimate; common in Yukon Delta
Buff-breasted Sandpiper	30,000	2	42,839	Most of range not surveyed
Short-billed Dowitcher	75,000	2	148,314	Our estimate exceeds the Morrison estimate; nearly all the range outside the arctic
Long-billed Dowitcher	400,000	1	704,387	Our estimate exceeds the Morrison estimate; common in Yukon Delta
Wilson's Snipe	2,000,000	1	316,255	>95% of range outside the arctic
Red-necked Phalarope	2,500,000	1	927,867	Abundant in Yukon Delta; >50% of range outside the arctic
Red Phalarope	1,250,000	1	1,617,678	Our estimate exceeds the Morrison estimate; common in the Yukon Delta, much of the arctic range not surveyed

[a] Estimated population size in the areas we have surveyed. Large areas of the arctic in Alaska and Canada remain to be surveyed, so these are *not* estimates of the range-wide population size.

Regarding the first question, our patterns of relative abundance are consistent with the Morrison estimates. The Morrison estimates are low for species that we rarely encountered and high for species that we often encountered. For example, among species restricted largely to the arctic, the five smallest Morrison estimates are for Purple Sandpiper, Buff-breasted Sandpiper, Pacific Golden-Plover, Whimbrel, and Hudsonian Godwit. The five largest estimates are for Western Sandpiper, Red-necked Phalarope, Semipalmated Sandpiper, Dunlin, and Red Phalarope. There is no question, from our surveys, that the population sizes for all five of the first group are markedly less than the sizes for all members of the second group. Doubt remains about some of the species whose ranges are largely outside the arctic such as Short-billed Dowitcher, Wilson's Snipe, and Least Sandpiper, but for the arctic species our results are consistent with their rank order list.

Second, their accuracy assessments seem realistic to us. Most species, as noted above, were given a rating of 2, meaning that Morrison et al. (2006) believe their estimate, say $\hat{Y}$, is within the "right order of magnitude." They do not define "right order of magnitude," but one interpretation is that the true value, Y, is closer to $\hat{Y}$ than either $0.1\hat{Y}$ or $10\hat{Y}$. This implies that Y lies (approximately) within the interval $(0.5\hat{Y}, 5\hat{Y})$. We believe their estimates meet this accuracy target in most if not all cases. For example, their estimate for Black-bellied Plovers is 200,000 and the accuracy rating is 2. The order of magnitude range would thus be 100,000 to 1,000,000. Our estimate for the areas we have surveyed is about 300,000. A large amount of high-quality range has not been surveyed, so we suspect the true number is substantially higher than 300,000, but we have no evidence to suggest it is greater than 1,000,000. The same argument can made for most if not all of the species where we have a substantial amount of information. We therefore conclude that their accuracy ratings for arctic species are plausible.

Third, we believe that most or all of the Morrison estimates that we can evaluate are low rather than high, which is not surprising since the Morrison estimates were presented as minimum population sizes. Our estimates for the regions surveyed to date exceeded the range-wide Morrison estimates for many species, including Black-bellied Plover, American Golden-Plover, Whimbrel, Dunlin, Pectoral Sandpiper, and Buff-breasted Sandpiper. For many other species, such as Pacific Golden-Plover, Greater Yellowlegs, Least Sandpiper, Wilson's Snipe, and Red Phalarope, conservative extrapolation across the range would produce estimates much larger than those provided by Morrison et al. (2006).

Access to the Data

An extensive data set, including the primary GIS products and all survey results, has been placed on the Coordinated Bird Monitoring home page at http://greatbasin.wr.usgs.gov/CBM/default.asp?PageID=1. All of the data described in this report can be downloaded at no charge. The data are also incorporated into the Coordinated Bird Monitoring Database, which provides a comprehensive method to sort and analyze the data. The GIS programs used to partition the regions into plots, clusters, zones, and strata are also available on the CBM site, as is the analytic program, DS, used to compute all of the estimates in this report. Anyone with questions is encouraged to contact the editors.

ACKNOWLEDGMENTS

This chapter has summarized results from all previous chapters and thus used the results from the hundreds of people who participated in developing Arctic PRISM. We particularly appreciate discussions with V. Johnston, J. Rausch, and B. J. McCaffery and reviews of a previous draft by L. Niles and D. Lank. The U.S. Geological Survey and the Canadian Wildlife Service provided support for preparing the chapter. This report is PCSP Contribution No. 01309.

ONLINE CONTENT

Abstracts are available in French and Spanish from www.ucpress.edu/go/sab.
Une traduction du résumé est disponible en français.
Una traducción del resumen está disponible en español.

CHAPTER FIFTEEN

Priorities for Future PRISM Surveys

Jonathan Bart, Victoria Johnston, Jennie Rausch, Paul A. Smith, and Brian J. McCaffery

Abstract. This chapter summarizes strengths of the current Arctic PRISM method, discusses how it can be improved, and suggests which regions should have highest priority for future surveys. Strengths of the method include its ability to be used in conjunction with intensive surveys to address specific management issues (as in the Mackenzie Delta; see Rausch and Johnston, chapter 5, this volume), its ability to vary plot selection methods and sample sizes across regions, and its unbiased estimation of density and population size through the use of detection ratios. Needed improvements include minor refinements in the plot selection algorithm. We also suggest a major effort be undertaken in collaboration with waterfowl, waterbird, and landbird specialists to refine the surveys so that they provide reliable estimates for a wider suite of non-shorebirds. We also believe that additional information, for example on phenology, habitat relationships, and variables related to climate change, could usefully be gathered on the intensive plots. Work should also continue on the analytic program, for example, to provide for analysis of aerial survey data. The chapter closes with suggestions for which regions should have the highest priority for future surveys. In Alaska, the regions are (1) Yukon Delta NWR and (2) other areas of western Alaska that have not yet been surveyed. In Canada, the regions are (1) Region 8, which includes the mainland from Kent to Boothia Peninsulas, King William Island, and Victoria Island; (2) Region 9, which includes Banks Island to Somerset Island; and (3) the high arctic.

Key Words: arctic, methodology, monitoring, PRISM, shorebirds, surveys.

EVALUATION OF THE ARCTIC PRISM DESIGN

Previous chapters have described the Arctic PRISM approach in detail, but we wish to add a few comments about it in this chapter. First, it should be noted that the design permits managers to concentrate effort in areas of special interest. For example, an oil and gas pipeline has been proposed in the Mackenzie Delta, one of the most important breeding and migration sites for birds in arctic Canada. The proposal led to detailed surveys in this area during the past few years,

Bart, J., V. Johnston, J. Rausch, P. A. Smith, and B. J. McCaffery. 2012. Priorities for future PRISM surveys. Pp. 239–244 *in* J. Bart and V. Johnston (editors). Arctic shorebirds in North America: a decade of monitoring. Studies in Avian Biology (no. 44), University of California Press, Berkeley, CA.

especially at the proposed gas production facilities and on the pipeline corridor. During PRISM surveys, the Mackenzie region was subdivided into more subregions than would normally have been the case for an area this size, and the pipeline corridor and proposed facilities were delineated as one small, non-contiguous subregion. This approach permitted biologists to conduct a greater number of surveys in the areas proposed for development than in other parts of the region, yet the results could both be integrated with, and compared to, results from surrounding subregions. Managers could put the survey results in perspective in ways that would not have been possible had the pipeline surveys been carried out in isolation. Other examples of integrating results from surveys in areas of high interest include surveys in the Arctic National Wildlife Refuge (Brown et al. 2007), surveys on the Colville Delta and the Teshekpuk Lake area in the National Petroleum Reserve-Alaska, and surveys in the Queen Maud Gulf Migratory Bird Sanctuary. The Arctic PRISM design thus accommodates the need for detailed surveys in certain areas while producing results that become part of the arctic-wide data set.

Second, the design has generally proven flexible enough to handle logistic issues that are particular to different parts of the arctic. The number of surveyors per rapid plot, plot size, distance between plots, and number of plots per cluster can all be varied. The size and number of intensive plots per camp can also be varied, as can the visitation schedule and number of surveyors per intensive plot—as long as the surveys yield unbiased estimates of the number of territorial birds present.

Third, the program includes sophisticated methods to estimate detection ratios. In developing Arctic PRISM, we first assumed that existing methods such as point counts or line transect counts, perhaps with distance estimation, would be suitable. A special method using two parallel line transects was even proposed. We found, however, that birds responded too strongly to the observers for these methods to work. Some birds sneak away from their nest and are only observed at substantial distances from it. Other birds are attracted to the observer but may only be seen after they have approached from great distances. Finally, we found that in some plots, we could best find the shorebirds by going to the locations (e.g., pond margins) where they were feeding. Unless we searched out these portions of the plot, we would miss a large fraction of the birds present. All of these realities led us to develop the area search methods. Recognizing that some nesting birds would not be present during surveys and others would not be detected led us to develop the double sampling approach. The method thus developed gradually, during a period of several years, in response to conditions encountered during hundreds of surveys.

The most critical assumption in double sampling is that the intensive surveys yield unbiased counts. We tested this assumption in a special study, involving four field camps widely dispersed across the arctic, to evaluate the detection rates on the intensive surveys (Smith et al. 2009). We found that when nests survive for several visits by surveyors, then detection probabilities are high. If the nest fails soon after initiation, so that it is only available for discovery on one or two visits, then detection rates may be much lower. If birds remain on the plot, however, then they may still be recorded by the intensive surveyors. Among all intensive plots, surveyors did not find nests for 26% of the 806 shorebirds recorded as being present.

FUTURE WORK: METHODOLOGY

Region-specific Sampling Plans

The approach for delineating plots and assigning them to clusters, zones, and strata is fairly well developed, but a few improvements are still needed. At present, there is no easy way to incorporate uncrossable linear barriers (usually river channels) into the process of plot delineation. As a result, plots often cross such features, making them difficult, and in some cases impossible, to survey. The Arctic PRISM ArcGIS extension (Bart et al., chapter 2, this volume), which is used to delineate plots and assign them to clusters, zones, and strata, was developed only recently and has not yet been extensively used. As experience is gained with it, other needed features will undoubtedly be identified and will need to be incorporated into the extension. It is thus important to plan for revisions of this tool during the next few years.

Recording Other Species

Perhaps the most important remaining innovation in the Arctic PRISM program is improving the methods and analytical capability for non-shorebird species. This must be done in cooperation with representatives of the waterfowl, waterbird, and landbird groups. We recognize that some of these species, especially waterfowl and waterbirds, are well covered in other surveys in most of Alaska and some areas near coastlines in Canada. Many other areas exist, however, in which little information is currently being collected, especially on landbirds. It would be optimal for PRISM surveys to be able to fill the gaps. While we feel that decisions about how best to use PRISM surveys for non-shorebirds must be made by the non-shorebird groups, we would like to indicate the potential for PRISM surveys to yield useful information on these other taxa.

In addition to the 34 species of shorebirds recorded on PRISM rapid surveys, we also recorded 75 other species (13,751 individuals). Selected non-shorebird species are shown in Table 15.1. We recorded large numbers of many species, and, perhaps of equal importance, the records come from many regions across the arctic, which greatly enhances their value. We have also computed region-specific density estimates for the most common non-shorebird species (see individual region chapters, chapters 3–8, this volume).

We hope that a careful analysis will be carried out during the coming few years of how the Arctic PRISM surveys can be most useful in surveying non-shorebirds. Questions include how to train surveyors to detect and record all species encountered, how to survey large water bodies (which at present are ignored in PRISM surveys), and how to estimate detection ratios where this is desirable. These questions should be answered by PRISM personnel and representatives from the other bird initiatives working together. We hope that such an effort will occur during the next few years.

TABLE 15.1

Total numbers of indicated pairs of selected non-shorebirds recorded on Arctic PRISM rapid surveys, and number of subregions in which each species was recorded.

Species	Number recorded	Number of subregions
Pacific Loon	168	17
King Eider	126	13
Willow Ptarmigan	675	18
Rock Ptarmigan	144	14
Sandhill Crane	95	13
Long-tailed Jaeger	150	17
Parasitic Jaeger	165	14
Glaucous Gull	251	11
Sabine's Gull	161	8
Arctic Tern	239	15
Horned Lark	120	10
Tree Swallow	254	4
Eastern Yellow Wagtail	100	7
American Pipit	224	5
Savannah Sparrow	1,499	17
American Tree Sparrow	378	14
White-crowned Sparrow	93	10
Lapland Longspur	3,293	23
Snow Bunting	91	10
Redpoll (spp.)	312	16

Intensive Plots

Much progress has been made in Canada during the past several years on using intensive plots to collect ancillary data. More work is needed, however, especially for collecting nesting success data when crews are present long enough to do that, other aspects of the species' natural history, and other information such as weather data and variables to document climate change. A database is being developed to maintain and record all of this information. If more attention is given to non-shorebirds, then intensive plot surveys will likely need to include non-shorebirds too, which will increase the effort needed in some areas. Methods may thus need to be revised, and additional people may be needed. Gathering this additional information, however, could substantially increase the overall value of Arctic PRISM.

Analytic Programs

The program to analyze our data, originally called Arctic PRISM, has now been renamed DS (for double sampling). Double sampling and DS are being used in a number of studies in the western United

States, and funding for these surveys has been used to improve the analysis in several ways. For example, the program now carries out numerous internal tests designed to identify errors in the data, and output has been substantially improved. Other improvements are planned, including a series of options for different sampling plans within strata, and it will be important to continue testing DS to ensure that it is easy to use with arctic data as improvements are made. For example, if other species and aerial surveys become more important in the Arctic PRISM program, additional changes may be needed to analyze these data.

FUTURE WORK: REGIONAL SURVEYS

In this section we identify the areas which we believe should receive the highest priority for new surveys in arctic Alaska and Canada.

Alaska

Yukon Delta NWR

This area has the highest densities, by far, of nesting shorebirds anywhere in arctic North America. Intensive studies have been conducted for years in a few areas, but most of the area has never been surveyed systematically. Knowledge of density for nearly all shorebirds across the refuge is so poor that biologists have been reluctant even to provide order of magnitude estimates of population size. Given the importance of the area and how poorly it is known, this is the top priority for additional shorebird surveys in arctic Alaska. We suggest using a design that will obtain rapid surveys from at least 300 widely distributed plots. The breeding season is longer there than in areas farther north, so we suspect a four-person field crew could survey approximately 100 plots per season (e.g., three plots/day per person; 12 plots/crew-day; 8–9 survey days/crew). This suggests that three such field crews doing rapid surveys would be sufficient. Ideally, each crew-year would occur in a different calendar year to sample year-to-year fluctuations. Given the high densities of shorebirds in at least portions of this area, it is possible that rapid surveys will be less accurate than elsewhere (although this was not our preliminary finding; see McCaffery et al., chapter 3, this volume) and that more effort should be allocated to intensive plots than has been the case elsewhere.

Other Western Alaska Areas

We conducted only preliminary surveys elsewhere in western Alaska. We suggest 1–2 field crew-years for rapid surveys on or near to each of four NWRs (Alaska Maritime, Alaska Peninsula, Togiak, Selawik). The fieldwork should cover all lowland shorebird habitat in western Alaska outside the Yukon Delta NWR. A total of four field crew-years would probably be sufficient for area-wide estimates (i.e., all of western Alaska). If refuge-specific estimates are desired, then at least two field crew-years should be expended on each NWR.

Northern Alaska

The Arctic NWR and the northeastern part of the NPRA (and Colville Delta) have been well sampled. Teshekpuk Lake has also recently been sampled intensively, though we did not include these plots in our analyses. They should be incorporated into the Arctic PRISM database once results have been analyzed and published by the principal investigators in those studies. Coverage was sparse in the western two-thirds of the NPRA and in the central region (Colville River to the Arctic NWR), and Region 15, southwest of the NPRA, has not been surveyed at all in the Arctic PRISM program. Given the high interest in these regions, especially the NPRA, expending 2–3 more field crew-years in these two areas would be desirable.

Canada

Region 8, Queen Maud Gulf[1]

Region 8 encompasses a large area of the mainland coast between the Kent and Boothia peninsulas, King William Island, and a portion of Victoria Island. Our surveys so far have covered 50% of the region's potential shorebird habitat, and we expect that remaining surveys could be completed with two field crew-years. Completing the surveys of this region is a priority, not only because they are already well under way, but also because the eastern portion of the region, where we have not yet conducted surveys, is thought to be important habitat for the endangered *rufa* subspecies of Red Knot. Our surveys would generate a quantitative estimate of abundance and determine the distribution of knots in this area. The area is also the site of some historic shorebird breeding ecology research (Parmelee et al. 1967, 1968), allowing us to compare timing of breeding and nest survival from our intensive plots with historical records. The portions of the region that remain to be surveyed are disjunct and separated by several hundred kilometers, so we plan to have one rapid crew in the east (King William Island) and one in the west (southwestern Victoria Island) during the same field season.

Region 9, Banks Island to Somerset Island

This region encompasses a large portion of the southern arctic archipelago and contains more than 340,000 km^2 of potential shorebird habitat. The vast majority (94%) of this habitat is uplands. We have surveyed a small area of Somerset Island, including Creswell Bay, and found moderate to high densities of shorebirds (8–60 birds/km^2). The Rasmussen lowlands in southeastern Region 9 has also been surveyed, was species rich compared to other similar sites across the arctic, and had moderate densities of shorebirds (average 15.2 birds/km^2; Johnston et al. 2000) across all habitat types. The large expanses of upland habitats may also contain significant numbers of shorebirds, although densities are probably low.

We think that the best-quality wetland habitats are concentrated in three distinct areas (see Fig. 2.1 in Bart et al., chapter 2, this volume), along the coasts of Banks Island and Victoria Island and throughout Prince of Wales Island[2]. We propose that these areas might be suitable for intensive camps and that covering this large region over three crew-years, based out of these camps, might be appropriate.

Regions 10 and 11, the High Arctic

Although vast, the high arctic of Canada has limited suitable shorebird habitat. Much of this habitat is concentrated in central Ellesmere and eastern Axel Heiberg Islands, which we have therefore defined as a separate region (Region 11). The remainder of the high-arctic islands are rugged and sparsely vegetated. Wetlands appear to be uncommon (less than 6,000 km^2) and the vast majority of the potential habitat (84 %) is sparsely vegetated uplands requiring a comparatively light sampling intensity. Based on work in several high-arctic locations, we believe that shorebirds generally nest near wet areas which they and their young can feed in, and that this behavior provides an effective basis for sampling the large barren areas (i.e., place barren areas close to wetlands in their own stratum). If this is done, then we expect that three crew-years would be sufficient to complete the surveys here. Our reconnaissance surveys (Bart et al., chapter 8, this volume) suggest that shorebird densities in most of this region are low. The distinct bird community (see Bart et al., chapter 8, this volume, Bart and Smith, chapter 14, this volume) and extremely limited historical data, however, make surveys here a priority.

These immediate priorities for surveys are presented here as short-term goals of Arctic PRISM. Surveys to date have provided a wealth of information on breeding ranges, densities, population size, and habitat associations, and the upcoming surveys will continue to provide new information in this vein. The full potential of the program will not be realized, however, until the surveys are repeated.

Current concerns over shorebird population declines come primarily from migration monitoring. Although the evidence suggests widespread declines, potential problems with bias and inconsistent trend estimates among surveys, analyses, and time periods reduce our ability to provide decision makers with conclusive results. Once completed, the Arctic PRISM surveys will provide the first ever rigorous, range-wide estimates of population status for many shorebird species. Moreover, regional estimates of population status may allow managers to highlight flyways or regions of particular concern. These data will provide a level of information previously unavailable.

Despite the pressing need for better information on shorebird population status, and the demonstrated ability of Arctic PRISM to provide this information, a stable source of funding for the program has not been identified in either Canada or the United States. This remains a limiting issue and thus can be a priority for the Arctic PRISM committee. To that end, we hope that the accounts provided in this monograph serve to demonstrate both the quality and the quantity of new information provided by range-wide arctic surveys.

CONCLUSION

Arctic PRISM has made much progress since ambitious plans to survey shorebirds were first made on the banks of the Colville River in the mid-1990s. The initial methods have been refined, modified, and subjected to numerous tests. The sampling plan has been improved repeatedly and is now, we believe, among the most rigorous approaches ever devised for sampling natural populations in large landscapes. The rapid and intensive surveys have been used by dozens of skilled ornithologists, many of whom have helped improve the methods. Extensive work on the analytic methods has been carried out, and a user-friendly program is now routinely used to analyze the data. Many sites have been surveyed, and this volume synthesizes the surveys conducted to date. Plans for future refinements and where to survey next have also been developed. Taken together, we believe that the plan for surveying arctic shorebirds is the equal of any wildlife survey in the world. We hope that the next decade will result in the first ever rigorous population estimates for a large suite of arctic shorebird species and will provide the foundation for detection of future trends, thus achieving the goal established by the Canadian and U.S. Shorebird Conservation Plan Initiatives.

NOTES

1. Surveys for Region 8 were completed in 2010.
2. Surveys were conducted on Prince of Wales Island in 2011; the remainder of Region 9 is slated for completion this season (2012).

ONLINE CONTENT

Abstracts are available in French and Spanish from www.ucpress.edu/go/sab.
Une traduction du résumé est disponible en français.
Una traducción del resumen está disponible en español.

APPENDIX A

Other Methods for Estimating Trends of Arctic Birds

Jonathan Bart, Stephen Brown, R. I. G. Morrison, and Paul A. Smith

Abstract. The Arctic PRISM was designed to determine shorebird population size and trend. During an extensive peer review of PRISM, some reviewers suggested that measuring demographic rates or monitoring shorebirds on migration would be more appropriate than estimating population size on the breeding grounds. However, each method has its own limitations. For demographic monitoring, an unbiased estimate based on a large sample of first-year survivorship would be extremely difficult for shorebirds in the arctic because the needed sample size would be unobtainable (in Canada at least) and the level of effort that would need to be expended (both financial and human resource-wise) would far exceed that of the current Arctic PRISM methodology. For migration monitoring, issues such as changes in use of monitored to non-monitored sites, residency times, and detection rates introduce bias that has not yet been resolved. While we believe demographic and migration monitoring are very valuable and are already components of the PRISM approach (e.g., Tier 2 sites focus on the collection of demographic data), we do not believe that either is likely to achieve the PRISM accuracy target of an 80% power to detect a 50% decline.

Key Words: arctic, demographics, migration, monitoring, PRISM, shorebirds.

Several years ago, a peer review of Arctic PRISM was carried out by six shorebird experts (Arctic PRISM Peer Review Committee 2010). Many constructive suggestions were received and led to useful changes in the design of the Arctic PRISM program. Some reviewers suggested that perhaps population trends in arctic-nesting shorebirds could be estimated by measuring demographic rates (both productivity and survival rates) and/or by monitoring shorebirds on migration rather than by estimating population size on the breeding grounds. We believe both of these activities are valuable for many reasons, and both are considered critical components of the overall PRISM strategy (Skagen et al. 2003, Johnston and Bart, chapter 1, this volume, Pirie et al., chapter 11, this volume, Armer et al., chapter 12, this volume), but we do not believe

Bart, J., S. Brown, R. I. G. Morrison, and P. A. Smith. 2012. Other methods for estimating trends of arctic birds. Pp. 245–252 *in* J. Bart and V. Johnston (editors). Arctic shorebirds in North America: a decade of monitoring. Studies in Avian Biology (no. 44), University of California Press, Berkeley, CA.

that either demographic studies or migration counts are likely to achieve the PRISM accuracy target. In this appendix, we briefly explain our rationale for this view. Other sections, especially Bart and Smith (chapters 13 and 14, this volume), explain why we believe Arctic PRISM will achieve the PRISM accuracy target. Although we focus on population trend in this appendix, it is worth noting that demographic modeling and migration surveys would probably not achieve another of the PRISM goals, estimation of population size, unless new methods, based on capture–recapture theory, were developed specially for such an effort.

DEMOGRAPHIC METHODS

Before commenting on the utility of demographic methods for trend estimation, we wish to emphasize our enthusiasm for demographic studies in general and, especially, for their use in combination with estimates of population size. While recent shorebird trend analyses indicate that many arctic-nesting shorebirds may be declining (Johnston and Bart, chapter 1, this volume), the lack of information about what is causing the declines currently limits our ability to design effective conservation actions. Demographic analyses for declining species would help determine factors causing declines by distinguishing whether survivorship or fecundity has the most influence on population change. PRISM surveys designed to measure population size and trend cannot provide information on the mechanisms behind declines (e.g., poor reproductive success, low juvenile or adult survival) and when (e.g., breeding, migration, non-breeding) shorebird population sizes are likely to be limited without the addition of targeted demographic analyses. Determining when shorebird populations are limited will have significant impacts on future conservation actions to address population declines, and demographic analyses will play an important role. PRISM explicitly incorporates demographic studies at the Tier 2 sites (Pirie et al., chapter 11, this volume), and the authors of this chapter are participating in the newly developed Arctic Shorebird Demographic Network of sites across the U.S. and Canadian arctic designed to further study shorebird demography. Our only point in this chapter is that we do not believe that monitoring demographic rates, with existing methods, is an efficient, or even feasible, way to achieve the PRISM accuracy target for estimating trend in population size.

A Simple Model

We begin with a simple model of how populations change in size. The model involves no assumptions or estimates; it is simply a way expressing how size changes. Let population size at the start of the breeding season in two years be N_1 and N_2. We may write $N_2 = N_1BS_0 + N_1S$, where B = the birth (i.e., hatching) rate in year one, S_0 is the proportion of birds hatched in year 1 that survive until the age of breeding (which may be more than one year), and S is the proportion of birds of adult age and alive at the start of year 1 that survive to the end of year 1. The change in population size between the start of year one and the start of year two (often referred to as the finite rate of change) thus may be written as

$$\lambda = \frac{N_2}{N_1} = BS_0 + S. \qquad (1)$$

If we further assume that λ is constant during a period of interest, then the problem is to estimate B, S_0, and S during the period. The main difficulty in estimating λ is that methods currently available are based on closed population models, whereas the populations studied are open populations. More specifically, it is virtually impossible, with shorebirds nesting in the arctic, to obtain accurate estimates of S_0. A few birds hatched at a study site return there in future years, but most do not (Schamel and Tracy 1991). There is no way to be certain whether the ones that did not return survived or died. An additional problem is that some birds breeding at the study site one year survive but do not return to the site the next year and thus are likely to be counted as mortalities. It should be added that the problem of estimating juvenile survival is a general one, not restricted to shorebirds or even to birds.

Obtaining a large enough sample size to estimate trends, however, would be quite difficult even if field methods to estimate first-year survival rates could be developed. At the range-wide scale, bird populations seldom change faster than about 3% per year (Bart et al. 2004). For example, among 141 species that occurred on at least 500 routes of the Breeding Bird Survey (so that trend estimates were fairly accurate), only 16 (11%)

changed at rates exceeding 3% per year during 1980–1999 (J. Bart, unpubl. data obtained from Sauer et al. 2007). This means that we can be moderately confident, even without collecting any data, that λ is between 0.97 and 1.03. In its simplest form (i.e., without getting into the intricacies of power analyses), the problem is thus to collect enough reliable data to show that an estimate of, say, 0.97 is significantly different from 1.0, which means that the standard error of $\hat{\lambda}$ must be no larger than about 0.015. This in turn means that $V(\hat{\lambda})$, the square of $SE(\hat{\lambda})$, must be no larger than about 0.000225. We therefore turn to the variance of $\hat{\lambda}$. If we assume that estimates of B, S_0, S (b, s_0, s) are independent (which also makes λ easier to estimate), then to a close approximation

$$V(\hat{\lambda}) = S_0^2 V(b) + B^2 V(s_0) + V(s). \quad (2)$$

Let us further assume, for this initial analysis, that the rates are constant across space and time. In this case, we may think of the problem as monitoring a large sample of eggs, first-year birds, and older birds and determining how many of them survive. For this initial analysis, let us further assume that all birds being monitored are re-sighted with a probability of 1. In each case, we are thus estimating a proportion, P. The variance of the estimate, assuming no spatial or temporal heterogeneity, is approximately $P(1 - P)/n$, where n is the number of individuals monitored. So we must obtain samples of sufficient size that $V(\hat{\lambda}) < 0.000225$. Let us assume that $B = S_0 = 0.5$ and $S = 0.72$. These values mean that $\lambda = 0.97$, the case we are interested in.

If we further assume that both parents and all young that hatch are monitored, then nests contribute an average of four eggs, two first-year birds (because we assume 50% hatching rate), and two adults to the monitoring program. The sample size can thus be expressed as the number of nests needed to achieve the accuracy target. With n = number of nests, and using the formula above for the variance of a proportion, we obtain $S_0^2 V(b) = 0.5^4/4n$, $B_0^2 V(s_0) = 0.5^4/2n$, and $V(s) = (0.72)(0.28)/2n$. Substituting these values into expression (2), with the left-hand side set equal to 0.000225, and solving for n, yields 656, the number of nests needed for each species. With 26 species covered by Arctic PRISM for which estimates are needed, the total number of nests would be 17,056. Under the assumption that λ is constant throughout the study period, the 17,056 nests could be observed throughout the study period, so for a 20-year program an average of 853 nests could be monitored per year.

Assumptions

The arctic, especially the high arctic, is notorious for spatial and temporal heterogeneity in productivity (Mayfield 1978). We know less about variation in adult survival rates and virtually nothing about variation in first-year survival rates, but it would not be surprising if these too exhibit substantial variation, as suggested by the wintering ground surveys for Red Knots (for scientific names, see Appendix C). The effect of this variation is substantial, as shown by the following argument. If we assume that sites vary in the proportion being estimated, then the variance of the estimate (assuming equal sample sizes at each site), instead of being $P(1 - P)/n$, becomes

$$V(p) = \frac{V(P_k)}{n} + \frac{ave.[P_k(1 - P_k)]}{nm} \quad (3)$$

where P_k is the proportion at the kth site, m is the number of individuals monitored per site, and n is the number of sites. To show the effect of variation in P_k, first imagine that all of ten sites have the same value of P_k [so $V(P_k) = 0$] and that $m = 100$ individuals are monitored per site, so $nm = 1000$. With $P = 0.5$, $V(p) = (0.5*0.5)/1000 = 0.00025$. Now suppose P is 0.25 at half the sites and 0.75 at the other half (so its overall value does not change). How many sites are required to maintain $V(p) = 0.00025$? Substituting these values and simplifying yields

$$V(p) = \frac{1}{n}(0.0625 + 0.0019). \quad (4)$$

Setting $V(p)$ to 0.00025 yields $n = 258$, a 26-fold increase in the needed sample size. It thus seems likely that heterogeneity could easily cause an order of magnitude increase in the needed sample size. Note, too, from expression (4) that the contribution to the overall variance from site-to-site variation (0.0625) is much larger than the contribution from within-site variation (0.0019). Increasing the number of nests per site would have hardly any effect on $V(p)$ under these assumptions.

Variation in productivity between sites, however, is only one of several factors that will increase the required sample size. Partial predation of nests is rare; typically, clutches either hatch or are lost completely. Further, reproductive output at a site varies significantly among years in relation to local predator abundance and timing of breeding (Smith et al. 2007a, 2009). Thus, the fate of eggs within nests and nests within sites is far from independent, further increasing the needed sample size. Even more important, re-sighting rates for adult birds will not be 100%, and thus must be estimated. Site fidelity is high for some monogamous species, and re-sighting rates could therefore be moderately high given adequate observer effort. For some polygamous species, however, annual redistribution to new breeding sites is thought to be common (Meltofte et al. 2007b and references therein).

In conclusion, we cannot say at present how large the needed sample would be, but it would almost certainly be at least 2,000 nests every year for 20 years (i.e., 40,000 nest attempts), and the effort would surely have to be distributed across at least 10–15 sites, perhaps more. We do not claim that this is impossible, but it would certainly be expensive, especially in Canada, where shorebird density is generally far lower than in Alaska. It is interesting to note, in this regard, that the largest current effort in Canada, at East Bay on Southampton Island, has monitored fewer than 1,000 nests *total* during its ten years of operation, with an annual budget of approximately $45,000. At Churchill, where shorebirds were historically numerous, sample sizes for the commonest species, the Semiplamated Sandpiper, reached 100 nests per year in a study by Gratto-Trevor (1994), whereas in the high arctic, at Alert on northern Ellesmere Island, total shorebird nests per year have varied from 0–30 over the past several decades (R. I. G. Morrison, unpubl. data).

Geographic Variation

Another major, and largely unacknowledged, problem is that trends are rarely the same throughout a species' range. BBS data, for example, show this clearly (Sauer et al. 2007). Populations go up in some areas, down in others, and remain constant in yet other areas. This reality has great importance for estimating population trends with demographic methods. It means that the results from each site at which demographic rates were measured would have to be weighted according to what fraction of the population that site "represented" (a term which would have to be defined). At present, however, we would have little basis for estimating these weights because we know too little about the distribution and abundance of shorebirds throughout their ranges in North America. Arctic PRISM is obtaining this information, but unless these surveys are completed, it would be difficult to decide what weights to give to different trends. The costs of demographic monitoring would thus be in addition to, not instead of, the costs for the first round of Arctic PRISM surveys. And if the Arctic PRISM surveys were not continued, we would have little basis for estimating the relative importance of site-specific demographic trends in the future, when populations size and distribution may have changed. Thus, a further dilemma for anyone trying to design a trend estimation program using demographic methods is how to do so without also conducting the PRISM surveys which would, by themselves, provide the needed trend estimates.

Measurement Errors

Finally, numerous practical problems could further increase the effort required to monitor shorebird population trends with demographic methods. For example, repeated visits to the nest (e.g., to catch and band both adults) can attract predators, which would cause negative bias in the estimated hatching rate. Similarly, if even a small amount of renesting occurred undetected, the estimated birth rates would be substantially underestimated. For example, suppose that 50% of the first nesting attempts fail but that one in eight of these birds renests successfully. Suppose, however, these birds move off-plot and renesters coming onto the surveyed plots are counted as late arrivers, not as renesters. Renesting is thus missed. In this case, using the values above for S_0 and S, $b = 0.5 + (0.5 \times 0.125) = 0.5625$. As a result, $\hat{\lambda} = 0.5625(0.5) + 0.72 = 1.00$. Instead of concluding that $\lambda = 0.97$, which implies a 20-year decline of nearly 50%, we would conclude that the population was stable. Other scenarios, in which the population is stable but the survey indicates a serious decline (e.g., due to visitor impacts on nest success), are also easy to construct.

Summary

The biggest obstacle to estimating population trends from demographic rates is obtaining an essentially unbiased estimate, based on a large sample, of first-year survivorship. For shorebirds in the arctic, we do not know of any way to do this. But even if it could be done, the required sample size would probably be more, and maybe much more, than 2,000 nest attempts per year or 40,000 attempts during 20 years. We suspect that at least 10–15 camps, widely distributed across the arctic, would be needed to collect the data and that the cost would substantially exceed the cost of the current methods being used in Arctic PRISM, especially since the current Arctic PRISM surveys would probably be needed to obtain the weights for combining estimates from different regions. Even if this large sample could be obtained, however, several sources of potential bias exist, particularly undetected renesting and visitor impacts, that could raise doubts about the validity of trend estimates. It might be noted that measurement of effective juvenile production may be more practically measured outside the arctic at some of the first stop-over areas during southward migration, though this would provide little information on reasons for annual success or origin of breeding areas of the birds involved. We return to the idea of coordinated programs inside and outside the arctic at the end of this appendix.

MIGRATION COUNTS

Migration counts provide valuable information about use of critical stop-over habitat, changes in phenology which may provide useful information about conditions on the breeding and wintering grounds, and they certainly provide an indication of trend in population size. They are also relatively inexpensive, especially compared to surveys in the arctic. In addition, several programs have already been in operation for over 30 years, providing a valuable historical database on which to build future sampling schemes more suited to trend analyses. Migration counts have always been viewed as a critical component of the overall PRISM program (Johnston and Bart, chapter 1, this volume). All of the authors of this appendix have devoted substantial time to conducting, refining, and using migration counts to estimate population trends, and their value for many purposes has been shown in many studies (Kingsford 1999, Kingsford et al. 1999, Baker et al. 2004, Nebel et al. 2008). Our only point in this appendix is that we do not believe migration counts—at least for most shorebird populations—are likely to yield the "gold standard" (i.e., estimates with sufficiently high precision and low bias to achieve the PRISM accuracy target of 80% power to detect a 50% decline in 20 years), which is usually needed before significant resources are committed to restoring populations. In the rest of this section, we elaborate on this view. See Bart and Smith (chapter 13, this volume) for a detailed discussion of why we believe the Arctic PRISM surveys will achieve the PRISM accuracy target.

Causes of Change in Migration Counts

A decline in the mean numbers counted at migration sites can arise from three causes other than a reduction in population size: a net shift of birds from monitored to non-monitored sites, a reduction in the residence times at monitored sites, and a change in the proportion of birds at the sites that are detected. It should be noted that while a net shift could also occur on sites surveyed on the breeding grounds, random selection of the sites makes a large net shift extremely unlikely, whereas it is much harder to predict how likely a net shift is on migration sites because they are almost never selected under a well-defined sampling plan. Among the three problems above, change in residence times may be especially significant. For example, Ydenberg et al. (2004) showed that an 83% decline in Western Sandpiper numbers on Sidney Island, British Columbia, was explained entirely by a four-fold decrease in length of stay (probably due to an increase in the number of Peregrine Falcons preying on the transient shorebirds at the site). Turnover times of Semipalmated Sandpipers on migration in the Bay of Fundy have been reported to have fallen by an estimated 50% (15 days to 7–8 days; P. Hicklin and J. Chardine, pers. comm.); declines in residence times of shorebirds on the Atlantic coast would not be surprising given the reintroduction and subsequent recovery of nesting Peregrine Falcons at key shorebird staging sites around the Bay of Fundy.

Improving Migration Counts

We now discuss three improvements in migration monitoring which could reduce the three sources of bias noted above, and the feasibility of implementing these changes. Suppose that a detailed description of the sites at which shorebirds occur during migration existed and that sites to survey were selected following a well-defined sampling plan. In this case, a change in distribution of shorebirds between periods being compared would not cause any bias in the estimated numbers present. The description of where migrants occur would involve specific locations, like wildlife refuges, and entire regions like the Prairie Potholes. Stratification would be used in selecting sites. For example "high-value" sites at which surveyors were available would probably be placed in one stratum, and all sites in it might be selected. Sites in other strata would be randomly selected. This approach would allow the survey to be focused on the best and most accessible areas while ensuring that a shift in distribution of the shorebirds would not cause any bias in the estimates of numbers present. This improvement in survey design has already been applied to migration counts being made for the PRISM migration program by the International Shorebird Survey throughout the Atlantic coastal states of the U.S., and plans to expand the revisions in survey design have been proposed for other regions as well.

The second improvement needed for migration counts to yield unbiased estimates of trend in population size is to extend the sampling frame so that the entire population being surveyed occurs within the survey area during the survey period. Many of the regions could be lightly sampled, but all sites would have to have positive probability of entering the sample. With such a sampling frame, movements by birds within the study area would not cause any bias in the estimate of population size. Note that the sampling frame would have to include the wintering area (Central and South America) unless a survey period could be defined in which all birds have left the breeding grounds but have not yet reached the wintering grounds.

The third needed improvement in migration counts is to place much more emphasis on getting unbiased counts of the number present during surveys. In many areas, especially those in coastal locations where the birds are found on open habitats, obtaining complete counts of shorebirds is not difficult. In others, however, poor access or visibility hinders observers. Such locations might be more lightly sampled, and professional surveyors—at times following complex plans such as double sampling or a combination of air and ground surveys—might be needed. An alternative approach, less satisfactory but still perhaps adequate, would be to standardize counts carefully in each area so that at least there were grounds for assuming that detection rates did not change within sites, between years or other survey periods being compared. The necessity for standardizing coverage at individual sites is emphasized for participants in surveys in eastern Canada and the U.S. Under this approach, an estimate of the detection ratio would still be needed in combining results from different sites, but even fairly large errors in these estimates would not seriously compromise the estimated trend in population size. Farmer and Durbian (2006) assessed the improvements in survey accuracy that would result from adjusting counts with detection ratios for vegetated wetlands at Squaw Creek National Wildlife Refuge, and developed techniques that could be widely applied in improvement of migration surveys.

These three improvements—a rigorous sampling plan to select sites, a study area covering all locations at which birds are present, and more accurate counts—provide a framework for assessing the current plan for making migration counts. At present, if numbers counted change, it is difficult to know whether that is because distribution of the birds changed, turnover rates changed, or detection ratios changed. The issue of biases and uncertainties in migration surveys was examined by Bart et al. (2007) using data from the International Shorebird Survey and the Maritimes Shorebird Survey. Despite large samples and a sophisticated analysis, they were not able to show conclusively whether declining reported numbers were due to population declines or to some combination of the potential problems identified above.

Conclusion

If the "three problems" described above could be solved, then we believe that migration counts could provide trend estimates that would achieve the PRISM accuracy target. Efforts have been made to reduce these problems in specific areas (e.g., the east coast of the U.S.) but extending this work throughout the U.S. and most or all of Central and South America does not currently seem feasible. We emphasize, however, that migration counts are highly valuable for a variety of reasons, summarized at the beginning of this section, and that they certainly provide strong indications of change in population size at a low cost relative to surveys on the breeding or wintering grounds. This is especially true when independent counts show similar patterns, as has often been the case (International Wader Study Group 2003).

FINAL COMMENT

Several people have suggested that perhaps a much more comprehensive (and better designed) program outside the arctic, along with an expanded set of Tier 2 (demographic) sites within the arctic, could (1) provide estimates of trend that meet the PRISM accuracy target but at lower cost and (2) provide additional information on demographic rates that could prove useful in understanding and reversing serious declines. This set of surveys might also lend itself to an "adaptive monitoring" approach in which competing hypotheses and models were constantly tested and refined using empirical data. Although we believe the obstacles to developing such a program are immense (for the reasons described above), so might be the potential values of such an approach. We would therefore encourage the formation of a working group to explore, in the same level of detail used to design the current Arctic PRISM surveys, how such a program might be developed. The ongoing first round of Arctic PRISM surveys will provide estimates of current population size as well as information on distribution, abundance, and habitat relationships in extensive areas that have never before been surveyed. If better methods are developed for estimating future population sizes (and thus trends), then they can be substituted for the current methods in round two. This ability to change the design (because we are not using index methods) is one of the great advantages of the Arctic PRISM approach.

Regional Density Estimates

CANADA: TABLE B.1

Region, species, and habitat-specific estimates of density (birds/km^2) and population size for shorebirds in surveyed regions of Canada.

Species shown with 4-letter codes. Refer to Appendix C for common and scientific names.

			Wetlands		Moist areas		Uplands		All habitats	
Species	Region ID	Region name	Dens.	SE	Dens.	SE	Dens.	SE	Pop. size	SE
AMGP	1.1	Québec	0.12	0.12	—	—	—	—	2,211	2,256
AMGP	3.1	PCI/AFI	0.55	0.55	—	—	2.12	1.25	10,927	5,694
AMGP	3.2	West Baffin	1.22	1.02	—	—	—	—	5,327	4,356
AMGP	4.1	SHI	2.97	2.45	4.21	2.70	—	—	65,039	27,177
AMGP	4.2	Coats	2.60	1.28	—	—	6.64	1.10	9,798	2,853
AMGP	8.1	QM-west	—	—	—	—	0.71	0.80	5,714	6,038
AMGP	8.2	QM-east	—	—	2.42	2.24	1.32	1.42	29,450	21,707
AMGP	8.3	Kent Penin	0.18	0.20	0.99	0.93	4.13	1.80	29,794	11,318
AMGP	9.1	C Bay	—	—	2.97	1.44	4.35	2.14	448	174
AMGP	9.2	Som Island	—	—	0.49	0.54	0.05	0.07	1,211	1,346
AMGP	12.1	Yukon NS	1.06	1.06	0.69	0.69	3.50	2.39	16,274	8,708
AMGP	12.2	MD-west	0.56	0.60	3.32	1.91	—	—	7,691	4,200
AMGP	12.3	MD-east	2.03	2.14	—	—	—	—	5,614	5,748
AMGP	12.4	MD-mgp	0.46	0.32	1.22	0.89	—	—	175	119
BASA	8.1	QM-west	—	—	—	—	1.41	1.60	11,428	12,076
BASA	8.3	Kent Penin	2.52	1.32	—	—	—	—	1,971	1,497
BASA	9.1	C Bay	—	—	1.78	1.08	3.73	1.64	348	128
BASA	9.2	Som Island	—	—	3.65	2.31	2.49	2.68	12,696	7,730

APPENDIX B.1 (*continued*)

Species	Region ID	Region name	Wetlands Dens.	Wetlands SE	Moist areas Dens.	Moist areas SE	Uplands Dens.	Uplands SE	All habitats Pop. size	All habitats SE
BASA	12.1	Yukon NS	0.29	0.30	0.69	0.69	—	—	3,553	3,257
BBPL	3.1	PCI/AFI	1.21	0.74	11.23	1.86	3.08	2.35	19,118	9,567
BBPL	3.2	West Baffin	0.46	0.45	0.55	0.58	—	—	3,698	2,694
BBPL	4.1	SHI	—	—	1.58	1.57	—	—	16,200	16,216
BBPL	4.2	Coats	2.19	0.99	—	—	—	—	2,247	1,129
BBPL	9.1	C Bay	—	—	—	—	3.73	1.45	262	103
BBPL	9.2	Som Island	—	—	—	—	4.53	3.27	7,989	5,060
BBPL	12.3	MD-east	1.26	1.30	—	—	—	—	3,477	3,583
BBSA	9.1	C Bay	—	—	2.97	1.64	—	—	142	76
DUNL	1.1	Québec	0.43	0.25	—	—	—	—	8,178	4,811
DUNL	3.1	PCI/AFI	3.95	2.77	—	—	1.18	0.79	24,766	14,609
DUNL	3.2	West Baffin	—	—	3.40	2.86	—	—	10,467	9,259
DUNL	4.1	SHI	12.53	5.77	7.38	3.42	—	—	167,907	62,468
DUNL	4.2	Coats	12.94	3.84	—	—	—	—	13,290	4,547
DUNL	8.1	QM-west	4.26	2.50	4.03	2.33	4.12	3.09	85,051	36,589
DUNL	8.2	QM-east	7.47	2.23	—	—	2.00	1.90	66,848	26,652
DUNL	8.3	Kent Penin	9.21	3.60	2.97	2.78	0.37	0.37	23,588	16,084
HUGO	12.4	MD-mgp	2.05	1.88	—	—	—	—	127	130
LBDO	12.1	Yukon NS	5.59	2.84	—	—	—	—	6,315	3,414
LBDO	12.2	MD-west	1.26	1.33	—	—	—	—	3,647	3,760
LBDO	12.4	MD-mgp	0.68	0.63	—	—	—	—	42	43
LESA	1.1	Québec	0.44	0.26	—	—	—	—	8,395	4,918
LESA	8.3	Kent Penin	—	—	1.34	1.03	2.29	1.95	20,239	13,117
LESA	12.1	Yukon NS	0.91	0.70	—	—	—	—	1,032	763
LESA	12.2	MD-west	1.80	1.46	—	—	—	—	5,190	4,063
LESA	12.3	MD-east	2.45	2.29	—	—	—	—	6,771	6,511
LESA	12.4	MD-mgp	0.86	0.80	0.28	0.31	—	—	87	62
PESA	3.1	PCI/AFI	0.71	0.50	—	—	0.06	0.07	3,891	2,617
PESA	4.1	SHI	—	—	1.93	1.90	—	—	19,809	20,119
PESA	8.1	QM-west	4.03	2.29	1.50	1.64	4.44	4.07	68,972	40,532
PESA	8.2	QM-east	14.02	5.81	18.00	10.35	1.32	1.42	201,815	82,836
PESA	8.3	Kent Penin	16.24	4.84	2.22	1.53	2.75	1.36	39,822	14,940
PESA	9.1	C Bay	2.79	1.86	2.38	1.12	—	—	170	70
PESA	12.1	Yukon NS	10.36	4.24	2.15	1.68	—	—	21,690	9,696
PESA	12.2	MD-west	5.16	3.34	1.66	1.60	—	—	17,904	10,358

APPENDIX B.1 *(continued)*

Species	Region ID	Region name	Wetlands		Moist areas		Uplands		All habitats	
			Dens.	SE	Dens.	SE	Dens.	SE	Pop. size	SE
PESA	12.3	MD-east	16.41	5.50	—	—	6.71	4.33	56,204	18,524
PESA	12.4	MD-mgp	2.73	2.50	1.84	1.90	3.20	3.60	772	807
PUSA	3.1	PCI/AFI	0.37	0.38	—	—	—	—	1,917	1,954
PUSA	4.2	Coats	0.57	0.57	—	—	—	—	585	603
REKN	3.1	PCI/AFI	1.20	1.21	—	—	—	—	6,158	6,303
REKN	3.2	West Baffin	0.91	0.96	—	—	—	—	3,971	4,145
REKN	9.2	Som Island	—	—	—	—	2.49	2.68	4,392	4,503
REKN	10.1	Alert	4.80	1.44	—	—	—	—	647	194
REPH	3.1	PCI/AFI	71.68	21.65	22.46	3.72	40.32	19.25	523,825	154,194
REPH	3.2	West Baffin	42.15	11.98	39.59	12.99	—	—	306,498	75,560
REPH	4.1	SHI	21.82	6.05	5.10	3.27	—	—	213,055	61,198
REPH	4.2	Coats	27.50	9.32	—	—	—	—	28,237	11,417
REPH	8.1	QM-west	2.71	2.52	—	—	—	—	15,391	16,123
REPH	8.3	Kent Penin	7.55	4.93	—	—	0.54	0.56	9,143	5,080
REPH	9.1	C Bay	44.68	13.58	25.58	7.11	13.67	4.46	3,092	695
REPH	9.2	Som Island	13.42	2.22	1.01	1.04	7.48	8.04	17,931	14,078
REPH	10.1	Alert	0.21	0.21	—	—	—	—	28	29
REPH	12.1	Yukon NS	—	—	0.30	0.31	—	—	1,377	1,426
RNPH	8.1	QM-west	1.37	0.96	1.50	1.64	3.22	2.36	44,068	22,751
RNPH	8.2	QM-east	10.05	9.21	—	—	—	—	58,469	53,738
RNPH	8.3	Kent Penin	3.47	3.46	1.25	1.28	—	—	8,673	7,279
RNPH	12.1	Yukon NS	14.04	5.07	2.92	1.68	—	—	29,422	10,504
RNPH	12.2	MD-west	17.33	6.83	1.66	1.60	3.14	3.24	59,916	21,688
RNPH	12.3	MD-east	11.35	5.44	—	—	14.52	8.66	54,907	21,038
RNPH	12.4	MD-mgp	16.60	7.54	2.27	1.94	35.11	33.62	5,485	4,264
RUTU	3.1	PCI/AFI	6.37	3.33	22.46	3.72	4.90	3.14	53,546	23,919
RUTU	4.2	Coats	2.00	1.40	—	—	—	—	2,052	1,491
RUTU	9.1	C Bay	—	—	—	—	1.24	0.91	87	63
RUTU	10.1	Alert	3.34	1.15	—	—	—	—	450	154
SAND	9.2	Som Island	—	—	2.27	1.48	9.97	10.72	22,721	18,593
SAND	10.1	Alert	2.30	0.98	—	—	—	—	309	131
SEPL	1.1	Québec	0.92	0.38	—	—	—	—	17,529	7,186
SEPL	8.3	Kent Penin	—	—	—	—	0.43	0.42	2,590	2,574
SESA	1.1	Québec	3.59	0.92	—	—	—	—	68,229	17,512
SESA	4.1	SHI	5.83	4.61	13.44	4.30	—	—	180,662	66,862

APPENDIX B.1 (*continued*)

Species	Region ID	Region name	Wetlands Dens.	Wetlands SE	Moist areas Dens.	Moist areas SE	Uplands Dens.	Uplands SE	All habitats Pop. size	All habitats SE
SESA	4.2	Coats	60.81	15.31	7.94	2.53	6.64	1.10	78,064	22,294
SESA	8.1	QM-west	3.57	2.51	4.52	3.47	5.24	3.56	93,511	41,779
SESA	8.2	QM-east	4.69	3.16	4.84	4.49	11.73	3.25	192,281	53,300
SESA	8.3	Kent Penin	10.37	6.58	1.25	1.28	1.75	1.32	24,628	16,460
SESA	12.1	Yukon NS	2.61	1.83	0.30	0.31	—	—	4,318	2,575
SESA	12.2	MD-west	3.15	2.14	12.98	7.06	3.00	3.12	39,435	16,578
SESA	12.3	MD-east	11.94	4.41	—	—	15.96	7.57	58,867	18,728
SESA	12.4	MD-mgp	3.58	2.17	1.23	1.26	—	—	369	199
STSA	8.1	QM-west	—	—	3.73	3.16	—	—	25,547	25,366
STSA	8.3	Kent Penin	3.47	1.48	—	—	1.86	1.26	13,964	8,351
STSA	12.1	Yukon NS	3.75	1.84	—	—	—	—	4,233	2,207
STSA	12.2	MD-west	1.41	0.92	1.66	1.60	—	—	7,121	4,014
STSA	12.3	MD-east	3.24	1.65	—	—	1.89	2.05	12,023	4,619
STSA	12.4	MD-mgp	2.40	1.74	0.61	0.63	—	—	222	121
WHIM	12.1	Yukon NS	1.42	0.75	2.68	2.68	—	—	14,036	12,586
WHIM	12.2	MD-west	1.78	1.37	—	—	—	—	5,135	3,761
WHIM	12.3	MD-east	0.69	0.62	—	—	—	—	1,918	1,628
WHIM	12.4	MD-mgp	5.37	2.85	6.76	4.03	—	—	1,143	645
WISN	1.1	Québec	0.97	0.46	—	—	—	—	18,353	8,822
WISN	8.1	QM-west	1.88	1.89	—	—	—	—	10,649	10,792
WISN	12.1	Yukon NS	1.89	1.22	0.72	0.71	—	—	5,486	3,652
WISN	12.2	MD-west	10.95	4.76	1.66	1.60	1.57	1.62	38,071	16,125
WISN	12.3	MD-east	1.69	1.68	—	—	3.84	1.92	10,894	5,926
WISN	12.4	MD-mgp	3.45	1.30	2.80	1.94	9.60	10.79	1,694	1,475
WRSA	3.1	PCI/AFI	47.30	20.41	112.31	18.60	58.48	36.83	477,220	188,304
WRSA	3.2	West Baffin	18.83	6.16	22.11	8.48	—	—	150,517	44,714
WRSA	4.1	SHI	10.53	5.68	—	—	—	—	77,639	43,565
WRSA	4.2	Coats	4.39	2.81	5.27	2.28	—	—	10,153	4,814
WRSA	8.1	QM-west	—	—	—	—	0.90	0.90	7,276	7,306
WRSA	8.3	Kent Penin	1.11	1.17	—	—	0.27	0.28	2,491	1,912
WRSA	9.1	C Bay	12.57	5.31	17.85	4.78	15.53	4.07	2,201	473
WRSA	9.2	Som Isl	13.42	2.22	0.50	0.52	—	—	3,611	1,581

ALASKA: TABLE B.2

Region, species, and habitat-specific estimates of density (birds/km²) and population size for shorebirds in surveyed regions of Alaska.

Species shown with 4-letter codes. Refer to Appendix C for common and scientific names.

Species	Region ID	Region name	Wetlands		Moist areas		Uplands		All habitats	
			Dens.	SE	Dens.	SE	Dens.	SE	Pop. size	SE
AMGP	13.1	Arctic NWR	2.46	0.77	4.30	1.13	0.41	0.32	11,888	2,578
AMGP	13.2	Central NS	1.06	1.11	3.37	1.22	—	—	24,230	7,517
AMGP	13.3	Foothills east	—	—	1.82	1.71	—	—	2,572	2,492
AMGP	14.1	NPRA NE	2.88	0.65	6.80	2.69	—	—	8,455	2,558
AMGP	14.2	NPRA coast	2.50	0.80	8.80	2.78	—	—	124,297	36,523
AMGP	14.3	Foothills west	—	—	—	—	5.50	4.33	112,796	88,885
AMGP	14.4	Colville River	—	—	—	—	1.78	1.41	20	14
BARG	13.2	Central NS	—	—	0.40	0.40	—	—	2,263	2,283
BARG	14.1	NPRA NE	2.26	0.62	1.59	1.07	—	—	3,209	1,060
BARG	14.2	NPRA coast	1.07	0.71	0.71	0.41	3.19	3.24	59,724	44,905
BARG	16.1	Selawik NWR	0.60	0.59	—	—	—	—	4,313	4,316
BARG	17.1	YKD central	2.79	1.56	4.69	2.44	—	—	3,121	1,114
BASA	13.1	Arctic NWR	14.36	14.40	0.25	0.25	—	—	16,194	15,698
BASA	13.2	Central NS	—	—	0.55	0.56	—	—	3,150	3,220
BASA	14.1	NPRA NE	0.02	0.02	—	—	—	—	15	15
BASA	14.2	NPRA coast	—	—	0.32	0.32	—	—	3,935	3,942
BBPL	13.2	Central NS	5.06	2.97	1.67	1.15	—	—	33,720	15,210
BBPL	13.3	Foothills east	—	—	2.64	2.61	—	—	3,725	3,700
BBPL	14.1	NPRA NE	6.40	1.24	4.38	2.42	—	—	8,973	2,506
BBPL	14.2	NPRA coast	11.05	2.29	3.20	1.33	1.33	1.43	134,313	34,499
BBPL	17.1	YKD central	2.78	2.23	7.16	4.23	—	—	4,081	1,668
BBPL	18.1	AK Pen	—	—	5.81	8.37	1.75	1.16	45,293	40,393
BBSA	13.1	Arctic NWR	1.10	0.79	0.04	0.05	—	—	1,293	895
BBSA	13.2	Central NS	2.30	2.32	0.91	0.65	—	—	16,194	11,761
BBSA	14.1	NPRA NE	1.62	1.07	0.04	0.04	—	—	1,292	846
BBSA	14.2	NPRA coast	1.35	1.32	0.92	0.66	—	—	20,649	12,372
BLTU	16.1	Selawik NWR	0.90	0.91	—	—	—	—	6,488	6,557
BLTU	17.1	YKD central	47.71	8.99	2.36	0.56	—	—	22,986	5,704
DUNL	13.1	Arctic NWR	2.89	1.20	0.04	0.05	—	—	3,255	1,441
DUNL	13.2	Central NS	8.27	3.55	3.92	1.87	—	—	61,883	21,430
DUNL	14.1	NPRA NE	16.63	3.94	14.45	4.95	—	—	26,111	5,813
DUNL	14.2	NPRA coast	32.05	5.57	5.12	1.99	4.87	3.48	352,900	75,482
DUNL	16.1	Selawik NWR	2.99	2.10	—	—	—	—	21,500	15,360
DUNL	17.1	YKD central	198.99	37.63	8.91	3.75	—	—	95,509	22,384

APPENDIX B.2 (*continued*)

APPENDIX B.2 (CONTINUED)

Species	Region ID	Region name	Wetlands		Moist areas		Uplands		All habitats	
			Dens.	SE	Dens.	SE	Dens.	SE	Pop. size	SE
DUNL	18.1	AK Pen	65.84	36.94	—	—	31.81	18.40	933,999	459,030
GRYE	18.1	AK Pen	5.69	4.15	—	—	—	—	48,656	29,894
LBDO	13.1	Arctic NWR	1.04	0.74	0.99	0.57	0.43	0.45	3,828	1,623
LBDO	13.2	Central NS	5.23	2.28	5.12	2.27	—	—	54,196	17,137
LBDO	13.3	Foothills east	—	—	2.64	2.61	—		3,725	3,700
LBDO	14.1	NPRA NE	8.92	1.50	6.93	2.32	—	—	13,258	2,623
LBDO	14.2	NPRA coast	29.10	5.46	11.62	2.65	10.59	5.15	489,274	96,575
LBDO	14.4	Colville River	—	—	—	—	0.99	1.18	11	12
LBDO	16.1	Selawik NWR	1.20	1.18	—	—	—	—	8,626	8,633
LBDO	17.1	YKD central	1.24	1.02	8.17	5.68	—	—	3,764	2,737
LBDO	18.1	AK Pen	—	—	—	—	1.75	1.16	20,408	18,724
LESA	18.1	AK Pen	18.18	3.91	—	—	9.55	2.87	266,915	57,869
LESA	19.1	AK Mar	—	—	0.48	0.46	—	—	4,450	4,292
MAGO	18.1	AK Pen	2.06	1.67	—	—	—	—	17,655	12,856
PAGP	16.1	Selawik NWR	3.71	2.05	—	—	—	—	26,649	15,455
PAGP	18.1	AK Pen	0.52	0.59	—	—	1.75	1.16	24,856	19,032
PESA	13.1	Arctic NWR	17.70	3.85	7.79	2.05	2.36	1.50	38,822	7,501
PESA	13.2	Central NS	22.60	6.06	19.09	5.18	—	—	216,783	45,729
PESA	13.3	Foothills east	—	—	8.86	5.90	1.57	1.56	14,394	8,784
PESA	14.1	NPRA NE	35.69	4.72	32.69	8.16	—	—	57,574	9,360
PESA	14.2	NPRA coast	37.09	6.81	9.44	2.13	23.32	10.02	692,835	159,501
PESA	14.4	Colville River	—	—	—	—	4.96	5.90	55	61
PESA	17.1	YKD central	4.75	4.72	8.60	1.92	—	—	5,554	1,923
REPH	13.1	Arctic NWR	6.42	2.65	1.45	0.72	—	—	9,894	3,688
REPH	13.2	Central NS	10.64	5.33	3.83	2.04	—	—	72,753	27,539
REPH	14.1	NPRA NE	23.88	5.99	20.92	7.08	—	—	37,652	9,685
REPH	14.2	NPRA coast	40.02	7.37	9.33	3.64	—	—	393,304	74,480
REPH	17.1	YKD central	4.86	3.26	3.49	3.27	—	—	3,609	1,922
RNPH	13.1	Arctic NWR	15.04	3.91	2.90	1.22	2.49	1.43	26,480	6,627
RNPH	13.2	Central NS	8.14	4.27	3.73	3.39	—	—	60,195	26,927
RNPH	13.3	Foothills east	—	—	3.11	2.97	1.57	1.56	6,285	4,649
RNPH	14.1	NPRA NE	11.99	2.08	8.16	2.10	—	—	16,764	2,752
RNPH	14.2	NPRA coast	18.49	4.15	9.77	3.16	10.28	6.72	388,552	105,335
RNPH	14.4	Colville River	—	—	—	—	2.98	3.54	33	37
RNPH	16.1	Selawik NWR	2.43	1.86	—	—	0.24	0.27	19,366	13,864

APPENDIX B.2 (*continued*)

Species	Region ID	Region name	Wetlands		Moist areas		Uplands		All habitats	
			Dens.	SE	Dens.	SE	Dens.	SE	Pop. size	SE
RNPH	17.1	YKD central	64.03	10.82	55.02	11.55	—	—	51,099	8,107
RNPH	18.1	AK Pen	8.66	1.27	—	—	—	—	74,038	22,351
ROSA	17.1	YKD central	0.89	0.64	16.03	4.84	—	—	6,674	2,285
ROSA	19.1	AK Mar	—	—	6.18	3.89	—	—	57,152	35,994
RUTU	13.1	Arctic NWR	0.40	0.23	0.60	0.49	—	—	1,613	1,020
RUTU	14.1	NPRA NE	1.09	0.63	1.54	1.17	—	—	2,256	1,215
RUTU	17.1	YKD central	0.58	0.43	2.36	0.56	—	—	1,193	321
SBDO	18.1	AK Pen	6.49	2.79	5.81	8.37	3.50	2.32	121,193	56,140
SEPL	13.1	Arctic NWR	0.36	0.26	0.12	0.12	0.25	0.26	1,060	585
SEPL	13.3	Foothills east	20.48	10.01	0.25	0.29	—	—	2,538	2,220
SEPL	14.1	NPRA NE	0.04	0.04	—	—	—	—	30	30
SEPL	14.4	Colville River	—	—	—	—	3.89	1.95	43	21
SEPL	18.1	AK Pen	0.52	0.59	4.05	2.30	—	—	21,777	20,901
SESA	13.1	Arctic NWR	15.32	3.43	8.15	2.37	0.18	0.20	33,172	6,955
SESA	13.2	Central NS	29.06	10.95	18.79	4.65	—	—	246,024	63,039
SESA	13.3	Foothills east	—	—	6.22	5.94	—	—	8,769	8,570
SESA	14.1	NPRA NE	49.73	6.83	29.25	6.07	—	—	65,307	8,757
SESA	14.2	NPRA coast	68.48	11.71	24.64	6.90	4.81	3.88	843,848	154,439
SESA	16.1	Selawik NWR	18.29	10.57	—	—	—	—	131,469	78,388
SESA	17.1	YKD central	69.74	14.03	2.95	3.78	—	—	33,405	11,111
STSA	13.1	Arctic NWR	4.25	1.54	1.39	0.79	—	—	7,382	2,525
STSA	13.2	Central NS	3.91	2.52	2.41	1.32	—	—	32,435	13,948
STSA	14.1	NPRA NE	3.27	0.78	1.19	0.51	—	—	3,630	829
STSA	14.2	NPRA coast	6.34	1.78	2.03	1.04	—	—	69,006	18,381
WESA	13.1	Arctic NWR	0.14	0.14	—	—	—	—	154	154
WESA	14.2	NPRA coast	6.26	2.98	5.68	2.98	15.36	6.53	322,700	101,477
WESA	16.1	Selawik NWR	4.10	2.48	—	—	—	—	29,458	18,283
WESA	17.1	YKD central	14.42	8.66	172.10	29.36	—	—	73,877	15,497
WESA	18.1	AK Pen	1.55	1.18	—	—	1.75	1.16	33,661	19,870
WHIM	13.1	Arctic NWR	—	—	0.54	0.42	—	—	1,064	850
WHIM	13.3	Foothills east	—	—	0.25	0.29	0.73	0.70	1,240	977
WHIM	14.1	NPRA NE	0.12	0.07	0.11	0.08	—	—	197	93
WHIM	14.2	NPRA coast	—	—	1.11	0.83	—	—	13,418	10,405
WHIM	14.4	Colville River	—	—	—	—	0.40	0.39	4	4
WHIM	16.1	Selawik NWR	5.44	1.95	—	—	5.13	4.34	80,052	37,616

APPENDIX B.2 (*continued*)

APPENDIX B.2 (CONTINUED)

Species	Region ID	Region name	Wetlands		Moist areas		Uplands		All habitats	
			Dens.	SE	Dens.	SE	Dens.	SE	Pop. size	SE
WISN	13.3	Foothills east	—	—	3.11	2.97	—	—	4,385	4,285
WISN	14.1	NPRA NE	0.25	0.14	0.21	0.13	—	—	384	165
WISN	14.2	NPRA coast	—	—	0.09	0.09	—	—	1,078	1,080
WISN	14.4	Colville River	—	—	79.07	63.80	7.22	1.86	779	600
WISN	16.1	Selawik NWR	0.17	0.18	—	—	—	—	1,241	1,275
WISN	17.1	YKD central	2.73	3.02	8.29	2.10	—	—	4,497	2,417
WISN	18.1	AK Pen	8.16	5.17	—	—	9.78	5.18	183,801	87,656
WRSA	14.1	NPRA NE	0.55	0.49	0.14	0.14	—	—	554	403
WRSA	14.2	NPRA coast	0.25	0.25	—	—	—	—	1,758	1,766

APPENDIX C

Common, Scientific, and Abbreviated Names for Species Included in the Volume

Common name	Scientific name	Abbreviation
Shorebirds		
Black Oystercatcher	*Haematopus bachmani*	BLOY
Black-bellied Plover	*Pluvialis squatarola*	BBPL
American Golden-Plover	*Pluvialis dominica*	AMGP
Pacific Golden-Plover	*Pluvialis fulva*	PAGP
Semipalmated Plover	*Charadrius semipalmatus*	SEPL
Common Ringed Pover	*Charadrius hiaticula*	CRPL
Lesser Yellowlegs	*Tringa flavipes*	LEYE
Greater Yellowlegs	*Tringa melanoleuca*	GRYE
Spotted Sandpiper	*Actitis macularius*	SPSA
Eskimo Curlew	*Numenius borealis*	ESCU
Whimbrel	*Numenius phaeopus*	WHIM
Hudsonian Godwit	*Limosa haemastica*	HUGO
Bar-tailed Godwit	*Limosa lapponica*	BARG
Marbled Godwit	*Limosa fedoa*	MAGO
Ruddy Turnstone	*Arenaria interpres*	RUTU
Black Turnstone	*Arenaria melanocephala*	BLTU
Rock Sandpiper	*Calidris ptilocnemis*	ROSA
Purple Sandpiper	*Calidris maritima*	PUSA
Red Knot	*Calidris canutus*	REKN
Sanderling	*Calidris alba*	SAND
Dunlin	*Calidris alpina*	DUNL

APPENDIX C (*continued*)

Common name	Scientific name	Abbreviation
Semipalmated Sandpiper	*Calidris pusilla*	SESA
Western Sandpiper	*Calidris mauri*	WESA
Least Sandpiper	*Calildris minutilla*	LESA
White-rumped Sandpiper	*Calidris fuscicollis*	WRSA
Baird's Sandpiper	*Calidris bairdii*	BASA
Pectoral Sandpiper	*Calidris melanotos*	PESA
Buff-breasted Sandpiper	*Tryngites subruficollis*	BBSA
Short-billed Dowitcher	*Limnodromus griseus*	SBDO
Long-billed Dowitcher	*Limnodromus scolopaceus*	LBDO
Stilt Sandpiper	*Calidris himantopus*	STSA
Wilson's Snipe	*Gallinago delicata*	WISN
Red-necked Phalarope	*Phalaropus lobatus*	RNPH
Red Phalarope	*Phalaropus fulicarius*	REPH
Non-shorebirds		
Greater White-fronted Goose	*Anser albifrons*	GWFG
Snow Goose	*Chen caerulescens*	SNGO
Cackling Goose	*Branta hutchinsii*	CACG
Canada Goose	*Branta canadensis*	CAGO
Brant	*Branta bernicla*	BRAN
Tundra Swan	*Cygnus columbianus*	TUSW
Mallard	*Anas platyrhynchos*	MALL
Green-winged Teal	*Anas crecca*	GWTE
American Wigeon	*Anas americana*	AMWI
Northern Pintail	*Anas acuta*	NOPI
Northern Shoveler	*Anas clypeata*	NSHO
Canvasback	*Aythya valisineria*	CANV
Redhead	*Aythya americana*	REDH
Greater Scaup	*Aythya marila*	GRSC
Common Eider	*Somateria mollissima*	COEI
King Eider	*Somateria spectabilis*	KIEI
Spectacled Eider	*Somateria fischeri*	SPEI
Steller's Eider	*Polysticta stelleri*	STEI
Black Scoter	*Melanitta nigra*	BLSC
White-winged Scoter	*Melanitta fusca*	WWSC
Surf Scoter	*Melanitta perspicillata*	SUSC
Long-tailed Duck	*Clangula hyemalis*	LTDU

APPENDIX C (*continued*)

Common name	Scientific name	Abbreviation
Common Goldeneye	*Bucephala clangula*	COGO
Red-breasted Merganser	*Mergus serrator*	RBME
Rock Ptarmigan	*Lagopus muta*	ROPT
Willow Ptarmigan	*Lagopus lagopus*	WIPT
Red-throated Loon	*Gavia stellata*	RTLO
Pacific Loon	*Gavia pacifica*	PALO
Common Loon	*Gavia immer*	COLO
Yellow-billed Loon	*Gavia adamsii*	YBLO
Red-necked Grebe	*Podiceps grisegena*	RNGR
Northern Harrier	*Circus cyaneus*	NOHA
Golden Eagle	*Aquila chrysaetos*	GOEA
Rough-legged Hawk	*Buteo lagopus*	RLHA
Peregrine Falcon	*Falco peregrinus*	PEFA
Gyrfalcon	*Falco rusticolus*	GYRF
Sandhill Crane	*Grus canadensis*	SACR
Mew Gull	*Larus canus*	MEGU
Herring Gull	*Larus argentatus*	HEGU
Glaucous Gull	*Larus hyperboreus*	GLGU
Sabine's Gull	*Xema sabini*	SAGU
Arctic Tern	*Sterna paradisaea*	ARTE
Pomarine Jaeger	*Stercorarius pomarinus*	POJA
Parasitic Jaeger	*Stercorarius parasiticus*	PAJA
Long-tailed Jaeger	*Stercorarius longicaudus*	LTJA
Short-eared Owl	*Asio flammeus*	SEOW
Snowy Owl	*Bubo scandiacus*	SNOW
Common Raven	*Corvus corax*	CORA
Horned Lark	*Eremophila alpestris*	HOLA
Tree Swallow	*Tachycineta bicolor*	TRES
Arctic Warbler	*Phylloscopus borealis*	ARWA
Bluethroat	*Luscinia svecica*	BLUE
Gray-cheeked Thrush	*Catharus minimus*	GCTH
American Robin	*Turdus migratorius*	AMRO
Yellow Wagtail	*Motacilla tschutschensis*	YWAG
American Pipit	*Anthus rubescens*	AMPI
American Tree Sparrow	*Spizella arborea*	ATSP
Fox Sparrow	*Passerella iliaca*	FOSP

APPENDIX C (*continued*)

APPENDIX C (CONTINUED)

Common name	Scientific name	Abbreviation
Savannah Sparrow	*Passerculus sandwichensis*	SAVS
White-crowned Sparrow	*Zonotrichia leucophrys*	WCSP
Lapland Longspur	*Calcarius lapponicus*	LALO
Snow Bunting	*Plectrophenax nivalis*	SNBU
Hoary Redpoll	*Carduelis hornemanni*	HORE
Common Redpoll	*Carduelis flammea*	CORE
Other species		
Arctic fox	*Vulpes lagopus*	—
Grizzly bear	*Ursus arctos*	—
Muskox	*Ovibus moschatus*	—

LITERATURE CITED

Abraham, K. F., and C. D. Ankney. 1986. Birds of East Bay, Southampton Island, Northwest Territories. Canadian Field-Naturalist 100:180–185.

Abraham, K. F., and R. L. Jefferies. 1997. High goose populations: causes, impacts, and implications. Pp. 7–72 *in* B. D. J. Batt (editor), Arctic ecosystems in peril: report of the Arctic Goose Habitat Working Group. Arctic Goose Joint Venture Special Publication. U.S. Fish and Wildlife Service, Washington, DC, and Canadian Wildlife Service, Ottawa, ON.

ACIA. 2005. Arctic climate impact assessment. Cambridge University Press, Cambridge, UK. <http://www.acia.uaf.edu> (15 February 2011).

Aiken, L. S., S. G. West, and R. R. Reno. 1991. Multiple regression: testing and interpreting interactions. Sage Publications, London, UK.

Alaska Shorebird Group. 2008. Alaska shorebird conservation plan. Version II. Alaska Shorebird Group, Anchorage, AK.

Alexander, S. A., D. M. Ealey, and S. J. Barry. 1988. Spring migration of eiders, Oldsquaws, and Glaucous Gulls along offshore leads of the Canadian Beaufort Sea. Canadian Wildlife Service Technical Report Series No. 56. Canadian Wildlife Service, Edmonton, AB.

AMEC Americas Limited. 2005. Mackenzie Gas Project: wildlife habitat suitability modelling: birds. Unpublished report.

Andres, B. A., J. A. Johnson, J. Valenzuela, R. I. G. Morrison, L. A. Espinosa, and R. K. Ross. 2009. Estimating eastern Pacific coast populations of Whimbrels and Hudsonian Godwits, with an emphasis on Chiloé Island, Chile. Waterbirds 32:216–224.

Angerbjörn, A., M. Tannerfeldt, and S. Erlinge. 1999. Predator-prey relationships: arctic foxes and lemmings. Journal of Animal Ecology 68:34–49.

Anonymous. 1948. Discovery of two islands in eastern Foxe Basin. Arctic Circular 1:73–74.

Anthony, R. G., M. G. Garrett, and F. B. Isaacs. 1999. Double-survey estimates of Bald Eagle populations in Oregon. Journal of Wildlife Management 63:794–802.

Arctic PRISM Peer Review Committee. 2010. A peer review of the Arctic PRISM program. U.S. Shorebird Conservation Plan, U.S. Fish and Wildlife Service, Division of Migratory Bird Management, Arlington, VA, USA. <http://www.fws.gov/shorebirdplan/Prism.htm>.

Armstrong, W. T. 1998. Predation and antipredator tactics of nesting Black Brant and Lesser Snow Geese. Ph.D. dissertation, University of Saskatchewan, Saskatoon, SK.

Ashenhurst, A. R. 2004. The effects of seismic lines and drill pads on breeding migratory birds in the Kendall Island Migratory Bird Sanctuary, NWT. M.Sc. thesis, University of Alberta, Edmonton, AB.

Auerbach, N. A., M. D. Walker, and D. A. Walker. 1997. Effects of roadside disturbance on substrate and vegetation properties in arctic tundra. Ecological Applications 7:218–235.

Austin, J. E., and M. R. Miller. 1995. Northern Pintail (*Anas acuta*). No. 163 *in* A. Poole (editor), The Birds of North America. Cornell Lab of Ornithology, Ithaca, NY. <http://bna.birds.cornell.edu/bna/species/163>.

Badyaev, A. V., B. Kessel, and D. D. Gibson. 1998. Yellow Wagtail (*Motacilla tschutschensis*). No. 382 *in* A. Poole (editor), The Birds of North America. Cornell

Lab of Ornithology, Ithaca, NY. <http://bna.birds.cornell.edu/bna/species/382>.

Baker, A. J., P. M. González, T. Piersma, L. J. Niles, I. de Lima Serrano do Nascimento, P. W. Atkinson, N. A. Clark, C. D. T. Minton, M. K. Peck, and G. Aarts. 2004. Rapid population decline in Red Knots: fitness consequences of decreased refuelling rates and late arrival in Delaware Bay. Proceedings of the Royal Society of London, Series B 271:875–882.

Baldwin, W. K. W. 1953. List of plants collected on Prince Charles and Air Force Islands in Foxe Basin, Northwest Territories. Annual Report of the Museum of Canada 123, Ottawa, ON.

Barnhill, R. A., D. Weyer, W. F. Young, K. G. Smith, and D. James. 2005. Breeding biology of Jabirus (*Jabiru mycteria*) in Belize. Wilson Bulletin 117: 142–153.

Barr, J. F., C. Eberl, and J. W. Mcintyre. 2000. Red-throated Loon (*Gavia stellata*). No. 513 *in* A. Poole (editor), The Birds of North America. Cornell Lab of Ornithology, Ithaca, NY. <http://bna.birds.cornell.edu/bna/species/513>.

Barry, T. W. 1967. The geese of the Anderson River delta, Northwest Territories. Ph.D. dissertation, University of Alberta, Edmonton, AB.

Bart, J. 2005. Monitoring the abundance of bird populations. Auk 122:15–25.

Bart, J., B. A. Andres, S. Brown, G. Donaldson, B. Harrington, H. Johnson, V. Johnston, S. Jones, R. I. G. Morrison, M. Sallaberry, S. K. Skagen, and N. Warnock. 2002. Program for Regional and International Shorebird Monitoring (PRISM), ver. 0.7. <http://www.fws.gov/shorebirdplan/downloads/PRISMOverview1_02.pdf> (24 September 2009).

Bart, J., B. A. Andres, S. Brown, G. Donaldson, B. Harrington, V. Johnston, S. Jones, R. I. G. Morrison, and S. Skagen. 2005. The Program for Regional and International Shorebird Monitoring (PRISM). *In* C. J. Ralph and T. D. Rich (editors), Bird conservation, implementation and integration in the Americas: Proceedings of the Third International Partners in Flight conference, Vol. 2. General Technical Report PSW-GTR-191. Pacific Southwest Research Station, Forest Service, U.S. Department of Agriculture, Albany, CA.

Bart, J., S. C. Brown, B. A. Harrington, and R. I. G. Morrison. 2007. Survey trend of North American shorebirds: population declines or shifting distributions? Journal of Avian Biology 38:73–82.

Bart, J., K. P. Burnham, E. A. Dunn, C. M. Francis, and C. J. Ralph. 2004. Goals and strategies for estimating trends in landbird abundance. Journal of Wildlife Management 68:611–626.

Bart, J., and S. L. Earnst. 2002. Double sampling to estimate density and population trends in birds. Auk 119:36–45.

Bart, J., and S. L. Earnst. 2005. Methods for shorebird surveys in the Arctic. *In* C. J. Ralph and T. D. Rich (editors), Bird conservation, implementation and integration in the Americas: proceedings of the Third International Partners in Flight conference, Vol. 2. General Technical Report PSW-GTR-191. Pacific Southwest Research Station, Forest Service, U.S. Department of Agriculture, Albany, CA.

Bart, J., M. J. Fligner, and W. Notz. 1998. Sampling and statistical methods for behavioral-ecologists. Cambridge University Press, Cambridge, UK.

Bechard, M. J., and T. R. Swem. 2002. Rough-legged Hawk (*Buteo lagopus*). No. 641 *in* A. Poole (editor), The Birds of North America. Cornell Lab of Ornithology, Ithaca, NY. <http://bna.birds.cornell.edu/bna/species/641>.

Béchet, A., A. Reed, N. Pante, J. F. Giroux, and G. Gauthier. 2004. Estimating the size of the Greater Snow Goose population. Journal of Wildlife Management 68:639–649.

Beveridge, B. 2007. Habitat associations and nest success of Red-necked Phalaropes (*Phalaropus lobatus*) in the Mackenzie Delta, Northwest Territories. M.Sc. thesis, Trent University, Peterborough, ON.

Blomqvist, S., N. Holmgren, S. Åkesson, A. Hedenström, and J. Pettersson. 2002. Indirect effects of lemming cycles on sandpiper dynamics: 50 years of counts from southern Sweden. Oecologia 133:146–158.

Boarman, W. I., and B. Heinrich. 1999. Common Raven (*Corvus corax*), No. 476 *in* A. Poole (editor), The Birds of North America. Cornell Lab of Ornithology, Ithaca, NY. <http://bna.birds.cornell.edu/bna/species/476>.

Booms, T. L., T. J. Cade, and N. J. Clum. 2008. Gyrfalcon (*Falco rusticolus*). No. 114 *in* A. Poole (editor), The Birds of North America. Cornell Lab of Ornithology, Ithaca, NY. <http://bna.birds.cornell.edu/bna/species/114>.

Bordage, D., and J. L. Savard. 1995. Black Scoter (*Melanitta nigra*). No. 177 *in* A. Poole (editor), The Birds of North America. Cornell Lab of Ornithology, Ithaca, NY. <http://bna.birds.cornell.edu/bna/species/177>.

Brandt, H. 1943. Alaskan bird trails. Bird Research Foundation, Cleveland, OH.

Braun, C. E. 2005. Techniques for wildlife investigations and management. 6th ed. The Wildlife Society, Bethesda, MD.

Bray, R. J. O. 1943. Notes on the birds of Southampton Island, Baffin Island and Melville Peninsula. Auk 60:504–536.

Brown, P. W., and L. H. Fredrickson. 1997. White-winged Scoter (*Melanitta fusca*). No. 274 *in* A. Poole (editor), The Birds of North America. Cornell Lab of Ornithology, Ithaca, NY. <http://bna.birds.cornell.edu/bna/species/274>.

Brown, S. C., J. Bart, R. Lanctot, J. Johnson, S. Kendall, D. Payer, and J. Johnson. 2007. Shorebird abundance and distribution on the coastal plain of the Arctic National Wildlife Refuge. Condor 109:1–14.

Brown, S. C., C. Hickey, B. A. Harrington, and R. E. Gill (editors). 2001. United States shorebird conservation plan. 2nd ed. Manomet Centre for Conservation Sciences, Manomet, MA.

Butcher, G. S., and D. K. Niven. 2007. Combining data from the Christmas Bird Count and the Breeding Bird Survey to determine the continental status and trends of North America birds. National Audubon Society, New York, NY. <http://www.audubon.org/bird/stateofthebirds/CBID/report.php>.

CAVM Team. 2003. Circumpolar arctic vegetation map (scale 1:7,500,000). Conservation of Arctic Flora and Fauna (CAFF) Map No. 1. U.S. Fish and Wildlife Service, Anchorage, AK.

Chilton, G., M. C. Baker, C. D. Barrentine, and M. A. Cunningham. 1995. White-crowned Sparrow (*Zonotrichia leucophrys*). No. 183 *in* A. Poole (editor), The Birds of North America. Cornell Lab of Ornithology, Ithaca, NY. <http://bna.birds.cornell.edu/bna/species/183>.

Clark, P. J., and F. C. Evans. 1954. Distance to nearest neighbor as a measure of spatial relationships in populations. Ecology 35:445–453.

Cochran, W. G. 1977. Sampling techniques. J. Wiley and Sons, New York, NY.

Collins, B. T. 2007. Guidelines for using double sampling in avian population monitoring. Auk 124:1373–1387.

Connors, P. G., B. J. McCaffery, and J. L. Maron. 1993. Speciation in golden-plovers, *Pluvialis dominica* and *P. fulva*: evidence from the breeding grounds. Auk 110:9–20.

Conroy, M. J., J. T. Peterson, O. L. Bass, C. J. Fonnesbeck, J. E. Howell, C. T. Moore, and J. P. Runge. 2008. Sources of variation in detection of wading birds from aerial surveys in the Florida Everglades. Auk 125:731–743.

Cornish, B. J., and D. L. Dickson. 1996. Distribution and abundance of birds on western Victoria Island, 1992 to 1994. Canadian Wildlife Service Technical Report No. 253.

COSEWIC. 2008. COSEWIC assessment and update status report on the Short-eared Owl *Asio flammeus* in Canada. Committee on the Status of Endangered Wildlife in Canada, Ottawa, ON. <http://www.sararegistry.gc.ca/species/speciesDetails_e.cfm?sid=60> (15 February 2011).

Day, R. H., I. J. Stenhouse, and H. G. Gilchrist. 2001. Sabine's Gull (*Xema sabini*). No. 593 *in* A. Poole (editor), The Birds of North America. Cornell Lab of Ornithology, Ithaca, NY. <http://bna.birds.cornell.edu/bna/species/593>.

Dickson, H. L., S. Jaques, S. Barry, E. S. Telfer, and A. R. Smith. 1989. Identification of nesting and staging shorebird areas in the Mackenzie River Delta and Richards Island area, Northwest Territories, using LANDSAT Thematic Mapper imagery, 1985–1987, NOGAP Project C7.3. Edmonton: Canadian Wildlife Service. Unpublished report. Available at the Departmental Library, Environment Canada, Place Vincent Massey, 351 St. Joseph Boulevard, Hull, Quebec K1A 0H3, Canada.

Dickson, H. L., and A. R. Smith. 1991. Use of Landsat Thematic Mapper and multi-spectral scanning imagery to identify habitats and shorebird nesting areas on the outer Mackenzie River Delta, NWT. Pp. 91–106 *in* P. Marsh and C. S. L. Ommanney (editors), Mackenzie Delta: environmental interactions and implications of development. National Hydrology Research Institute (NHRI) Fourth Symposium, Saskatoon, SK, Canada. NHRI, Environment Canada, Ottawa, ON.

Donaldson, G., P. Hyslop, G. Morrison, L. Dickson, and I. Davidson. 2000. Canadian shorebird conservation plan. Canadian Wildlife Service, Environment Canada, Ottawa, ON.

Donnelly, K. 1978. Simulations to determine the variance and edge-effect of total nearest neighbor distance. Pp. 91–95 *in* I. Hodder (editor), Simulation models in archaeology, Cambridge University Press, London, UK.

Drilling, N., R. Titman, and F. McKinney. 2002. Mallard (*Anas platyrhynchos*). No. 658 *in* A. Poole (editor), The Birds of North America. Cornell Lab of Ornithology, Ithaca, NY. <http://bna.birds.cornell.edu/bna/species/658>.

Droege, S., A. Syr, and J. Larivée. 1998. Checklists: an under-used tool for the inventory and monitoring of plants and animals. Conservation Biology 12:1134–1138.

Dubowy, P. J. 1996. Northern Shoveler (*Anas clypeata*). No. 217 *in* A. Poole (editor), The Birds of North America. Cornell Lab of Ornithology, Ithaca, NY. <http://bna.birds.cornell.edu/bna/species/217>.

Dunn, E. H., J. Larivée, and A. Cyr. 2001. Site-specific observation in the breeding season improves the ability of checklist data to track population trends. Journal of Field Ornithology 72:547–555.

Dunn, J., and J. H. Alderfer. 2006. National Geographic field guide to North American birds. National Geographic Society, Washington, DC.

Durant, J. M., D. O. Hjermann, G. Ottersen, and N. C. Stenseth. 2007. Climate and the match or mismatch between predator requirements and resource availability. Climate Research 33:271–283.

Eadie, J. M., M. L. Mallory, and H. G. Lumsden. 1995. Common Goldeneye (*Bucephala clangula*). No. 170

in A. Poole (editor), The Birds of North America. Cornell Lab of Ornithology, Ithaca, NY. <http://bna.birds.cornell.edu/bna/species/170>.

Eberhardt, L. E., R. A. Garrott, and W. C. Hanson. 1983. Den use by arctic foxes in northern Alaska. Journal of Mammalogy 64:97–102.

Edlund, S. A. 1982. Vegetation of Melville Island, District of Franklin: eastern Melville Island and Dundas peninsula. Geological Survey of Canada, Open File Report 852.

Edlund, S. A. 1990. Bioclimatic zones in the Canadian arctic archipelago. Pp. 421–441 *in* C. R. Harington (editor), Canada's missing dimension: science and history in the Canadian arctic islands. Canadian Museum of Nature, Ottawa, ON.

Elliott, K. H., P. A. Smith, and V. H. Johnston. 2010. Aerial surveys do not reliably census boreal-nesting shorebirds. Canadian Field-Naturalist 124:145–150.

Elphick, C. S., and J. Klima. 2002. Hudsonian Godwit (*Limosa haemastica*). No. 629 *in* A. Poole (editor), The Birds of North America. Cornell Lab of Ornithology, Ithaca, NY. <http://bna.birds.cornell.edu/bna/species/629>.

Elphick, C. S., and T. L. Tibbitts. 1998. Greater Yellowlegs (*Tringa melanoleuca*). No. 355 *in* A. Poole and F. Gill (editors), The Birds of North America. The Birds of North America, Inc., Philadelphia, PA.

Ely, C. R., and A. X. Dzubin. 1994. Greater White-fronted Goose (*Anser albifrons*). No. 131 *in* A. Poole (editor), The Birds of North America. Cornell Lab of Ornithology, Ithaca, NY. <http://bna.birds.cornell.edu/bna/species/131>.

Environment Canada. 2006. Mackenzie Gas Project environmental assessment review written submission, Joint Review Panel topic specific hearing, theme 3: biological environment—fish wildlife and their habitats; topic 7: wildlife and wildlife habitat, migratory birds including Kendall Island Bird Sanctuary, November 15–16, 2006. Public Registry for the Review of the Mackenzie Gas Project Exhibit Number J-EC-00136.

Environment Canada. 2009. Climate data online for Tuktoyaktuk A and Inuvik A Stations. <http://climate.weatheroffice.ec.gc.ca/climateData/canada_e.html>.

Farmer, A. H., and F. Durbian. 2006. Estimating shorebird numbers at migration stopover sites. Condor 108:792–807.

Farmer, A. H. 2008. "Anchoring" and research priorities: factors that depress bird population estimates? Auk 125:980–983.

Fontaine, A. J., and M. L. Mallory. 2011. Detection and classification of land cover classes of Southampton Island, Nunavut, using Landsat ETM+ data. Canadian Wildlife Service Occasional Paper Number 119. Environment Canada, Ottawa, ON.

Forbes, G., K. Robertson, C. Ogilvie, and L. Seddon. 1992. Breeding densities, biogeography, and nest depredation of birds on Igloolik Island, NWT. Arctic 45:295–303.

Fredrickson, L. H. 2001. Steller's Eider (*Polysticta stelleri*), No. 571 *in* A. Poole (editor), The Birds of North America. Cornell Lab of Ornithology, Ithaca, NY. <http://bna.birds.cornell.edu/bna/species/571>.

Gabor, T. S., T. R. Gadawski, R. K. Ross, R. S. Rempl, and D. W. Kroeker. 1995. Visibility bias of waterfowl brood surveys using helicopters in the Great Clay Belt of Northern Ontario. Journal of Field Ornithology 66:81–87.

Gabrielson, I. N., and F. C. Lincoln. 1959. The birds of Alaska. Stackpole Co. and the Wildlife Management Institute, Washington, DC.

Ganter, B., F. Cooke, and P. Mineau. 1996. Long-term vegetation changes in a Snow Goose nesting habitat. Canadian Journal of Zoology 74:965–969.

Gaston, A. J., R. Decker, F. G. Cooch, and A. Reed. 1986. The distribution of larger species of birds breeding on the coasts of Foxe Basin and Northern Hudson Bay, Canada. Arctic 39:285–296.

Gaston, A. J., M. L. Mallory, H. G. Gilchrist, and K. O'Donovan. 2006. Status, trends and attendance patterns of the Northern Fulmar (*Fulmarus glacialis*) in Nunavut, Canada. Arctic 59:165–178.

Gaston, A. J., and H. Ouellet. 1997. Birds and mammals of Coats Island, N.W.T. Arctic 50:101–118.

Gaston, A. J., S. A. Smith, R. Saunders, G. I. Storm, and J. A. Whitney. 2007. Birds and marine mammals in southwestern Foxe Basin, Canada. Polar Record 43:33–47.

Gauthier, G., P. Besbeas, J. D. Lebreton, and B. J. T. Morgan. 2007. Population growth in Greater Snow Geese: a modeling approach integrating demographic and population survey information. Ecology 88:1420–1429.

Gibson, D. D., and G. V. Byrd. 2007. Birds of the Aleutian Islands, Alaska. Series in Ornithology No. 1. Nuttal Ornithological Club, Cambridge, MA, and the American Ornithologists Union, Washington, DC.

Gilchrist, H. G. 2001. Glaucous Gull (*Larus hyperboreus*). No. 573 *in* A. Poole (editor), The Birds of North America. Cornell Lab of Ornithology, Ithaca, NY. <http://bna.birds.cornell.edu/bna/species/573>.

Gilchrist, H. G., and M. L. Mallory. 2005. Declines in abundance and distribution of the Ivory Gull (*Pagophila eburnea*) in arctic Canada. Biological Conservation 121:303–309.

Gilders, M. A., and M. A. Cronin. 2000. North Slope oil field development. Pp. 15–33 *in* J. C. Truett, S. R. Johnson (editors), The natural history of an arctic oil field: development and the biota. Academic Press, San Diego, CA.

Gill, R. E., Jr., and C. M. Handel. 1981. Shorebirds of the eastern Bering Sea. Pp. 719–738 *in* D. W. Hood and J. A. Calder (editors), Eastern Bering Sea shelf: oceanography and resources, Vol. 2. Office of Marine Pollution Assessment, NOAA. Distributed by University of Washington Press, Seattle, WA.

Gill, R. E., Jr., and C. M. Handel. 1990. The importance of subarctic intertidal habitats to shorebirds: a study of the central Yukon–Kuskokwim Delta, Alaska. Condor 92:709–725.

Gill, R.E. Jr., M. R. Petersen, and P. D. Jorgensen. 1981. Birds of the northcentral Alaska Peninsula, 1976–1980. Arctic 34:286–306.

Gill, R. E., P. S. Tomkovich, and B. J. McCaffery. 2002. Rock Sandpiper (*Calidris ptilocnemis*). No. 686 *in* A. Poole (editor), The Birds of North America. Cornell Lab of Ornithology, Ithaca, NY. <http://bna.birds.cornell.edu/bna/species/686doi:10.2173/bna.686>.

Godfrey, W. E. 1986. The birds of Canada. Rev. ed. National Museum of Natural Science. Ottawa, ON.

Goudie, R. I., G. J. Robertson, and A. Reed. 2000. Common Eider (*Somateria mollissima*). No. 546 *in* A. Poole (editor), The Birds of North America. Cornell Lab of Ornithology, Ithaca, NY. <http://bna.birds.cornell.edu/bna/species/546>.

Government of Canada. 2009. Species profile: Short-eared Owl. Species at Risk Public Registry. <http://www.sararegistry.gc.ca/species/speciesDetails_e.cfm?sid=60> (2 June 2009).

Government of Canada. 2010. Species profile: Eskimo Curlew. Species at Risk Public Registry. <http://www.sararegistry.gc.ca/species/speciesDetails_e.cfm?sid=21> (5 January 2010).

Gratto-Trevor, C. L. 1992. Semipalmated Sandpiper (*Calidris pusilla*), No. 6 *in* A. Poole (editor), The Birds of North America. Cornell Lab of Ornithology, Ithaca, NY. <http://bna.birds.cornell.edu/bna/species/006>.

Gratto-Trevor, C. L. 1993. Monitoring shorebird populations in the arctic. Bird Trends 3:10–12.

Gratto-Trevor, C. L. 1994. Use of Landsat TM imagery in determining priority shorebird habitat in the outer Mackenzie Delta, N.W.T. (NOGAP subproject C.24). Unpublished report. Available at Canadian Wildlife Service, Environment Canada, Prairie and Northern Wildlife Research Centre, 115 Perimeter Road, Saskatoon, SK S7N 0X4, Canada.

Gratto-Trevor, C. L. 1996. Use of Landsat TM imagery in determining important shorebird habitat in the outer Mackenzie Delta, N.W.T. Arctic 49:11–22.

Gratto-Trevor, C. L. 1997. Climate change: proposed effects on shorebird habitat, prey, and numbers in the outer Mackenzie Delta. Pp. 205–210 *in* C. Stewart (editor), Mackenzie Basin Impact Study final report. Environment Canada, Ottawa, ON.

Gratto-Trevor, C. L., V. H. Johnston, and S. T. Pepper. 1998. Changes in shorebird and eider abundance in the Rasmussen Lowlands, NWT. Wilson Bulletin 110:316–325.

Groves, D. J., E. J. Mallek, R. Macdonald, and T. J. Moser. 2009. Migratory bird surveys in the Canadian arctic 2007/2008. Unpublished Report. U.S. Fish and Widlife Service.

Guzy, M. J., and B. J. McCaffery. 2002. Bluethroat (*Luscinia svecica*). No. 670 *in* A. Poole (editor), The Birds of North America. Cornell Lab of Ornithology, Ithaca, NY. <http://bna.birds.cornell.edu/bna/species/670>

Halkette, A. 1904. List of birds. Pp. 314–319 *in* A.P. Lowe (editor), Cruise of the Neptune. Government Printing Bureau, Ottawa, ON.

Handel, C. M., and C. P. Dau. 1988. Seasonal occurrence of migrant Whimbrels and Bristle-thighed Curlews on the Yukon–Kuskokwim Delta. Condor 90:782–790.

Handel, C. M., and R. E. Gill, Jr. 1992. Breeding distribution of the Black Turnstone. Wilson Bulletin 104:122–135.

Handel, C. M., and R. E. Gill. 2001. Black Turnstone (*Arenaria melanocephala*). No. 585 *in* A. Poole and F. Gill (editors), The Birds of North America. Cornell Lab of Ornithology, Ithaca, NY. <bna.birds.cornell.edu/bna/species/585>.

Hannon, S. J., P. K. Eason, and K. Martin. 1998. Willow Ptarmigan (*Lagopus lagopus*). No. 369 *in* A. Poole (editor), The Birds of North America. Cornell Lab of Ornithology, Ithaca, NY. <bna.birds.cornell.edu/bna/species/369>.

Harrington, B. A. 2001. Red Knot (*Calidris canutus*). No. 563 *in* A. Poole (editor), The Birds of North America. Cornell Lab of Ornithology, Ithaca, NY. <http//bna.birds.cornell.edu/bna/species/563>.

Hatch, J. J. 2002. Arctic Tern (*Sterna paradisaea*). No. 707 *in* A. Poole (editor), The Birds of North America. Cornell Lab of Ornithology, Ithaca, NY. <http://bna.birds.cornell.edu/bna/species/707>.

Haven W. R., and D. S. Lee. 2000. Pomarine Jaeger (*Stercorarius pomarinus*). No. 483 *in* A. Poole (editor), The Birds of North America. Cornell Lab of Ornithology, Ithaca, NY. <http://bna.birds.cornell.edu/bna/species/483>.

Heywood, W. W., and B. V. Sanford. 1976. Geology of Southampton, Coats and Mansel Islands, District of Keewatin, Northwest Territories. Memoir No. 382. Geological Survey of Canada, Ottawa, ON.

Hines, J. E., P. Latour, and C. Machtans. 2010. The effects on lowland habitat, breeding shorebirds and songbirds in the Banks Island Migratory Bird

Sanctuary Number 1 by the growing colony of Lesser Snow Goose (*Chen caerulescens caerulescens*). Occasional Paper No. 118. Canadian Wildlife Service, Ottawa, ON.

Hines, J. E., and M. O. Wiebe Robertson (editors). 2006. Surveys of geese and swans in the Inuvialuit Settlement Region, western Canadian arctic, 1989–2001. Occasional Paper No. 112. Canadian Wildlife Service, Ottawa, ON.

Hochachka, W. M., K. Martin, F. Doyle, and C. J. Krebs. 2000. Monitoring vertebrate populations using observational data. Canadian Journal of Zoology 78:521–529.

Holmes, R. T. 1970. Differences in population density, territoriality, and food supply of Dunlin on arctic and subarctic tundra. Pp. 303–319 *in* A. Watson (editor), Animal populations in relation to their food resources. Oxford University Press, Oxford, UK.

Holmes, R. T. 1971. Density, habitat, and the mating system of the Western Sandpiper (*Calidris mauri*). Oecologia 7:191–208.

Holmes, R. T., and C. P. Black. 1973. Ecological distribution of birds in the Kolomak River-Askinuk Mountain region, Yukon–Kuskokwim Delta, Alaska. Condor 75:150–163.

Holmes, R. T., and F. A. Pitelka. 1968. Food overlap among coexisting sandpipers on northern Alaska tundra. Systematic Zoology 17:305–318.

Holmes, R. T., and F. A. Pitelka. 1998. Pectoral Sandpiper (*Calidris melanotos*). No. 348 *in* A. Poole (editor), The Birds of North America. Cornell Lab of Ornithology, Ithaca, NY. <http://bna.birds.cornell.edu/bna/species/348>.

Howe, M. A., P. H. Geissler, and B. A. Harrington. 1989. Population trends of North American shorebirds based on the International Shorebird Survey. Biological Conservation 49:185–199.

Hussell, D. J., and R. Montgomerie. 2002. Lapland Longspur (*Calcarius lapponicus*). No. 656 *in* A. Poole (editor), The Birds of North America. Cornell Lab of Ornithology, Ithaca, NY. <http://bna.birds.cornell.edu/bna/species/656>.

Innes, M. J. G., A. K. Goodacre, A. A. Weston, and J. R. Weber. 1968. Gravity and isostacy in the Hudson Bay region. Pp. 703–728 *in* C. S. Beals (editor), Science, history and Hudson Bay, Vol. 2. Department of Energy Mines and Resources, Ottawa, ON.

International Wader Study Group (IWSG). 2003. Waders are declining worldwide: conclusions from the 2003 International Wader Study Group Conference. Cadiz, Spain: International Wader Study Group.

Jaques, D. 1987a. Landsat imagery for identification and mapping of shorebird habitat in the Mackenzie Delta, Northwest Territories, Canada. Unpublished report prepared by Ecosat Geobotanical Surveys, Inc., for the Canadian Wildlife Service, Edmonton, AB.

Jaques, D. 1987b. Identification of potential shorebird nesting and staging habitats using Landsat Thematic Mapper imagery in the outer Mackenzie Delta, NWT, Canada. Unpublished report prepared by Ecosat Geobotanical Surveys, Inc., for the Canadian Wildlife Service, Edmonton, AB.

Jehl, J. R. 2007. Disappearance of breeding Semipalmated Sandpipers from Churchill, Manitoba: more than a local phenomenon. Condor 109:351–360.

Jehl, J. R., and W. Lin. 2001. Population status of shorebirds nesting at Churchill, Manitoba. Canadian Field-Naturalist 115:487–494.

Johnson, J. A., R. B. Lanctot, B. A. Andres, J. R. Bart, S. C. Brown, S. J. Kendall, and D. C. Payer. 2007. Distribution of breeding shorebirds on the arctic Coastal Plain of Alaska. Arctic 60:277–293.

Johnson, J. M., and B. J. McCaffery. 2004. Use of upland tundra habitats by Western and Rock Sandpipers during brood-rearing on the Yukon–Kuskokwim Delta, Alaska. Wader Study Group Bulletin 103:36–39.

Johnson, K. 1995. Green-winged Teal (*Anas crecca*). No. 193 *in* A. Poole (editor), The Birds of North America. Cornell Lab of Ornithology, Ithaca, NY. <http://bna.birds.cornell.edu/bna/species/193>.

Johnson, M., J. R. Conklin, B. L. Johnson, B. J. McCaffrey, S. M. Haig, and J. R. Walters. 2009. Behavior and reproductive success of Rock Sandpipers breeding on the Yukon-Kuskokwim River Delta, Alaska. Wilson Journal of Ornithology 121:328–337.

Johnson, O. W., A. J. Bennett, L. Alsworth III, L. A. Bennett, P. M. Johnson, J. R. Morgart, and R. J. Kienholz. 2001. Radio-tagged Pacific Golden-Plovers: the Hawaii–Alaska link, spring destinations, and breeding season survival. Journal of Field Ornithology 72:537–546.

Johnson, O. W., and P. G. Connors. 1996a. American Golden-Plover (*Pluvialis dominica*). No. 201 *in* A. Poole (editor), The Birds of North America. Cornell Lab of Ornithology, Ithaca, NY. <http://bna.birds.cornell.edu/bna/species/201>.

Johnson, O. W., and P. G. Connors. 1996b. Pacific Golden-Plover (*Pluvialis fulva*). No. 202 *in* A. Poole (editor), The Birds of North America. Cornell Lab of Ornithology, Ithaca, NY. <http://bna.birds.cornell.edu/bna/species/202>.

Johnson, O. W., R. Goodwill, and P. M. Johnson. 2006. Wintering ground fidelity and other features of Pacific Golden-Plovers (*Pluvialis fulva*) on Saipan, Mariana Islands, with comparative observations from Oahu, Hawaiian Islands. Wader Study Group Bulletin 109:67–72.

Johnston, V. H., C. L. Gratto-Trevor, and S. T. Pepper. 2000. Assessment of bird populations in the

Rasmussen Lowlands, Nunavut. Occasional Paper 101. Canadian Wildlife Service, Ottawa, ON.

Johnston, V. H., and S. T. Pepper. 2009. The birds of Prince Charles Island and Air Force Island, Foxe Basin, Nunavut. Occasional Paper 117. Canadian Wildlife Service, Yellowknife, NT.

Jorgensen, T., and C. Ely. 2001. Topography and flooding of coastal ecosystems on the Yukon–Kuskokwim Delta, Alaska: implications for sea-level rise. Journal of Coastal Research 17:124–136.

Kemeny, J. G., and T. E. Kurtz. 1992. True Basic. True Basic, West Lebanon, NH.

Kemper, J. T. 2006. Vegetation changes on seismic lines from recent (2000–2001) and historic (1970–1986) seismic programs in the Mackenzie Delta area. Report No. 159. Environmental Studies Research Funds, Calgary, AB.

Kerbes, R. H. 1969. Biology and distribution of nesting blue geese on Koukdjuak Plain, Baffin Island, NWT. M.Sc. thesis, University of Western Ontario, Guelph, ON.

Kerbes, R. H., P. N. Kotanen, and R. L. Jeffries. 1990. Destruction of wetland habitats by Lesser Snow Geese: a keystone species on the west coast of Hudson Bay. Journal of Applied Ecology 27:242–258.

Kessel, B. 1989. Birds of the Seward Peninsula. University of Alaska Press, Fairbanks, AK.

Kessel, B., D. A. Rocque, and J. S. Barclay. 2002. Greater Scaup (*Aythya marila*). No. 650 *in* A. Poole (editor), The Birds of North America. Cornell Lab of Ornithology, Ithaca, NY. <http://bna.birds.cornell.edu/bna/species/650>.

Kingsford, R. T. 1999. Aerial survey of waterbirds on wetlands as a measure of river and floodplain health. Freshwater Biology 41:425–438.

Kingsford, R. T., P. S. Wong, L. W. Braithwaite, and M. T. Maher. 1999. Waterbird abundance in eastern Australia, 1983–92. Wildlife Research 26:351–366.

Kirk, W. L., and C. J. Markon. 1989. Land cover mapping of the Selawik National Wildlife Refuge: a user's guide. U.S. Fish and Wildlife Service, Anchorage, AK.

Klima, J., and J. R. Jehl. 1998. Stilt Sandpiper (*Calidris himantopus*). No. 341 *in* A. Poole (editor), The Birds of North America. Cornell Lab of Ornithology, Ithaca, NY. <http://bna.birds.cornell.edu/bna/species/341doi: 10.2173/bna.341>.

Knox, A. G., and P. E. Lowther. 2000a. Common Redpoll (*Carduelis flammea*). No. 543 *in* A. Poole (editor), The Birds of North America. Cornell Lab of Ornithology, Ithaca, NY. <http://bna.birds.cornell.edu/bna/species/543>.

Knox, A. G., and P. E. Lowther. 2000b. Hoary Redpoll (*Carduelis hornemanni*). No. 544 *in* A. Poole (editor), The Birds of North America. Cornell Lab of Ornithology, Ithaca, NY. <http://bna.birds.cornell.edu/bna/species/544>.

Kochert, M. N., K. Steenhof, C. L. Mcintyre, and E. H. Craig. 2002. Golden Eagle (*Aquila chrysaetos*). No. 684 *in* A. Poole (editor), The Birds of North America. Cornell Lab of Ornithology, Ithaca, NY. <http://bna.birds.cornell.edu/bna/species/684>.

Koski, W. R. 1975. A study of the distribution and movements of Snow Geese, other geese, and Whistling Swans on the Mackenzie Delta, Yukon north slope, and Alaskan north slope in August and September, 1974, including a comparison with similar data from 1973. Arctic Gas Biological Report Series No. 30. Canadian Arctic Gas Study Ltd., Calgary, AB.

Kotanen, P., and R. L. Jefferies. 1997. Long-term destruction of wetland vegetation by Lesser Snow Geese. Ecoscience 4:179–182.

Krebs, C. J. 1989. Ecological methodology. Harper Collins, New York, NY.

Krebs, C. J., D. Reid, A. Kenney, D. Morris, and S. Gilbert. 2008. Small mammal population monitoring: arctic WOLVES monitoring protocols. <http://www.cen.ulaval.ca/arcticwolves/en_project_descrip_CAN_method.htm> (14 April 2009).

Lanctot, R. B., and C. D. Laredo. 1994. Buff-breasted Sandpiper (*Tryngites subruficollis*). No. 91 *in* A. Poole (editor), The Birds of North America. Cornell Lab of Ornithology, Ithaca, NY. <http://bna.birds.cornell.edu/bna/species/091>.

Larned, W., R. Stehn, and R. Platte. 2005. Eider breeding population survey, Arctic Coastal Plain, Alaska. U.S. Fish and Wildlife Service, Division of Migratory Bird Management. Soldotna, AK. <http//www.arlis.org>.

Latour, P. B., C. Machtans, and G. Beyersbergen. 2005. Shorebird and passerine abundance and habitat use at a high arctic breeding site: Creswell Bay, Nunavut. Arctic 58:55–65.

Latour, P. B., J. Leger, J. E. Hines, M. L. Mallory, D. L. Mulders, H. G. Gilchrist, P. A. Smith, and D. L. Dickson. 2008. Key migratory bird terrestrial habitat sites in the Northwest Territories and Nunavut. Occasional Paper 114. Canadian Wildlife Service, Yellowknife, NT.

Laursen, K., J. Frikke, and J. Kahlert. 2008. Accuracy of "total counts" of waterbirds from aircraft in coastal waters. Wildlife Biology 14:165–174.

Liebezeit, J. R., S. Kendall, S. Brown, C. Johnson, P. Martin, T. McDonald, D. Payer, C. Rea, B. Streever, A. Wildman, and S. Zack. 2009. Influence of human development and predators on nest survival of tundra birds, Arctic Coastal Plain, Alaska. Ecological Applications 19:1628–1644.

Liebezeit, J. R., P. A. Smith, R. B. Lanctot, H. Schekkerman, I. Tulp, S. J. Kendall, D. Tracy,

R. J. Rodrigues, H. Meltofte, J. A. R. Robinson, C. Gratto-Trevor, B. J. McCaffery, J. Morse, and S. W. Zack. 2007. Assessing the development of shorebird eggs using the flotation method: species-specific and generalized regression models. Condor 109:32–47.

Limpert, R. J., and S. L. Earnst. 1994. Tundra Swan (*Cygnus columbianus*). No. 89 *in* A. Poole (editor), The Birds of North America. Cornell Lab of Ornithology, Ithaca, NY. <http://bna.birds.cornell.edu/bna/species/089>.

Littell, R. C., G. A. Milliken, W. W. Stroup, R. D. Wolfinger, and O. Schabenberger. 2006. SAS for Mixed Models, 2nd edition. SAS Institute Inc., Cary, NC.

Lowther, P. E., C. Celada, N. K. Klein, C. C. Rimmer, and D. A. Spector. 1999. Yellow Warbler (*Setophaga petechia*). No. 454 *in* A. Poole (editor), The Birds of North America. Cornell Lab of Ornithology, Ithaca, NY. <http://bna.birds.cornell.edu/bna/species/454>.

Lowther, P. E., C. C. Rimmer, B. Kessel, S. L. Johnson, and W. G. Ellison. 2001. Gray-cheeked Thrush (*Catharus minimus*). No. 591 *in* A. Poole (editor), The Birds of North America. Cornell Lab of Ornithology, Ithaca, NY. <http://bna.birds.cornell.edu/bna/species/591>.

Lowther, P. E., and S. Sharbaugh. 2008. Arctic Warbler (*Phylloscopus borealis*). No. 590 *in* A. Poole (editor), The Birds of North America. Cornell Lab of Ornithology, Ithaca, NY. <http://bna.birds.cornell.edu/bna/species/590>.

Lyon, B., and R. Montgomerie. 1995. Snow Bunting (*Plectrophenax nivalis*). No. 198 *in* A. Poole (editor), The Birds of North America. Cornell Lab of Ornithology, Ithaca, NY. <http://bna.birds.cornell.edu/bna/species/198>.

Mackenzie Gas Project. 2004. Project overview. Mackenzie Gas Project, Calgary, AB. <http://www.mackenziegasproject.com/index.asp>.

Macwhirter, R. B., P. Austin-Smith, Jr., and D. Kroodsma. 2002. Sanderling (*Calidris alba*). No. 653 *in* A. Poole (editor), The Birds of North America. Cornell Lab of Ornithology, Ithaca, NY. <http://bna.birds.cornell.edu/bna/species/653>.

Macwhirter, R. B., and K. L. Bildstein. 1996. Northern Harrier (*Circus cyaneus*). No. 210 *in* A. Poole (editor), The Birds of North America. Cornell Lab of Ornithology, Ithaca, NY. <http://bna.birds.cornell.edu/bna/species/210>.

Mallek, E. J., R. Platte, and R. Stehn. 2004. Aerial breeding pair surveys of the arctic Coastal Plain of Alaska—2004. U.S. Fish and Wildlife Service, Waterfowl Management. Fairbanks, AK. <http://www.arlis.org>.

Markham, A. 1996. Potential impacts of climate change on ecosystems: a review of implications for policymakers and conservation biologists. Climate Research 6:179–191.

Markham, I. 1986. Ice cover. Pp. 101–116 *in* I. P. Martini (editor), Canadian inland seas. Elsevier Oceanography Series No. 44. Elsevier, New York, NY.

Marshall, I. B., and P. H. Schut. 1999. A national ecological framework for Canada. Ecosystems Science Directorate, Environment Canada and Agriculture, and Agri-Food Canada, Ottawa, ON. <http://sis.agr.gc.ca/cansis/nsdb/ecostrat/intro.html>.

Martin, J. L., A. Clamens, and S. Blangy. 1988. First breeding record of the Dunlin, *Calidris alpina*, on Baffin Island, Northwest Territories. Canadian Field-Naturalist 102:257–258.

Mayfield, H. F. 1961. Nesting success calculated from exposure. Wilson Bulletin 73:255–261.

Mayfield, H. F. 1978. Undependable breeding conditions in the Red Phalarope. Auk 95:590–592.

McCaffery, B. J. 1996. The status of Alaska's large shorebirds: a review and an example. International Wader Studies 8:28–32.

McCaffery, B. J., and R. Gill. 2001. Bar-tailed Godwit (*Limosa lapponica*). No. 581 *in* A. Poole (editor), The Birds of North America. Cornell Lab of Ornithology, Ithaca, NY. <http://bna.birds.cornell.edu/bna/species/581>.

McCaffery, B. J., R. E. Gill and D. R. Ruthrauff. 2006. Bar-tailed Godwits staging in western Alaska: evidence of steep population decline. Wader Study Group Bulletin 109:39.

McCaffery, B. J., and C. M. Harwood. 1997. Results of the 1996 avifaunal inventory, Cape Romanzof Long Range Radar Site, Alaska. Final report for Department of Defense Legacy Project No. 9510096. Prepared for U.S. Air Force, 611th Air Support Group, Civil Engineer Squadron/Environmental Flight, Elmendorf Air Force Base, AK. U.S. Fish and Wildlife Service, Bethel, AK.

McCaffery, B. J., and C. M. Harwood. 2000. Status of Hudsonian Godwits on the Yukon–Kuskokwim Delta, Alaska. Western Birds 31:165–177.

McCaffery, B. J., C. M. Harwood, and H. Moore. 1998. Results of the 1997 avifaunal inventory, Cape Romanzof Long Range Radar Site, Alaska. Final report for Department of Defense Legacy Project No. 97611CES001. Prepared for U.S. Air Force, 611th Air Support Group, Civil Engineer Squadron/Environmental Flight, Elmendorf Air Force Base, AK. U.S. Fish and Wildlife Service, Bethel, AK.

McCaffery, B. J., and D. R. Ruthrauff. 2004. How intensive is intensive enough? Limitations of intensive searching for estimating shorebird nest numbers. Wader Study Group Bulletin 103:63–66.

McCaffery, B. J., J. van de Kam, K. Woodley. 2008. Observations of Red Knots *Calidris canutus* at Old Chevak, Yukon–Kuskokwim Delta, Alaska, spring 2008. Wader Study Group Bulletin 115:167–170.

Mcintyre, J. W., and J. F. Barr. 1997. Common Loon (*Gavia immer*). No. 313 *in* A. Poole (editor), The Birds of North America. Cornell Lab of Ornithology, Ithaca, NY. <http://bna.birds.cornell.edu/bna/species/313>.

McKinnon, L., J. Bêty, P. Smith, G. Morrison, and E. Bolduc. 2008. Arthropod monitoring: seasonal variation in the abundance of surface active, sub-surface and low flying arthropods: arctic-WOLVES monitoring protocols. <http://www.cen.ulaval.ca/arcticwolves/en_project_descrip_CAN_method. htm> (14 April 2009).

Meltofte, H. 1985. Populations and breeding schedules of waders, Charadrii, in high arctic Greenland. Meddelelser om Grønland, Bioscience 16:1–43.

Meltofte, H. 2001. Wader population censuses in the arctic: getting the timing right. Arctic 54:367–376.

Meltofte, H., T. T. Høye, N. M. Schmidt, and M. C. Forchhammer. 2007a. Differences in food abundance cause inter-annual variation in the breeding phenology of high arctic waders. Polar Biology 30:601–606.

Meltofte, H., T. Piersma, H. Boyd, B. McCaffery, B. Ganter, V. V. Golovnyuk, K. Graham, C. L. Gratto-Trevor, R. I. G. Morrison, E. Nol, H. U. Rösner, D. Schamel, H. Schekkerman, M. Y. Soloviev, P. S. Tomkovich, D. M. Tracy, I. Tulp, and L. Wennerberg. 2007b. Effects of climate variation on the breeding ecology of arctic shorebirds. Meddelelser om Grønland Bioscience No. 59. Danish Polar Center, Copenhagen.

Mlodinow, S. G., P. F. Springer, B. Deuel, L. S. Semo, T. Leukering, T. D. Schonewald, W. Tweit, and J. H. Barry. 2008. Distribution and identification of Cackling Goose (*Branta hutchinsii*) subspecies. North American birds 63:344–360.

Montgomerie, R. D., R. V. Cartar, B. E. Lyon, and R. L. McLaughlin. 1982. Shorebird studies at Sarcpa Lake, Melville Peninsula, Northwest Territories. Wader Study Group Bulletin 34:30–32.

Montgomerie, R. D., R. V. Cartar, R. L. McLaughlin, and B. Lyon. 1983. Birds of Sarcpa Lake, Melville Peninsula, Northwest Territories: breeding phenologies, densities and biogeography. Arctic 36:65–75.

Montgomerie, R., and K. Holder. 2008. Rock Ptarmigan (*Lagopus muta*). No. 51 *in* A. Poole (editor), The Birds of North America. Cornell Lab of Ornithology, Ithaca, NY. <http://bna.birds.cornell.edu/bna/species/051>.

Morrison, R. I. G. 1997. The use of remote sensing to evaluate shorebird habitats and populations on Prince Charles Island, Foxe Basin, Canada. Arctic 50:55–75.

Morrison, R. I. G., Y. Aubry, R. W. Butler, G. W. Beyersbergen, G. M. Donaldson, C. L. Gratto-Trevor, P. W. Hicklin, V. H. Johnston, and R. K. Ross. 2001a. Declines in North American shorebird populations. Wader Study Group Bulletin 94:34–38.

Morrison, R. I. G., C. Downes, and B. Collins. 1994. Population trends of shorebirds on fall migration in eastern Canada 1974–1991. Wilson Bulletin 106:431–447.

Morrison, R. I. G., R. E. Gill, Jr., B. A. Harrington, S. Skagen, G. W. Page, C. L. Gratto-Trevor, and S. M. Haig. 2001b. Estimates of shorebird populations in North America. Occasional Paper No. 104. Canadian Wildlife Service, Ottawa, ON.

Morrison, R. I. G., and P. Hicklin. 2001. Shorebird population trends and issues in Canada: an overview. Bird Trends 8:1–5.

Morrison, R. I. G., B. J. McCaffery, R. E. Gill, S. K. Skagen, S. L. Jones, G. W. Page, C. L. Gratto-Trevor, and B. A. Andres. 2006. Population estimates of North American shorebirds, 2006. Wader Study Group Bulletin 111:67–85.

Moskoff, W., and L. R. Bevier. 2002. Mew Gull (*Larus canus*). No. 687 *in* A. Poole (editor), The Birds of North America. Cornell Lab of Ornithology, Ithaca, NY. <http://bna.birds.cornell.edu/bna/species/687>.

Moskoff, W., and R. Montgomerie. 2002. Baird's Sandpiper (*Calidris bairdii*). No. 661 *in* A. Poole (editor), The Birds of North America. Cornell Lab of Ornithology, Ithaca, NY. <http://bna.birds.cornell.edu/bna/species/661>.

Mowbray, T. B. 1999. American Wigeon (*Anas americana*). No. 401 *in* A. Poole (editor), The Birds of North America. Cornell Lab of Ornithology, Ithaca, NY. <http://bna.birds.cornell.edu/bna/species/401>.

Mowbray, T. B. 2002. Canvasback (*Aythya valisineria*). No. 659 *in* A. Poole (editor), The Birds of North America. Cornell Lab of Ornithology, Ithaca, NY. <http://bna.birds.cornell.edu/bna/species/659>.

Mowbray, T. B., F. Cooke, and B. Ganter. 2000. Snow Goose (*Chen caerulescens*). No. 514 *in* A. Poole (editor), The Birds of North America. Cornell Lab of Ornithology, Ithaca, NY. <http://bna.birds.cornell.edu/bna/species/514>.

Mowbray, T. B., C. R. Ely, J. S. Sedinger, and R. E. Trost. 2002. Canada Goose (*Branta canadensis*). No. 682 *in* A. Poole (editor), The Birds of North America. Cornell Lab of Ornithology, Ithaca, NY. <http://bna.birds.cornell.edu/bna/species/682>.

Mueller, H. 1999. Wilson's Snipe (*Gallinago delicata*). No. 417 *in* A. Poole (editor), The Birds of North America. Cornell Lab of Ornithology, Ithaca, NY. <http://bna.birds.cornell.edu/bna/species/417>.

Muller, S. V., A. E. Racoviteanu, and D. A. Walker. 1999. Landsat MSS-derived land-cover map of northern Alaska: extrapolation methods and a comparison with photo-interpreted and AVHRR-derived maps. International Journal of Remote Sensing 20:2921–2946.

NABCI. 1998. Workshop on the North American Bird Conservation Initiative. November 4–6, Puebla, Mexico.

NABCI. 2004. Map of North American Bird Conservation Initiative (NABCI) defined bird conservation regions. <http://www.nabci-us.org/map.html> (26 May 2009).

National Geographic Society. 2006. National Geographic field guide to the birds of North America. National Geographic Society, Washington, DC.

National Research Council. 2003. Cumulative environmental effects of oil and gas activities on Alaska's North Slope. National Academies Press, Washington, DC.

NatureWatch. 2009. NatureWatch. <http://www.naturewatch.ca> (2 May 2009).

Naugle, D. E., R. R. Johnson, T. R. Cooper, M. M. Holland, and K. F. Higgins. 2000. Temporal distribution of waterfowl in eastern South Dakota: implications for aerial surveys. Waterbirds 20:177–183.

Naugler, C. T. 1993. American Tree Sparrow (*Spizella arborea*). No. 37 *in* A. Poole (editor), The Birds of North America. Cornell Lab of Ornithology, Ithaca, NY. <http://bna.birds.cornell.edu/bna/species/037>.

Nebel, S., and J. M. Cooper. 2008. Least Sandpiper (*Calidris minutilla*). No. 115 *in* A. Poole (editor), The Birds of North America. Cornell Lab of Ornithology, Ithaca, NY. <http://bna.birds.cornell.edu/bna/species/115>.

Nebel, S., and B. J. McCaffery. 2003. Vocalization activity of breeding shorebirds: documentation of its seasonal decline and applications for breeding bird surveys. Canadian Journal of Zoology 81:1702–1708.

Nebel, S., J. L. Porter, and J. T. Kingsford. 2008. Long-term trends of shorebird populations in eastern Australia and impacts of freshwater extraction. Biological Conservation 141:971–980.

Nettleship, D. N. 2000. Ruddy Turnstone (*Arenaria interpres*). No. 537 *in* A. Poole (editor), The Birds of North America. Cornell Lab of Ornithology, Ithaca, NY. <http://bna.birds.cornell.edu/bna/species/537>.

Niles, L. J., J. Bart, H. P. Sitters, A. D. Dey, D. A. Clark, P. W. Atkinson, A.J. Baker, K. A. Bennett, K. S. Kalasz, N. A. Clark, J. Clark, S. Gillings, A. S. Gates, P. M. Gonzaílez, D. Hernandez, C. D. T. Minton, R. I. G. Morrison, R. R. Porter, K. Ross, and R.Veitch. 2009. Effects of horseshoe crab harvest in Delaware Bay on Red Knots: are harvest restrictions working? Bioscience 59:153–164.

Niles, L. J., H. P. Sitters, A. D. Dey, P. W. Atkinson, A. J. Baker, K. A. Bennett, R. Carmona, K. E. Clark, N. A. Clark, C. Espoz, P. M. González, B. A. Harrington, D. E. Hernández, K. S. Kalasz, R. G. Lathrop, R. N. Matus, C. D. T. Minton, R. I. G. Morrison, M. K. Peck, W. Pitts, R. A. Robinson, and I. L. Serrano. 2008. Status of the Red Knot (*Calidris canutus rufa*) in the Western Hemisphere. Studies in Avian Biology 38.

Nol, E., and M. S. Blanken. 1999. Semipalmated Plover (*Charadrius semipalmatus*). No. 444 *in* A. Poole (editor), The Birds of North America. Cornell Lab of Ornithology, Ithaca, NY. <http://bna.birds.cornell.edu/bna/species/444doi:10.2173/bna.444>.

North, M. R. 1994. Yellow-billed Loon (*Gavia adamsii*). No. 121 *in* A. Poole (editor), The Birds of North America. Cornell Lab of Ornithology, Ithaca, NY. <http://bna.birds.cornell.edu/bna/species/121>.

Northwest Territories/Nunavut Bird Checklist Survey. 2009. Northwest Territories/Nunavut bird checklist survey. Canadian Wildlife Service, Yellowknife NT. <http://www.pnr-rpn.ec.gc.ca/checklist>.

Obst, J., and A. Spaulding. 1994. Annual report on the Eskimo Curlew (*Numenius borealis*) recovery plan project in the Northwest Territories, Canada in 1994: results of the investigations into an unconfirmed Eskimo Curlew nest site in the Northwest Territories. Unpublished report. Wildlife Management Division, Department of Renewable Resources, Government of the Northwest Territories, Yellowknife, NT.

Oring, L. W., E. M. Gray, and J. M. Reed. 1997. Spotted Sandpiper (*Actitis macularius*). No. 289 *in* A. Poole (editor), The Birds of North America. Cornell Lab of Ornithology, Ithaca, NY. <http://bna.birds.cornell.edu/bna/species/289>.

Parker, G. R., and R. J. Ross. 1973. Notes on the birds of Southampton Island, Northwest Territories. Arctic 26:123–129.

Parmelee, D. F., D. W. Greiner, and W. D. Graul. 1968. Summer schedule and breeding biology of the white-rumped sandpiper in the Central Canadian arctic. Wilson Bulletin 80:5–29.

Parmelee, D. F., H. A. Stephens, and R. H. Schmidt. 1967. The birds of southeastern Victoria Island and adjacent small islands. National Museum of Canada Bulletin 222.

Parmelee, D. F. 1992a. Snowy Owl (*Bubo scandiacus*). No. 10 *in* A. Poole (editor), The Birds of North

America. Cornell Lab of Ornithology, Ithaca, NY. <http://bna.birds.cornell.edu/bna/species/010>.

Parmelee, D. F. 1992b. White-rumped Sandpiper (*Calidris fuscicollis*). No. 29 *in* A. Poole (editor), The Birds of North America. Cornell Lab of Ornithology, Ithaca, NY. <http://bna.birds.cornell.edu/bna/species/029doi: 10.2173/bna.29>.

Patterson, L. A., and W. G. Alliston. 1978. Breeding bird surveys at selected sites on southern Somerset Island and Boothia Peninsula, July 1977. LGL Limited for Polar Gas Project, Toronto, ON.

Paulson, D. R. 1995. Black-bellied Plover (*Pluvialis squatarola*). No. 186 *in* A. Poole (editor), The Birds of North America. Cornell Lab of Ornithology, Ithaca, NY. <http://bna.birds.cornell.edu/bna/species/186doi: 10.2173/bna.186>.

Payne, L. X., and E. P. Pierce. 2002. Purple Sandpiper (*Calidris maritima*). No. 706 *in* A. Poole (editor), The Birds of North America. Cornell Lab of Ornithology, Ithaca, NY. <http://bna.birds.cornell.edu/bna/species/706>.

Perkins, D. E. 2004. The breeding ecology and behavioral endocrinology of Ruddy Turnstones (*Arenaria interpres*) in the eastern Canadian arctic. M.Sc. thesis, University of Maine, Orono, ME.

Perkins, D. E., P. A. Smith, and H. G. Gilchrist. 2007. The breeding ecology of Ruddy Turnstones in the eastern Canadian arctic. Polar Record 43:135–142.

Petersen, M. R., J. B. Grand, and C. P. Dau. 2000. Spectacled Eider (*Somateria fischeri*). No. 547 *in* A. Poole (editor), The Birds of North America. Cornell Lab of Ornithology, Ithaca, NY. <http://bna.birds.cornell.edu/bna/species/547>.

Petersen, M. R., D. N. Weir, and M. H. Dick. 1991. Birds of the Kilbuck and Ahklun Mountain Region, Alaska. North American Fauna 76.

Pierotti, R. J., and T. P. Good. 1994. Herring Gull (*Larus argentatus*). No. 124 *in* A. Poole (editor), The Birds of North America. Cornell Lab of Ornithology, Ithaca, NY. <http://bna.birds.cornell.edu/bna/species/124>.

Pirie, L. D. 2009. Identifying and modeling Whimbrel (*Numenius phaeopus*) breeding habitat in the outer Mackenzie Delta, Northwest Territories. M.Sc. thesis, Carleton University, Ottawa, ON.

Pirie, L. D., V. Johnston, J. Rausch, and C. Dockrill. 2009. 2008 Final Report. Whimbrel of the Mackenzie Delta: A study of survey techniques, distribution and breeding biology. Unpublished Report prepared for the Canadian Wildlife Service, Yellowknife, NT.

Pirie, L. D., C. M. Francis, and V. H. Johnston. 2009. Evaluating the potential impact of a gas pipeline on Whimbrel breeding habitat in the outer Mackenzie Delta, Northwest Territories. Avian Conservation and Ecology – Ecologie et conservation des oiseaux 4(2):2. [online] <http://www.ace-eco.org/vol4/iss2/art2/>.

Pitelka, F. A. 1979. Introduction: the Pacific coast shorebird scene. Studies in Avian Biology 2:1–11.

Pitelka, F. A., P. Q. Tomich, and G. W. Treichel. 1955. Ecological relations of jaegers and owls as lemming predators near Barrow, Alaska. Ecological Monographs 25:85–117.

Pollock, K. H., and W. L. Kendall. 1987. Visibility bias in aerial surveys: a review of estimation procedures. Journal of Wildlife Management 51:502–510.

Raven, G. H., and D. L. Dickson. 2009. Surveys of pacific Common Eiders (*Somateria mollissima v.-nigra*) in the Bathurst Inlet area of Nunavut, 2006–2008. Canadian Wildlife Service Technical Report No. 503.

Reed, A., D. H. Ward, D. V. Derksen, and J. S. Sedinger. 1998. Brant (*Branta bernicla*). No. 337 *in* A. Poole (editor), The Birds of North America. Cornell Lab of Ornithology, Ithaca, NY. <http://bna.birds.cornell.edu/bna/species/337>.

Rehfisch, M., and H. Crick. 2003. Predicting the impact of climate change on arctic-breeding waders. Wader Study Group Bulletin 100:86–95.

Richardson, W. J., and S. R. Johnson. 1981. Waterbird migration near the Yukon and Alaskan coast of the Beaufort Sea. Arctic 34:106–121.

Richardson, W. J., M. R. Morrell, and S. R. Johnson. 1975. Bird migration along the Beaufort Sea coast: radar and visual observations in 1975. Beaufort Sea Technical Report No. 3c. Environment Canada, Victoria, BC.

Ridgely, R. S., T. F. Allnutt, T. Brooks, D. K. McNicol, D. W. Mehlman, B. E. Young, and J. R. Zook. 2007. Digital distribution maps of the birds of the Western Hemisphere, ver. 3.0. NatureServe, Arlington, VA.

Riordan, B., D. Verbyla, and A. D. McGuire. 2006. Shrinking ponds in subarctic Alaska based on 1950–2002 remotely sensed images. Journal of Geophysical Research 111:G04002.

Robertson, G. J., and J. L. Savard. 2002. Long-tailed Duck (*Clangula hyemalis*). No. 651 *in* A. Poole (editor), The Birds of North America. Cornell Lab of Ornithology, Ithaca, NY. <http://bna.birds.cornell.edu/bna/species/651>.

Robertson, R. J., B. J. Stutchbury, and R. R. Cohen. 1992. Tree Swallow (*Tachycineta bicolor*). No. 11 *in* A. Poole (editor), The Birds of North America. Cornell Lab of Ornithology, Ithaca, NY. <http://bna.birds.cornell.edu/bna/species/011>.

Rodrigues, R. 1994. Microhabitat variables influencing nest-site selection by tundra birds. Ecological Applications 4:110–116.

Rubega, M. A., D. Schamel, and D. M. Tracy. 2000. Red-necked Phalarope (*Phalaropus lobatus*). No. 538 *in* A. Poole (editor), The Birds of North America. Cornell Lab of Ornithology, Ithaca, NY. <http://bna.birds.cornell.edu/bna/species/538>.

Rumble, M. A., and L. D. Flake. 1982. A comparison of two waterfowl brood survey techniques. Journal of Wildlife Management 46:1048–1053.

Russell, R. W. 2002. Pacific Loon (*Gavia pacifica*). No. 657 *in* A. Poole (editor), The Birds of North America. Cornell Lab of Ornithology, Ithaca, NY. <http://bna.birds.cornell.edu/bna/species/657>.

Ruthrauff, D. R., and T. L. Tibbitts. 2009. Inventory of breeding birds in Aniakchak National Monument and Preserve. Unpublished final report for National Park Service. U.S. Geological Survey, Alaska Science Center, Anchorage, AK.

Ruthrauff, D. R., T. L. Tibbitts, R. E. Gill, Jr., and C. M. Handel. 2007. Inventory of montane-nesting birds in Katmai and Lake Clark National Parks and Preserves. Unpublished final report for National Park Service. U.S. Geological Survey, Alaska Science Center, Anchorage, AK.

Sallabanks, R., and F. C. James. 1999. American Robin (*Turdus migratorius*). No. 462 *in* A. Poole (editor), The Birds of North America. Cornell Lab of Ornithology, Ithaca, NY. <http://bna.birds.cornell.edu/bna/species/462>.

Salter, R., and R. A. Davis. 1974. Surveys of terrestrial bird populations in Alaska, Yukon Territory, Northwest Territories and northern Alberta, May, June, July, 1972. *In* W. W. H. Gunn, and J. A. Livingston (editors), Bird distribution and populations ascertained through ground survey techniques, 1972. Arctic Gas Biological Report Series, Vol. 12. LGL Limited, Calgary, AB.

Sauer, J. R., J. E. Hines, and J. Fallon. 2007. The North American Breeding Bird Survey: results and analysis 1966–2006, ver. 10.13.2007. USGS Patuxent Wildlife Research Center, Laurel, MD.

Savage, S. E., and O.W. Johnson. 2005. Breeding range extensions for the Pacific Golden-Plover and Black-bellied Plover on the Alaska Peninsula. Wader Study Group Bulletin 108:63–65.

Savard, J. L., D. Bordage, and A. Reed. 1998. Surf Scoter (*Melanitta perspicillata*). No. 363 *in* A. Poole (editor), The Birds of North America. Cornell Lab of Ornithology, Ithaca, NY. <http://bna.birds.cornell.edu/bna/species/363>.

Schamel, D., and D. M. Tracy. 1991. Breeding site fidelity and natal philopatry in the sex role-reversed Red and Red-necked Phalaropes. Journal of Field Ornithology 62:390–398.

Schamel, D., D. M. Tracy, P. G. Mickelson, and A. Seguin. 1979. Avian community ecology at two sites on Espenberg Peninsula in Kotzebue Sound, Alaska. Pp. 289–607 *in* Final reports of principal investigators, Environmental Assessment of the Alaska Continental Shelf. Biological Studies, Vol. 5. NOAA-BLM, Boulder, CO.

Schrader, S., and R. Pouncey. 1997. Classification: ERDAS field guide. 4th ed. Erdas Inc., Atlanta, GA. <http: //support.erdas.com/documentation/files/FieldGuide.pdf>.

Schweinsberg, R. E. 1974. An ornithological study of proposed gas pipeline routes in Alaska, Yukon Territory and the Northwest Territories, 1971. Arctic Gas Biological Report Series, Vol. 10. LGL Limited, Calgary, AB.

Shur, Y., M. Jorgenson, and E. Pullman. 2003. Widespread degradation of ice wedges on the arctic Coastal Plain in northern Alaska in response to the recent warmer climate. American Geophysical Union, Fall Meeting 2003, Abstract C11A-05. <http://adsabs.harvard.edu/abs/2003AGUFM.C11A..05S>.

Sinclair, P., Y. Aubry, J. Bart, V. Johnston, R. Lanctot, B. McCaffrey, K. Ross, P. Smith, and L. Tibbitts. 2004. Boreal shorebirds: an assessment of conservation status and potential for population monitoring. Unpublished PRISM report.

Skagen, S., J. Bart, B. Andres, S. Brown, G. Donaldson, B. Harrington, V. Johnston, S. L. Jones, and R. I. G. Morrison. 2003. Monitoring the shorebirds of North America: towards a unified approach. Wader Study Group Bulletin 100:102–104.

Skalski, J. R., K. E. Ryding, and J. J. Millspaugh. 2005. Wildlife demography: analysis of sex, age, and count data. Elsevier, New York, NY.

Skeel, M. A. 1983. Nesting success, density, philopatry, and nest-site selection of the Whimbrel (*Numenius phaeopus*) in different habitats. Canadian Journal of Zoology 61:218–225.

Skeel, M. A., and E. P. Mallory. 1996. Whimbrel (*Numenius phaeopus*). No. 219 *in* A. Poole (editor), The Birds of North America. Cornell Lab of Ornithology, Ithaca, NY. <http://bna.birds.cornell.edu/bna/species/219>.

Smith, G. 1995. A critical review of the aerial and ground surveys of breeding waterfowl in North America. U.S. Fish and Wildlife Service Technical Report. <http://www.dtic.mil/cgi-bin/GetTRDoc?AD=ADA322667&Location=U2&doc=GetTRDoc.pdf>.

Smith, L. C., Y. Sheng, G. M. MacDonald, and L. D. Hinzman. 2005. Disappearing arctic lakes. Science 208:1429.

Smith, P. A. 2003. Factors affecting nest site selection and reproductive success of tundra nesting shorebirds. M.Sc. thesis, University of British Columbia, Vancouver, BC.

Smith, P. A. 2009. Variation in shorebird nest survival: proximate pressures and ultimate constraints. Ph.D. dissertation, Carleton University, Ottawa, ON.

Smith, P. A., J. Bart, R. B. Lanctot, B. J. McCaffery, and S. Brown. 2009. Probability of detection of nests and implications for survey design. Condor 111:414–423.

Smith, P. A., H. G. Gilchrist, M. R. Forbes, J. L. Martin, and K. Allard. 2010. Inter-annual variation in the breeding chronology of arctic shorebirds: effects of weather, snow melt and predators. Journal of Avian Biology 41:292–304.

Smith, P. A., H. G. Gilchrist, and J. N. M. Smith. 2007a. Effects of nest habitat, food, and parental behaviour on shorebird nest success. Condor 109:15–31.

Smith, P. A., H. G. Gilchrist, J. N. M. Smith, and E. Nol. 2007b. Annual variation in the benefits of a nesting association between Red Phalaropes (*Phalaropus fulicarius*) and Sabine's Gulls (*Xema sabini*). Auk 124:276–290.

Smith, P. A., and S. D. Wilson. 2010. Intraseasonal patterns in shorebird nest survival are related to nest age and defence behaviour. Oecologia 163:613–624.

Soper, J. D. 1940. Local distribution of some eastern Canadian arctic birds. Auk 57:13–21.

Stenhouse, I. J., H. G. Gilchrist, and W. A. Montevecchi. 2001. Reproductive biology of Sabine's Gull in the Canadian arctic. Condor 103:98–107.

Stout, B. E., and G. L. Nuechterlein. 1999. Red-necked Grebe (*Podiceps grisegena*). No. 465 *in* A. Poole (editor), The Birds of North America. Cornell Lab of Ornithology, Ithaca, NY. <http://bna.birds.cornell.edu/bna/species/465>.

Sturm, M., C. Racine, and K. Tape. 2001. Increasing shrub abundance in the arctic. Nature 411:546–547.

Sutton, G. M. 1932. The exploration of Southampton Island, Hudson Bay, Part II: Zoology, Section 2: The birds of Southampton Island. Memoirs of the Carnegie Museum 12, II, 2.

Suydam, R. S. 2000. King Eider (*Somateria spectabilis*). No. 491 *in* A. Poole (editor), The Birds of North America. Cornell Lab of Ornithology, Ithaca, NY. <http://bna.birds.cornell.edu/bna/species/491>.

Tacha, T. C., S. A. Nesbitt, and P. A. Vohs. 1992. Sandhill Crane (*Grus canadensis*). No. 31 *in* A. Poole (editor), The Birds of North America. Cornell Lab of Ornithology, Ithaca, NY. <http://bna.birds.cornell.edu/bna/species/031>.

Takekawa, J. Y., and N. Warnock. 2000. Long-billed Dowitcher (*Limnodromus scolopaceus*). No. 493 *in* A. Poole (editor), The Birds of North America. Cornell Lab of Ornithology, Ithaca, NY. <http://bna.birds.cornell.edu/bna/species/493>.

Tande, G. F., and T. W. Jennings. 1986. Classification and mapping of tundra near Hazen Bay, Yukon Delta National Wildlife Refuge, Alaska. Report AI 87/01. U.S. Fish and Wildlife Service, Anchorage, AK.

Tarves, T. L. 1987. Shorebirds of the outer Mackenzie Delta, N.W.T./Fish Island area habitat and nest data of the Whimbrel and the Semipalmated Plover. B.Sc. thesis, Northern Alberta Institute of Technology, Edmonton, AB.

Tibbitts, T. L., and W. Moskoff. 1999. Lesser Yellowlegs (*Tringa flavipes*). No. 427 *in* A. Poole (editor), The Birds of North America. Cornell Lab of Ornithology, Ithaca, NY. <http://bna.birds.cornell.edu/bna/species/427>.

Tibbitts, T. L., D. R. Ruthrauff, R. E. Gill, Jr., and C. M. Handel. 2005. Inventory of montane-nesting birds in the Arctic Network of National Parks, Alaska. NPS/AKARCN/NRTR-2006/02. Arctic Network Inventory and Monitoring Program, USDI National Park Service, Fairbanks, AK.

Titman, R. D. 1999. Red-breasted Merganser (*Mergus serrator*). No. 443 *in* A. Poole (editor), The Birds of North America. Cornell Lab of Ornithology, Ithaca, NY. <http://bna.birds.cornell.edu/bna/species/443>.

Tracy, D. M., D. Schamel, and J. Dale. 2002. Red Phalarope (*Phalaropus fulicarius*). No. 698 *in* A. Poole (editor), The Birds of North America. Cornell Lab of Ornithology, Ithaca, NY. <http://bna.birds.cornell.edu/bna/species/698>.

Underhill, L. G., M. Waltner, and R. W. Summers. 1989. Three-year cycles in breeding productivity of Knots, (*Calidris canutus*) wintering in southern Africa suggest Taimyr Peninsula provenance. Bird Study 36:83–87.

U.S. Fish and Wildlife Service. 2002. Waterfowl breeding population survey for northern Alberta, northeastern British Columbia, and the Northwest Territories (Mackenzie District). U.S. Fish and Wildlife Service, Laurel, MD. <http://www.fws.gov/migratorybirds/reports/wps02/May/nwt.pdf>.

U.S. Shorebird Conservation Plan. 2004. High priority shorebirds—2004. Unpublished report. U.S. Fish and Wildlife Service, Arlington, VA.

Van Belle, G. 2002. Statistical rules of thumb. John Wiley & Sons, New York, NY.

Verbeek, N. A., and P. Hendricks. 1994. American Pipit (*Anthus rubescens*). No. 95 *in* A. Poole (editor), The Birds of North America. Cornell Lab of Ornithology, Ithaca, NY. <http://bna.birds.cornell.edu/bna/species/095>.

Vermeer, K., and G. G. Anweiler. 1975. Oil threat to aquatic birds along the Yukon coast. Wilson Bulletin 87:467–480.

Visser, M. E., C. Both, and M. M. Lambrechts. 2004. Global climate change leads to mistimed avian reproduction. Advances in Ecological Research 35:89–110.

Walpole, B., E. Nol, and V. Johnston. 2008a. Breeding habitat preference and nest success of Red-necked Phalaropes on Niglintgak Island, Northwest Territories. Canadian Journal of Zoology 86:1346–1357.

Walpole, B., E. Nol, and V. Johnston. 2008b. Pond characteristics and occupancy by Red-necked Phalaropes in the Mackenzie Delta, Northwest Territories, Canada. Arctic 61:426–432.

Warnock, N. D., and R. E. Gill. 1996. Dunlin (*Calidris alpina*). No. 203 *in* A. Poole (editor), The Birds of North America. Cornell Lab of Ornithology, Ithaca, NY. <http://bna.birds.cornell.edu/bna/species/203>.

Watkinson, A. R., J. A. Gill, and M. Hulme. 2004. Flying in the face of climate change: a review of climate change, past, present and future. Ibis 146:4–10.

Weckstein, J. D., D. E. Kroodsma, and R. C. Faucett. 2002. Fox Sparrow (*Passerella iliaca*). No. 715 *in* A. Poole (editor), The Birds of North America. Cornell Lab of Ornithology, Ithaca, NY. <http://bna.birds.cornell.edu/bna/species/715>.

Wheelwright, N. T., and J. D. Rising. 2008. Savannah Sparrow (*Passerculus sandwichensis*). No. 45 *in* A. Poole (editor), The Birds of North America. Cornell Lab of Ornithology, Ithaca, NY. <http://bna.birds.cornell.edu/bna/species/045>.

White, C. M., N. J. Clum, T. J. Cade, and W. G. Hunt. 2002. Peregrine Falcon (*Falco peregrinus*). No. 660 *in* A. Poole (editor), The Birds of North America. Cornell Lab of Ornithology, Ithaca, NY. <http://bna.birds.cornell.edu/bna/species/660>.

White, C. M., F. S. L. Williamson, and W. B. Emison. 1977. Avifaunal investigations. Pp. 227–260 *in* M. L. Merritt and R. G. Fuller (editors), The environment of Amchitka Island, Alaska. Technical Information Center, Energy Research and Development Administration, Springfield, VA.

Wibbenmeyer, M. W., J. G. Grunblatt, and L. S. Shea. 1982. User's guide for Bristol Bay land cover maps. Alaska Department of Natural Resources, Anchorage, AK.

Wiens, J. A. 2009. Is "monitoring" a dirty word? Bulletin of the British Ecological Society 40:39–40.

Wiggins, D. A., D. W. Holt, and S. M. Leasure. 2006. Short-eared Owl (*Asio flammeus*). No. 62 *in* A. Poole (editor), The Birds of North America. Cornell Lab of Ornithology, Ithaca, NY. <http://bna.birds.cornell.edu/bna/species/062>.

Wiley, R. H., and D. S. Lee. 1998. Long-tailed Jaeger (*Stercorarius longicaudus*). No. 365 *in* A. Poole (editor), The Birds of North America. Cornell Lab of Ornithology, Ithaca, NY. <http://bna.birds.cornell.edu/bna/species/365>.

Wiley, R. H., and D. S. Lee. 1999. Parasitic Jaeger (*Stercorarius parasiticus*). No. 445 *in* A. Poole (editor), The Birds of North America. Cornell Lab of Ornithology, Ithaca, NY. <http://bna.birds.cornell.edu/bna/species/445>.

Wiley, R. H., and D. S. Lee. 2000. Pomarine Jaeger (*Stercorarius pomarinus*). No. 483 *in* A. Poole (editor), The Birds of North America. Cornell Lab of Ornithology, Ithaca, NY. <http://bna.birds.cornell.edu/bna/species/483>.

Williams, B. K., J. D. Nichols, and M. J. Conroy. 2005. Analysis and management of animal populations. Academic Press, New York, NY.

Wilson, W. H. 1994. Western Sandpiper (*Calidris mauri*). No. 90 *in* A. Poole (editor), The Birds of North America. Cornell Lab of Ornithology, Ithaca, NY. <http://bna.birds.cornell.edu/bna/species/090>.

Woodin, M. C., and T. C. Michot. 2002. Redhead (*Aythya americana*). No. 695 *in* A. Poole (editor), The Birds of North America. Cornell Lab of Ornithology, Ithaca, NY. <http://bna.birds.cornell.edu/bna/species/695>.

Ydenberg, R. C., R. W. Butler, D. B. Lank, B. D. Smith, and J. Ireland. 2004. Western sandpipers have altered migration tactics as Peregrine Falcon populations have recovered. Proceedings of the Royal Society of London B 271:1263–1269.

INDEX

STUDIES IN AVIAN BIOLOGY

1. Kessel, B., and D. D. Gibson. 1978. *Status and Distribution of Alaska Birds.*
2. Pitelka, F. A., editor. 1979. *Shorebirds in Marine Environments.*
3. Szaro, R. C., and R. P. Balda. 1979. *Bird Community Dynamics in a Ponderosa Pine Forest.*
4. DeSante, D. F., and D. G. Ainley. 1980. *The Avifauna of the South Farallon Islands, California.*
5. Mugaas, J. N., and J. R. King. 1981. *Annual Variation of Daily Energy Expenditure by the Black-billed Magpie: A Study of Thermal and Behavioral Energetics.*
6. Ralph, C. J., and J. M. Scott, editors. 1981. *Estimating Numbers of Terrestrial Birds.*
7. Price, F. E., and C. E. Bock. 1983. *Population Ecology of the Dipper* (Cinclus mexicanus) *in the Front Range of Colorado.*
8. Schreiber, R. W., editor. 1984. *Tropical Seabird Biology.*
9. Scott, J. M., S. Mountainspring, F. L. Ramsey, and C. B. Kepler. 1986. *Forest Bird Communities of the Hawaiian Islands: Their Dynamics, Ecology, and Conservation.*
10. Hand, J. L., W. E. Southern, and K. Vermeer, editors. 1987. *Ecology and Behavior of Gulls.*
11. Briggs, K. T., W. B. Tyler, D. B. Lewis, and D. R. Carlson. 1987. *Bird Communities at Sea off California: 1975 to 1983.*
12. Jehl, J. R., Jr. 1988. *Biology of the Eared Grebe and Wilson's Phalarope in the Nonbreeding Season: A Study of Adaptations to Saline Lakes.*
13. Morrison, M. L., C. J. Ralph, J. Verner, and J. R. Jehl, Jr., editors. 1990. *Avian Foraging: Theory, Methodology, and Applications.*
14. Sealy, S. G., editor. 1990. *Auks at Sea.*
15. Jehl, J. R., Jr., and N. K. Johnson, editors. 1994. *A Century of Avifaunal Change in Western North America.*
16. Block, W. M., M. L. Morrison, and M. H. Reiser, editors. 1994. *The Northern Goshawk: Ecology and Management.*
17. Forsman, E. D., S. DeStefano, M. G. Raphael, and R. J. Gutiérrez, editors. 1996. *Demography of the Northern Spotted Owl.*
18. Morrison, M. L., L. S. Hall, S. K. Robinson, S. I. Rothstein, D. C. Hahn, and T. D. Rich, editors. 1999. *Research and Management of the Brown-headed Cowbird in Western Landscapes.*
19. Vickery, P. D., and J. R. Herkert, editors. 1999. *Ecology and Conservation of Grassland Birds of the Western Hemisphere.*
20. Moore, F. R., editor. 2000. *Stopover Ecology of Nearctic–Neotropical Landbird Migrants: Habitat Relations and Conservation Implications.*
21. Dunning, J. B., Jr., and J. C. Kilgo, editors. 2000. *Avian Research at the Savannah River Site: A Model for Integrating Basic Research and Long-Term Management.*
22. Scott, J. M., S. Conant, and C. van Riper, II, editors. 2001. *Evolution, Ecology, Conservation, and Management of Hawaiian Birds: A Vanishing Avifauna.*
23. Rising, J. D. 2001. *Geographic Variation in Size and Shape of Savannah Sparrows* (Passerculus sandwichensis).
24. Morton, M. L. 2002. *The Mountain White-crowned Sparrow: Migration and Reproduction at High Altitude.*

25. George, T. L., and D. S. Dobkin, editors. 2002. *Effects of Habitat Fragmentation on Birds in Western Landscapes: Contrasts with Paradigms from the Eastern United States.*

26. Sogge, M. K., B. E. Kus, S. J. Sferra, and M. J. Whitfield, editors. 2003. *Ecology and Conservation of the Willow Flycatcher.*

27. Shuford, W. D., and K. C. Molina, editors. 2004. *Ecology and Conservation of Birds of the Salton Sink: An Endangered Ecosystem.*

28. Carmen, W. J. 2004. *Noncooperative Breeding in the California Scrub-Jay.*

29. Ralph, C. J., and E. H. Dunn, editors. 2004. *Monitoring Bird Populations Using Mist Nets.*

30. Saab, V. A., and H. D. W. Powell, editors. 2005. *Fire and Avian Ecology in North America.*

31. Morrison, M. L., editor. 2006. *The Northern Goshawk: A Technical Assessment of Its Status, Ecology, and Management.*

32. Greenberg, R., J. E. Maldonado, S. Droege, and M. V. McDonald, editors. 2006. *Terrestrial Vertebrates of Tidal Marshes: Evolution, Ecology, and Conservation.*

33. Mason, J. W., G. J. McChesney, W. R. McIver, H. R. Carter, J. Y. Takekawa, R. T. Golightly, J. T. Ackerman, D. L. Orthmeyer, W. M. Perry, J. L. Yee, M. O. Pierson, and M. D. McCrary. 2007. *At-Sea Distribution and Abundance of Seabirds off Southern California: A 20-Year Comparison.*

34. Jones, S. L., and G. R. Geupel, editors. 2007. *Beyond Mayfield: Measurements of Nest-Survival Data.*

35. Spear, L. B., D. G. Ainley, and W. A. Walker. 2007. *Foraging Dynamics of Seabirds in the Eastern Tropical Pacific Ocean.*

36. Niles, L. J., H. P. Sitters, A. D. Dey, P. W. Atkinson, A. J. Baker, K. A. Bennett, R. Carmona, K. E. Clark, N. A. Clark, C. Espoz, P. M. González, B. A. Harrington, D. E. Hernández, K. S. Kalasz, R. G. Lathrop, R. N. Matus, C. D. T. Minton, R. I. G. Morrison, M. K. Peck, W. Pitts, R. A. Robinson, and I. L. Serrano. 2008. *Status of the Red Knot* (Calidris canutus rufa) *in the Western Hemisphere.*

37. Ruth, J. M., T. Brush, and D. J. Krueper, editors. 2008. *Birds of the US–Mexico Borderland: Distribution, Ecology, and Conservation.*

38. Knick, S. T., and J. W. Connelly, editors. 2011. *Greater Sage-Grouse: Ecology and Conservation of a Landscape Species and Its Habitats.*

39. Sandercock, B. K., K. Martin, and G. Segelbacher, editors. 2011. *Ecology, Conservation, and Management of Grouse.*

40. Forsman, E.D., et al. 2011. *Population Demography of Northern Spotted Owls.*

41. Wells, J. V., editor. 2011. *Boreal Birds of North America: A Hemispheric View of Their Conservation Links and Significance.*

42. Paul, E., editor. 2012. *Emerging Avian Disease.*

43. Ribic, C. A., F. R. Thompson III, and P. J. Pietz, editors. 2012. *Video Surveillance of Nesting Birds.*

44. Bart, J. R., and V. H. Johnston, editors. 2012. *Arctic Shorebirds in North America: A Decade of Monitoring.*